Fiber Optic Networks

Fiber Optic Networks

Paul E. Green, Jr.

Computer Science Department
IBM T.J. Watson Research Center
Hawthorne, New York

Prentice Hall
Englewood Cliffs, New Jersey 07632

Library of Congress Cataloging-in-Publication Data

Green, Paul E.
    Fiber optic networks / P.E. Green.
        p. cm.
    Includes bibliographical references and index.
    ISBN 0-13-319492-2
    1. Optical communications. 2. Fiber optics. I. Title.
TK5103.59.G74   1993
621.382'75--dc20   92-4846
                                                     CIP

Acquisitions editor: Peter Janzow
Production editor: Irwin Zucker
Copy editor: Bob Lentz
Prepress buyer: Linda Behrens
Manufacturing buyer: David Dickey
Supplements editor: Alice Dworkin
Cover design: Dorrit Green
Editorial assistant: Phyllis Morgan

© 1993 by Prentice-Hall, Inc.
A Simon & Schuster Company
Englewood Cliffs, New Jersey 07632

The author and publisher of this book have used their best efforts in preparing this book. These efforts include the development, research, and testing of the theories and programs to determine their effectiveness. The author and publisher make no warranty of any kind, expressed or implied, with regard to these programs or the documentation contained in this book. The author and publisher shall not be liable in any event for incidental or consequential damages in connection with, or arising out of, the furnishing, performance, or use of these programs.

IBM® is a registered trademark of International Business Machines Corporation.

Printed in the United States of America

10 9 8 7 6 5 4 3 2 1

ISBN   0-13-319492-2

PRENTICE-HALL INTERNATIONAL (UK) LIMITED, *London*
PRENTICE-HALL OF AUSTRALIA PTY. LIMITED, *Sydney*
PRENTICE-HALL CANADA INC., *Toronto*
PRENTICE-HALL HISPANOAMERICANA, S.A., *Mexico*
PRENTICE-HALL OF INDIA PRIVATE LIMITED, *New Delhi*
PRENTICE-HALL OF JAPAN, INC., *Tokyo*
SIMON & SCHUSTER ASIA PTE. LTD., *Singapore*
EDITORA PRENTICE-HALL DO BRASIL, LTDA., *Rio de Janeiro*

To my inspiration

Skip

and to my teachers

*Rajiv, Kumar, Karen, Dave, Michael, and Frank ** 2*

Contents

PART II - BUILDING BLOCKS   33

PART IV - REALIZATION   459

Foreword

Many years ago I used to teach a course on data communications with Paul Green. From time to time Paul would chide me about the relative economics of computing and communications. He would point out the exponential progress over the years in getting ever more computing for a given cost. Then he would show a similar curve for the amount of communications for a given cost. This curve, however, was almost flat with time. Perhaps the minimal progress in communications was due to a lack of innovation in the communications industry, he would say.

I used to argue with Paul, saying that a computer was after all only a little box, and that it could take advantage of the ever-growing economies that were a fallout of the evolution of the integrated circuit. In contrast, I would say, communications takes place on a different scale, often covering thousands of kilometers on a given connection. How was it possible to make such a large scale system inexpensive? But Paul would not concede the point. Even though he did not know how it might be done, he felt that there might be some technological way to change the nature of communications.

Those arguments were before optical fiber came along. So there **was** something magical waiting to be applied to communications after all; something utterly fantastic in its potential. Here was a hair-thin thread of glass with a bandwidth of 25,000 gigahertz, so transparent that if the oceans were filled with this material we could see the bottom in even the deepest trenches. That was also before we had realized the potential of the semiconductor laser–a tiny flashlight the size of a grain of salt that was capable of emitting billions of pulses of light each second, pulses whose meager power could be still detected more than a hundred kilometers down the optical fiber.

Miracles have a way of sneaking up on us, until when they actually arrive they are taken for granted. Paul and I would never have dreamed of gigabits and terabits then, but every year has brought us nearer to those astronomical rates of communications. The pace of the integrated circuit evolution that I had once so envied has been slow in comparison to that of optical communications technology.

Paul Green and I are examples of what I think of as "classically trained" communications theorists. We and others of our generation were all followers of Claude Shannon and his beautiful theory of information. We lived in a world where communications channels were constrained by limited bandwidth and random noise. The bandwidth and the noise were determined by nature, while the communications designer tried with utmost cleverness to squeeze the maximum possible information rate from this limited resource. Shannon's famous bounds on channel capacity showed what might be obtained. We could only get so much information; more was impossible.

It has been almost impossible to fit optical fiber communications into that classical model of Shannon. The communications channel is not an intrinsic gift of nature, but a calculated design of material scientists. For all practical purposes it has no bandwidth limitation whatsoever, and the noise level is negligible compared with deterministic impairments and limitations. The allowable signaling rate at present has nothing to do with Shannon's capacity, but seems solely to be a function of the current state of the physical devices used for generating and detecting lightwave pulses.

Thus progress in optical communications has not been in the hands of the signal processing experts, who traditionally worry about such things as efficiency of modulation format, adaptive equalization, maximum likelihood sequence estimation, coded modulation, error correcting codes, and the like. These are techniques of great beauty and sophistication, but they have been conceived to extract the maximum data rate from precious, limited bandwidth in the face of the ancient enemy—random noise. Moreover, they require a great amount of processing per bit. No one could even consider applying these techniques in the gigabit regime—nor has there been any need to do so.

Instead, the increasing data rate of lightwave transmission has been largely due to the efforts of physicists and materials scientists—people who traditionally have known little of the communications systems methodologies developed during the last twenty-five years. Physicists publish papers on information-rate limited transmission when the bit rate equals the bandwidth. When dispersion limits the rate of transmission they move on to other problems; the channel has yielded its maximum capacity. Communications systems people know better.

The traditional communications theorist, however, has been reticent to attack the problems of optical fiber communications. On the one hand, it is too easy. A laser emits a pulse of light; a detector decides whether or not a pulse is present—like flashing a semaphore across the dark sea between swaying boats. By such a simple concept gigabits upon gigabits can be transmitted without sophistication. What is the point of analysis, and who could ever need improvement on such data rates? But on the other hand, the problem is too hard. Few communications theorists understand laser technology, solid state physics, and the quantum mechanical concepts that determine ultimate behavior in the world of photonics.

The design of an optical communications system cuts across many disciplines. Materials scientists grow wafers, using the formulations of device physicists, who then cut and test the prototype devices. The transmissions system itself, including especially the loss budget that accounts for various attenuation and signal impairments, is studied by systems researchers, who may be typically either physicists or electrical engineers. The transmission system itself fits into a progressively high level framework of architecture that may often be the province of computer scientists.

Because of the specialization required at each stage and level of this work, seldom do the personal skills and knowledge of the participants cut across boundaries. This book effectively does much of that. For a communications systems person like myself it gives a guided tour and explanation of photonics devices. For a device physicist it gives insight into systems issues. Frankly, I learned a great deal in reading it myself, and I believe there are few readers who will not be able to expand their personal inventory of knowledge by studying the chapters of this book.

It might seem that there is little left to do in optical communications systems now that the vast regions of gigabit transmission have been reached. However, in some respects we have not yet come to grips with the changed nature of the relationship of communications and computation. Many of us have been conditioned to think that transmission is inherently expensive; that we should use switching and processing wherever possible to minimize transmission. The limitless bandwidth of optical fiber changes these assumptions. Perhaps we should transmit signals thousands of kilometers to avoid even the simplest processing function. The opportunity that beckons us is to trade the bandwidth of the fiber for systems simplicity elsewhere. That is what this book is all about. We have a new resource; let us learn how best to use it.

Robert W. Lucky
Executive Director
Communication Sciences
AT&T Bell Laboratories

Preface

The purpose of this book is twofold: first, to present an up-to-date account of fiber optic communication in an accessible style using undergraduate-level physical reasoning, and second, to portray the "all-optical" revolution in networking that fiber optic technology is producing.

Lightwave communication is only 25 years old, but it would be difficult to exaggerate the impact it has already had on all branches of information transmission, from spans of a few centimeters to intercontinental distances. This strange technology, involving, of all things, the transmission of messages as pulses of light along an almost invisible thread of glass, is effectively taking over the role of guided transmission by copper and to a modest extent the role of unguided free-space radio and infrared transmission.

And yet, if one looks a bit more closely at how this supposedly revolutionary technology has been used to date, one comes away with a sense of vast new unexploited opportunities. The bandwidth is so much greater, the attenuation so much less, the bit error rate so much lower, and the potential for low costs so much more promising than anything communication engineers have ever had at their disposal before that it seems perfectly clear that much more can be done with lightwave communication than is being done today. One sometimes gets the impression that the only real applications found for this remarkable medium are those that substitute fiber for copper within the framework of some existing architecture. The performance is improved, the cost goes down, but the system is basically what it was before fiber optics came along.

A principal theme of this book is to emphasize the rapidly evolving field of *nonclassical* applications of fiber optic communication. This term requires a few words of elaboration.

We shall treat here the system architecture and technology issues for three different kinds of structure: (i) The point-to-point *link*–for example, an intercity telephone trunk, (ii) the single station-to-multistation *multipoint*, as exemplified by "fiber to the home" distribution systems, and (iii) the any-to-any connected *network*, as typified by the modern computer network, in which any node has equal access to any other node. The third of these embodies each of the first two as a special case, and therefore networks will receive the most emphasis.

A classical approach to the use of lightwave technologies to solve these problems begins by considering a detailed design already in existence and asking how it could be improved by substituting fiber. For example, we might try to improve point-to-point trunks to keep up with growing traffic demands by upgrading single copper connections to fiber connections and developing faster and faster electronic time-division multiplex, demultiplex and switch electronics, and faster and faster lasers and photodetectors. As another example of the classical approach, improved token ring computer networks can be built by substituting fiber and then upgrading the speed of the time-division electronics. It is clear that such systems do not exploit the full potential of fiber optic technology.

A nonclassical approach to these system designs asks, "What properties of fiber optic communication will allow us to do things very differently?" The nonclassical viewpoint about fiber-optic transmission and the higher-level functions that such transmission can support will be seen to offer new possibilities with respect to system capacity, reliability, and flexibility of use.

This challenge to exploit the intrinsic optical fiber parameters with imagination is being widely addressed these days, and there is not only a growing need for the practicing engineer to keep track of the evolving understanding, but there is an equally strong need to present it in the classroom.

In this book, we shall discuss the recent advances in links, multipoints, and networks justifying this definition. The advances are particularly exciting in the case of full networks, which have earned the name *all-optical networks*, because the routes through them are optical throughout, with the performance limitations imposed by electronics occurring only at the ends of the network connection.

Although the emphasis is on digital transmission, fiber optics has been used successfully in several analog applications, notably television distribution, so occasional mention will be made of analog techniques also.

This book is intended for the senior or first-year graduate level, and a course using the entire book could be expected to take two semesters. A standard undergraduate background in communication engineering and physics is assumed.

The fast-breaking field of optical communication systems is becoming so important that an elective course at this level should be considered for any university curriculum preparing students for a career to involve a heavy content of communication technology. Students of communications in either universities or mid-career professional programs need an up-to-date understanding of how fiber optic communication is evolving in order to best do their job, whether as planner, engineer, researcher, or manager. I have aimed to provide in this book a unified and easily accessible review of that understanding.

To give some coverage of all the important aspects of fiber optic communication networks has required discussing a variety of topics from several disciplines. I had to portray how the capabilities of photonic devices constrain the world of network protocols and architectures, and how the network world defines the requirements on the components. To carry this off within the bounds of a single 500-page volume and a single one-year course has required that some detail be sacrificed at many points. Where a proper understanding requires a derivation so complex as to consume an inordinate amount of space, I have tried to hit just the high spots and have referred the reader to the most readable complete treatment I could find. For example, I have skimped on the application of queuing theory to calculating the performance of third-generation networks, having found that answers that are accurate enough for realism can be gotten more simply. The lengthy solutions of the wave equation in cylindrical coordinates needed to completely understand propagation in single- and multimode fibers has been compressed to simply the highlights of the analysis, the details being relegated to selected references that are particularly clear on the subject.

I hope I have provided the student with a book that will require a minimum of prerequisite material, one from which he or she will bring away a good quantitative understanding. To me, the important thing has been to give a physical feel for the principles at work. I would like the reader to learn from my book *why* a given thing works, not just *that* it does so. I have attempted to be as quantitatively correct as a given topic required when viewed in the world of real networks. I have rarely attempted an exact quantitative treatment when I felt that that particular aspect of real networks did not need to be computed to any greater degree of accuracy in order to be practically applicable.

The volume is divided into four parts, *Challenge, Building Blocks, Architecture,* and *Realization.* Part I tells what all-optical (third-generation) lightwave networks are and what may be expected from them in support of various applications. In Part II, the physical details of fiber propagation, couplers, filters, tunable lasers, modulators, and so forth are discussed along with the detection theory relevant to various forms of photonic receiver. The aim is to present the way in which the internal physics of each device leads to its external appearance as a "black box" that can be placed at the appropriate point in the entire system. This is the longest part of the book. Part III deals with the system issues, taking as the framework two viewpoints, the standard protocol layering model of communication networks, and the time-sequential network-control viewpoint, both suitably reinterpreted to take into account the new building blocks. Finally, in Part IV, a number of third-generation links, multipoints, and networks that have actually been built and used are discussed.

In teaching a course from this book, the instructor must assume that the student has had undergraduate electrical engineering courses in electromagnetism, some aspects of modulation and detection, at least some exposure to multilevel quantum devices, and is not dismayed by the simpler ideas from vector analysis, probability theory, and Fourier transforms.

For a course for students who have a good background in communication networks and protocols, but need to understand how the latest photonic technology is affecting communication systems, much of Chapters 1, 11, 12, and 13 will be sufficiently easy that a quick pass through them will suffice. Conversely, those students who have a good background in wave propagation, quantum devices such as lasers and photodectors, and threshold detection of lightwave signals, but need to understand how these phenomena are organized into a system, will not have much difficulty with Chapters 3, 7, 8, and much of 5.

To aid in teaching and learning the material, selected problems have been provided with each chapter. A solution manual is available from the publisher. Some of the problems for each chapter are designed to start the student thinking about issues before they are gone into extensively in later chapters.

Of necessity, writing this book has been for me an exercise in going back to school. At the beginning, I possessed a level of ignorance that is quite appalling to recollect. However, I think that the incompleteness of my knowledge allowed me to play the role of nonexpert and look for explanations that seemed to work for me. I hope that those I have chosen will work equally well for you.

While the responsibility for technical correctness is mine alone, it was the education I received from others that made this book possible. For providing me many hours of enlightenment in this fascinating field, I would particularly like to thank the members of my group at IBM Research: Rajiv Ramaswami, Kumar Sivarajan, Karen Liu, Frank Tong, Dave Steinberg and Michael Choy, and Kumar most of all. In addition, Profs. Larry Coldren of the University of California at Santa Barbara, Emanuel Desurvire of Columbia University, and Pierre Humblet of M.I.T. and Drs. Harold Stone and Chung-Sheng Li of IBM Research were patient and generous friends of this book.

I am also grateful to the following individuals who cleared up many matters for me, both technical and logistical: Greg Abbas, Tony Acampora, Josh Auerbach, Nigel Baker, Alan Baratz, Richard Bates, Ian Brackenbury, Chuck Brackett, Mike Brain, Jim Chiddix, Peter Cochrane, Frank Corr, John Crow, Bob Gallager, Rafael Gidron, Barry Goldstein, Inder Gopal, Walid Hamdy, Paul Henry, Goff Hill, Ken Jackson, Jeff Jaffe, Kristina Johnson, Ivan Kaminow, Leonid Kazovsky, Jeff Kravitz, Betty McCarthy, Jim Meditch, Cal Miller, Tran Muoi, Mike O'Mahoney, Bob Olshanksy, Rajesh Pankaj, Abe Peled, Paul Prucnal, Ken Reay, Tom Rowbotham, Marty Sachs, Dave A. Smith, Vince Tekippe, Pat Trischitta, and Jean Walrand. My guardian angel at Prentice Hall, Irwin Zucker, was generous beyond belief with my many demands on his time and patience.

Paul E. Green, Jr.
Mt. Kisco, NY

Fiber Optic Networks

Part I
CHALLENGE

CHAPTER 1

Fiber Optic Networks

1.1 Some Perspective

This is a book about fiber optic communication systems, what they are made of, and where they seem to be going as we approach the end of the century.

The field of fiber optic transmission [1-3] is still young and dynamic. It can be considered to have be only 25 years ago when Charles Kao, then a young engineer at the ITT Laboratory at Harlow in England, realized that one should be able to send high-speed messages down a narrow filament of glass, and did an extensive series of experiments to prove that this was so [4]. At about the same time, in a seemingly unrelated series of events, the semiconductor laser diode was invented [5].

The attenuation in Kao's early experiments was about 1 dB per foot and the light source was a crude light-emitting diode at visible wavelength. Today, nobody seems to find it remarkable that fiber being installed widely has 1 dB of attenuation per 5 kilometers of length and that commercially available near-infrared laser diodes can be modulated at rates exceeding several gigabits per second. Figure 1-1 shows the year-to-year scorecard that has been kept for some years on the bitrate-repeater spacing product for a single link [6].

The span of lightwave systems already installed ranges from the short distances within a computer or computer room to intercontinental undersea cable connections. Since the mid-60s, the lightwave communication story has been one of steady and exciting progress on many fronts. One invention after another propelled fiber optics into its present position as **the** technology of choice for all communication systems using fixed paths longer than a few meters. One can see the emerging outlines of a

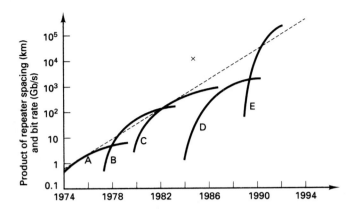

Figure 1-1. Growth history of single link fiber optic transmission capability expressed in product of inter-repeater spacing and bitrate. (From [7], © 1992 by *Scientific American*). (A) Multimode fiber at 0.8 μ wavelength, (B) Single-mode fiber at 1.3 μ, (C) Single-mode fiber, single-frequency laser, 1.5 μ, (D) Coherent detection substituted for direct detection, and (E) Erbium-doped fiber amplifiers substituted. The point marked × is an early laboratory result ([8]) using wavelength-division multiplexing.

two-layer future communication world in which essentially all communication between fixed points takes place by fiber rather than copper, while working off this backbone infrastructure there is a second world of *untethered* radio and infrared communication involving phones and terminals located in the car, airplane, boat, briefcase, purse and shirt pocket, or on one's wrist.

As we progress in the technologies of communication and computing, many of the people involved feel that something is out of balance; large increases in speed of electronics are becoming increasingly harder to get, while the capacity of fiber optic links is barely being touched. By straining very hard, silicon digital circuitry can be made to run at around one or two gigabits per second and gallium arsenide at from two to five times that speed. On the other hand, each fiber being installed today has three transmission passbands in the near-infrared (0.85, 1.3 and 1.55 μ). The letter μ stands for micron, 10^{-6} meter; one μ is 1000 nanometers (nm), or 10,000 angstroms (Å).

There are around 25,000 gigahertz of capacity in each of the three wavelength bands. This number is the equivalent of all the telephone calls in the United States today at the peak busy hour of the year. It is also roughly equal to one thousand times the entire RF spectrum used for radio communication in free space. And this capacity is there in each fiber that is going into the ground, being laid in ducts or strung on poles, potentially available for use at any time. Until very recently, few people seemed to be aware of this unused capacity, and even fewer were laying the system underpinnings to make use of it.

The relative abundance of computation resources and communication resources used to be the other way around. The communication link was a technological bottleneck between switches, terminals and computers, each of which internally had the speeds and low bit error rates that were unachievable on the connections between them. But in the last several years, as Figure 1-2 shows, there has been a reversal in the historical trend according to which costs of logic and memory were dropping much faster than those of communication. The lowered costs of doing the transmission by fiber have now presented the system architect and implementer with a rate of improvement of his communication costs that is faster than the rate of improvement of the cost of a given amount of compute power. Communication, the traditional bottleneck in system design, is now the enabler. An enabler of what? The next chapter attempts a partial answer by dealing with a few of the obvious areas in which new and better high-speed applications are being opened up by fiber optic systems. After these preliminaries we shall see how the different pieces of such a network function individually and then how they can function collectively in a complete network architecture.

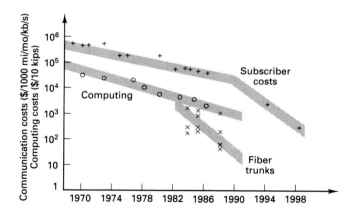

Figure 1-2. Cost history comparison for computation and for communication. The direct effect of the introduction of fiber technology can be seen with fiber trunk costs (lower curve), and the indirect effect on end user costs can also be seen (upper curve). (Courtesy of Brian Carr, IBM).

1.2 Link - Multipoint - Network

We can divide communication systems, from the topological viewpoint, into three general classes, as shown in Figure 1-3. These systems interconnect *nodes*, which in general could be switches, terminals, computers, workstations, telephones, etc. At the left is indicated a single point-to-point link, either unidirectional (*simplex*) or bidirectional (*duplex*). A transmitter at one end sends messages to a receiver at the other, and there are only two parties on the network. In the middle is indicated a one-to-many situation, which we shall arbitrarily call a *multipoint* network. Now there are N parties, but they are not all equal. One of them can send to a subset of all the other $N-1$, and perhaps these can reply. A special case of a multipoint is the *broadcast* situation in which the one station can send to **all** $N-1$ others.

These two topologies, the link and the multipoint, are subsets of the *network*, indicated at the right in the figure, a system in which each of the N nodes can reach each of the others. To be maximally useful, the network must provide this *any-to-any* connectivity, rather than a situation in which some nodes can communicate with only a subset of the $N-1$ others. We shall assume in the chapters ahead that this any-to-any pattern is the requirement. Sometimes a distinction will be made between links, multipoints and networks, and sometimes the catchall word *system* will be used to mean all three.

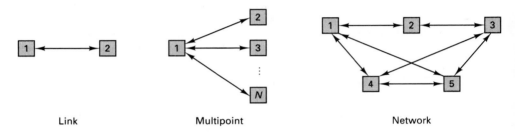

Link   Multipoint   Network

Figure 1-3. Three basic topologies, link, multipoint, and network.

An interesting thing seems to be happening now in the world of fiber optic communication. The preoccupation with the single long-distance link is shifting to one with multipoints and networks [9, 10]. The driving application for multipoint systems seems to be local fiber to the home or office, while that for networks is the "local area" or "metropolitan area" computer network.

For two and a half decades, the scorecard of Figure 1-1 has sufficed fairly well in capturing all the interesting things going on in lightwave communication. The ordinate represents the product of distance between repeater stations and the bitrate sent between them. The dominant challenge was to build point-to-point interoffice trunks for handling the growing number of telephones that needed to be connected. It wasn't the digital rate per telephone that was increasing, it was their number. Today, 1.7 Gb/s is a standard fiber interconnection speed between telephone company central offices, representing some 25,000 telephone conversations at 64 kb/s each.

The issues that are currently important in lightwave systems research and development, are focussed less on how to increase the bitrate and distance between repeaters on a single link, and more on how many bitstreams at different wavelengths (frequencies) can be supported, how broad the bandwidth of photonic amplifiers can be made, how fast a tunable receiver or transmitter can tune, how to interconnect purely optical networks by purely optical means, how fast a protocol can operate, how efficient and invisible to the user the communication resources can be made, and a

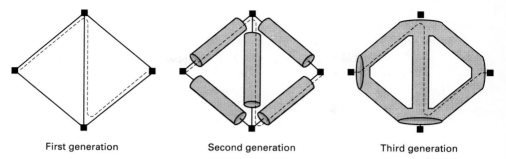

First generation   Second generation   Third generation

Figure   1-4. The basic idea behind the three generations of fiber usage in networks. For the first generation, fiber is not used. For the second generation fiber is used as a replacement for copper. For the third generation, unique fiber properties are exploited.

number of other such issues that relate to the multinode connection problems of multi-points and networks. The issue of cost is becoming important, as lightwave communication engineers strive mightily to bring about the same cost evolution that occurred with the photonic components in the compact disk player.

This recent change can be attributed to at least two things. First, it has long been the case that a new transmission technology is exploited first for long-distance communication and only later "trickled down" to the local-connection environment. This has been the case with microwave radio, coaxial cable, satellites, and now fiber optics. In the long-haul world, there is a preoccupation with getting large aggregations of traffic from one point to another: from central office to central office, or from one side of an ocean to the other. In the local world, connectivity is the important parameter, and the quantity of interest is the number of parties that can economically be connected, and not so much the bitrate per node.

Second, until recently the technology has not really been ready for aggressive new attacks on the problems of connecting users together in multipoints or networks. As will be seen in Part II, where we discuss these technologies, a great deal of usable art exists, and much of it is less than five years old.

Most of the pieces are now in place for a take-off of economical fiber interconnection involving many more nodes at higher speed per node than could be achieved by more classical approaches.

1.3   From Classical Use of Fiber to Nonclassical

Techniques for sending bits from one place to another constitute a *physical level*, an underpinning for all the other functions that a network does for its users. These other functions will be described in Part III as communication protocol layers. For the present discussion, we can say that the physical level has undergone a three-stage evolution. The first two stages represent the orderly or *classical* kind of evolution, but the third represents the *nonclassical* break with the past.   Figure 1-4 is an attempt to schematize this three-step progression.

At the left is seen an example of a first-generation network having a certain physical topology. The nodes are interconnected with copper links, drawn as narrow lines to indicate that their capabilities are limited. A logical connection between two nodes is shown as a dotted line. In the middle is schematized what happens to the same physical topology when the copper links are replaced one-for-one by fiber. At the right is shown what happens when a continuous physical fiber connection is available between all nodes. The way the links are diagrammed brings out the three principal things about a link that matter in designing the system:

- Its bandwidth,

- Its error rate, and

- Its delay.

The links are shown fatter in the center and right-hand diagrams in Figure 1-4 than in the left one to indicate that the bandwidth and error rate have improved, but are shown with the same length as before to indicate that the propagation delay has not changed.

The 25,000-GHz width of each of the three fiber passbands (at 0.85, 1.3, and 1.55 μ wavelength) is some ten orders of magnitude greater than that of voice-grade phone lines. The error rate, too, can be made to improve by about ten orders of magnitude, from around 10^{-5}, a standard figure for voice-grade modems fifteen years ago, to 10^{-15}, a working target today for computer-room fiber optic links. This improvement occurs because there are physically so many ways that noise can occur on a copper link, whereas on a fiber link we can control errors by keeping out extraneous light, by controlling transmitter and receiver internal noise levels, and by never letting the light signal get so weak that its random nature as discrete photons introduces more uncertainty into the detection process. From the physics point of view, copper links are more vulnerable to unwanted outside influences because moving electrons influence each other (and copper transmission links operate in a sea of moving electrons from other channels, ground loops, impulse noise, and so forth), whereas the moving photons of light transmission do not interact with other moving photons.

Another property of fibers will prove to be of some interest: their small size, which allows them to be bundled very tightly, in some cases with density as high as 10^6 fibers per cm^2. As we shall see, this can allow some interesting choices of the network physical topology.

Actually, the most important benefits of going to third-generation networks may not lie in such measurable parameters as bandwidth, error rate, attenuation, or packing density. There is an important user-oriented parameter that is being emphasized more and more today, especially in computer networks: *openness*. The term is usually taken to mean that the system is easy to use, easy to change, and can serve a wide variety of resources that are in use concurrently. As we shall see in succeeding chapters, many third-generation networks are essentially protocol-transparent − many different bitrates and formats may be accommodated at the same time. The networks can be very flexible topologically, so that adding a whole new set of nodes involves adding a few passive splitters and combiners and some short runs of fiber, rather than installing another switching node or making many new long cable runs. Such measures would be required if the network were based on space-division switches, to be discussed shortly. Perhaps reasons such as these explain why local-area networks such as Ethernet or token rings never suffered much competition from central switches such as PBXs (private branch exchanges).

It is customary to classify networks (of any generation) into three categories with respect to physical size. Thinking of the first generation at the left of Figure 1-4 as the traditional copper-based networks, one can categorize these as:

- *Local area networks*, LANs (up to, say, two kilometers total span), such as Ethernets, token rings, and token busses,

- *Metropolitan area networks*, MANs (up to, say, 100 kilometers), such as the telephone local exchange environment or cable television distribution systems, and

- *Wide area networks* WANs (up to thousands of kilometers), such as ARPAnet, the Research Internet, CSnet, such value-added networks as Tymnet and Telenet, and the thousands of commercial networks built on such architectures as System Network Architecture (IBM), Digital Network Architecture (DEC), and the Open System Interconnect (OSI) international standard.

The first-generation bitrate ranges from a few kilobits per second for older commercial networks based on voice-grade service up to a few megabits per second for coax-based local area networks. One particularly interesting first-generation system is the High Performance Parallel Interface (HIPPI) [11], which can be used as a short (25 meters) point-to-point 1 Gb/s parallel link, or can be switched.

In view of the better bandwidth and error rate of fiber optic transmission, the second generation of systems (middle of Figure 1-4) has evolved and deployment is proceeding . This generation is typified by such systems as the Fiber Digital Distributed Interface (FDDI) [12], which is essentially a large 100 Mb/s LAN, such MAN designs as IEEE 802.6 (the distributed queue dual bus − DQDB) [13], the Cyclic Reservation Multiple Access (CRMA) double bus [14], the Metaring/Orbit buffer insertion ring [15] and the National Research and Engineering Network (NREN), a wide area network [16]. FDDI is a slightly modified token ring network, DQDB is a bus, and NREN will use a faster version of packet switching. While all these techniques involve some detailed architectural improvements on their predecessors, they are essentially what one gets by the direct substitution of a fiber optic medium for copper. The speed has gone up, the error rate has gone down, there are a few protocol changes, but otherwise they are doing what they did before with copper.

In point of fact, fiber optics is not just a slightly better version of copper; it is so different quantitatively as to constitute a difference in kind. The medium is so remarkably different from its predecessors that to use it most effectively one must rethink much of the entire subject of networks.

The right-hand diagram in Figure 1-4 schematizes this third generation. One obvious difference between it and the second generation is that there is a minimum number of *electrical-to-optical* and *optical-to-electrical conversions*. Even more important is the notion that no node is obliged to carry traffic on behalf of other nodes. In the past this never made much difference. Once bitstreams got inside a node, whether it was an ARPAnet packet switch, an SNA communication controller, a token ring or FDDI node or Ethernet node, there was always plenty of electronic speed not only to handle the node's incoming and outgoing bits, but to look at and in some cases buffer and do some protocol handling of the traffic from any and all other stations in the network that wanted to use that node as a relay point. In the case of networks in which the physical medium was shared, such as Ethernets and token rings, each station had to see the bits from *all* other stations. As applications emerged that required high bitrates, the lowest possible delay (hopefully not much more than propagation time), and absolute integrity, the insertion of intermediate nodes running at hundreds or thousands of times the average bitrate of a single connection not only became unacceptable, it became technically infeasible. This is the famous *electronic bottleneck* that is being widely discussed today.

The electronic bottleneck occurs because the speed available from silicon and gallium arsenide electronics is disproportionately small compared to the total network throughput demanded by the new applications. This bandwidth is in turn small compared to that of the fiber.

Therefore, the third *nonclassical* generation discussed in this book envisions that the bandwidth, error performance, and propagation delay of fiber are to be made available along the entire path from one end of a connection to the other, as shown at the right in Figure 1-4. Such networks have been dubbed "all-optical" networks. There are two conversions between electrons and photons, but only two: at the transmitter and at the receiver. And the electronics involved can be responsible for the traffic needs of that node alone, without regard to the traffic of other nodes. This new freedom from the speed constraints of silicon and gallium arsenide will be seen, later in this chapter, to allow whole networks of extremely high throughput to be built using only modest speed electronics, and without large amounts of logical complexity and parallelism.

1.4   What an All-Optical Network Looks Like

The fiber optic networks that most naturally exploit the fiber's bandwidth and low error rate are those in which the different connections are made over a single physical medium, uninterrupted by optical-to-electronic conversions. Examples of such arrangements are those that use a central passive *star coupler*, as in Figure 1-5(A), or a serial bus, as in (B), or a tree (C). A star coupler is a device that broadcasts the merge of all the inputs to each output. The bus uses a tap on transmission and another on reception. It is possible to tie such individual networks together to form larger networks.

In the geometries shown, any transmission can be received by any receiver. Since there are no topological loops, there are no undesired multipath effects due to the existence of multiple routes by which light can travel from transmitter to receiver.

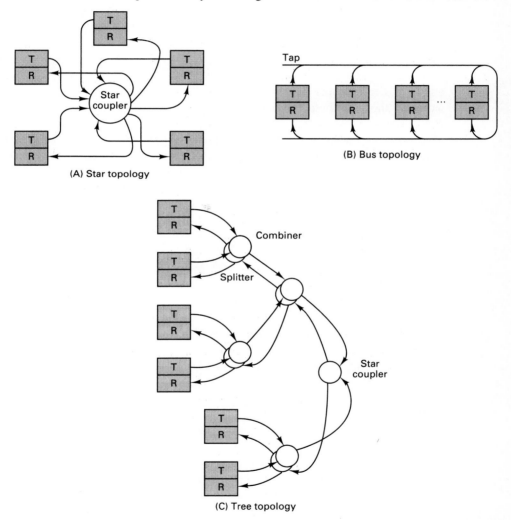

Figure 1-5. Some physical topologies of typical optical networks. (T = transmitter and R = receiver).

A significant price must be paid for escaping from the speed limitations of having electronics at points along the connection, and that is the limited amount of

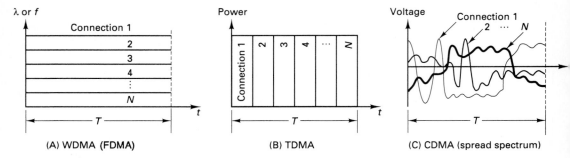

(A) WDMA (FDMA)   (B) TDMA   (C) CDMA (spread spectrum)

Figure   1-6. Three basic transmission formats. T is the bit duration.

optical power available. When the optical transmitter power is made available to all receivers, the *splitting loss* incurred in the process becomes a dominant issue. In principle, it should not be necessary to waste transmitted energy by directing it to all receivers, only one of which may make use of it. However, it has proved difficult to devise components that will allow this to be done, and the usual scheme is to suffer the splitting losses and try to recover them by photonic amplification.

With no loss of generality, the diagrams of Figure 1-5 show only one transmitter and one receiver per node, but it is clear that in principle, we may want to have several concurrently active connections into and out of any given node.

The available technology makes it possible to fit a very large number of simultaneous connections within the large available bandwidth of a single fiber. One can build a *multiple-access* network in which *dynamically* available connections between nodes use different optical wavelengths (wavelength division multiaccess − WDMA), different time slots (time-division multiaccess − TDMA), or by different signal waveforms (code division multiple access − CDMA). These options are shown in Figure 1-6. By "dynamic" we mean that the connectivity can be rapidly changed. In this book we shall not distinguish between the terms WDMA and FDMA (frequency division multiple access), instead using "WDMA" to refer to both. To date, most of the work on third-generation optical networks involves either putting all the different channels on different wavelengths or putting several on each of a number of wavelengths as RF microwave subcarriers.

Depending on the technology available, the dynamic association of one transmitter with a particular one or more receivers can be done using tunable transmitters, tunable receivers, or both. Or, if wavelength tunability must be avoided, we may disregard the fact that a given transmitter's signal is heard immediately by all receivers (including the intended one) and deliberately do "multihop" store and forward through a series of intermediate nodes carefully arranged so that every node can eventually reach every other node [17, 18]. Since each of these nodes performs a conversion from the optical domain to electronic and back to optical, such a network is not truly "all-optical."

There are several component classes that collectively realize the networks of Figure 1-5. Fiber links, couplers, and splitters distribute the light signals about the network.   The transmitters are usually semiconductor laser diodes, since each such device provides usably high output power over a sufficiently narrow optical bandwidth and can be conveniently modulated with the information bitstream.   More complex laser diodes can even tune over a significant wavelength span.   There are several options for digitally modulating the transmitter laser, notably amplitude-shift keying (ASK), often called on-off keying (OOK), frequency-shift keying (FSK), and phase-shift keying (PSK).   The choice of receiver designs is quite wide; there can be coherent detectors that observe the frequency and sometimes the phase of the incoming light signal, or direct detection in which only the received power level is observed. Because, unlike coherent detection, direct detection is essentially non-wavelength selective, the light must first be passed through some sort of narrowband optical filter that passes the desired bitstream but rejects signals from unwanted transmitters. Tunable filters of many kinds exist, and because these have become commercially available earlier than either tunable lasers or coherent receivers, many early practical lightwave networks use fixed-tuned laser diode transmitters, and receivers consisting of tunable optical filters followed by direct detection.

As mentioned, a significant challenge with optical networks is the splitting loss suffered in the couplers and splitters that distribute the light around the network.   For the LAN and MAN distances, path attenuation is a minor factor compared to splitting loss.   For example, an $N \times N$ star coupler splits the power coming into any one of its input ports equally among the N output ports.   This is serious if the number of nodes is too large and the bitrate too high, because then the amount of received energy per bit may not be sufficient to provide reliable enough detection to achieve the desired bit-error probability.   This splitting-loss problem existed, of course, in the earliest days of radio, but was soon forgotten once the triode amplifier tube was invented.   Similarly, one of the enabling technologies for the third generation of lightwave networks is the recently invented reliable and economical lightwave amplifier, which increases the power of a light input over a wide bandwidth, with negligible distortion, and without introducing too much noise of their own − the very problems that were met eighty years ago at the dawn of electronic communication.

The advent of cheap, reliable, purely photonic amplifiers suggests that it should be possible eventually not only to build LANs and MANs in which the paths are entirely optical, but to interconnect these to form WANs.   How easy is it to build third-generation networks as large as desired, extending wherever desired?   We can avoid intervening optical-to-electronic conversions only if we have a *dark fiber* path available.   This term refers to a fiber link whose ends are accessible to the system owner for him to send and receive his own light.   At LAN distances, this is usually no problem, since the fiber lies within the owner's own premises.   Dark fiber links for MAN distances (up to, say, 50 kilometers) can be gotten from local telephone companies, cable installers, or special contractors, depending on the practices in the local area.   For WAN distances, however, dark fiber has not been offered yet.   Thanks to amplifiers, the problems are not technical.

The network discussions of this book deal predominantly with situations that are most easily realized in LAN and MAN situations. How to achieve a third-generation WAN by interconnecting more localized subnetworks that are in the form of third-generation LANs and MANs is a very new research issue. We shall discuss what little is known. It will be observed that the third generation arrives with LANs first, then MANs, then WANs.

1.5   The Space-Division Switching Alternative

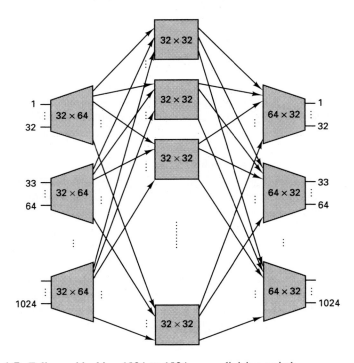

Figure   1-7. Fully nonblocking 1024 × 1024 space-division switch.

As the reader may have perceived, there is another way to avoid optical-to-electronic conversions except at the two ends, and that is to use spatial switching of the light path rather than the dynamic wavelength-division, time-division, or code-division approach. This *photonic switching* approach has been the subject of much research, but unfortunately no commercially viable technology has been found for building a low-attenuation low-crosstalk purely photonic switch module of size greater than about 16 × 16 [19, 20]. This is true even for that subclass of photonic switches in which the switch is thrown not by optical inputs but by electrical inputs, with the path through the switch still remaining optical. Perhaps some day practical photonic switches with hundreds or thousands of ports will compete with the optical networks described in this volume, but at this writing other technologies are so much farther advanced than the purely optical space-switching technology that the latter will not be discussed. However, we shall discuss later the interesting possibility that small optical space-division switches can be combined into a hybrid space- and wavelength-switching system.

Electronic subgigabit- or gigabit-per-second switch designs have been proposed and built [21-23]. Conversion from photons to electrons and back again is made at the switch in the middle of the network so that the switching can be done electronically. This artifice does a lot to avoid the electronic bottleneck, since the electronics in each node handles only that node's bit rate.

As is well known [24], we can achieve the same *fully nonblocking* character with a switch that third-generation lightwave networks achieve by using designs such as the three-stage arrangement illustrated in Figure 1-7. "Fully nonblocking" means that, no matter what pattern of connections is already set up in the switch, any new request for a connection between an unused input port and an unused output port can be satisfied without rearranging any of the connections already in place.

The 1024 $\times$ 1024 switch in the figure achieves this with an input stage of thirty-two 32 $\times$ 64 crosspoint switches (shown in a column), followed by a stage of sixty-four 32 $\times$ 32 switches, followed by an output stage of thirty-two 64 $\times$ 32 modules. In general, such an arrangement requires $6N^{3/2}$ crosspoints, which is about 200,000 for a 1024 $\times$ 1024 switch. Because high-speed silicon or gallium arsenide technology is so expensive, it is customary to make each of the many interconnection links ($4N$ for full nonblocking) in the form of a highly parallel bus, so that low-speed electronics may be used everywhere within the switch except at the input and output ports where conversions between serial and parallel are made. This combinatorial explosion with N, its cost consequences, and problems such as connection-request queuing in the shared controller that sets up the connections, and constraints on topological flexibility of the network, are all in contrast to the structural simplicity and flexibility of the networks to be described, in which, in effect, the switching takes place in small, single-threaded pieces, one in each node.

1.6 Throughput Capacity of Fiber Optic Networks

We close this introduction by presenting Figure 1-8, which quantifies the capacity of fiber optic networks using passive star-coupler distribution of the signals and any-to-any connectivity. This figure provides some perspective on the great increase in capacity that is possible by moving from the second to the third generation. By *capacity* is meant the total throughput of the network, the sum of all the bitrates of all the connections that can be active at any one time. The figure focusses on the limits on total capacity that are imposed by various technologies to be discussed in the next few chapters.

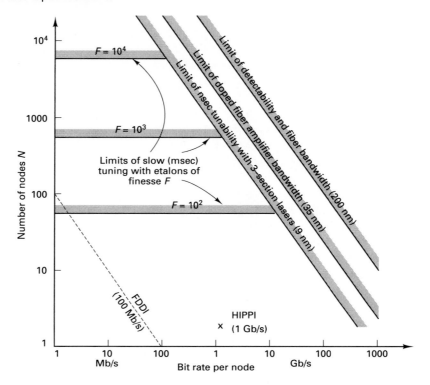

Figure   1-8. Limits on total throughput capacity of a fiber optic communication network. The dashed line shows FDDI, a contemporary second generation fiber optic network. The × shows HIPPI, an all-copper parallel gigabit interface.

It is interesting to note that we get about the same answer for capacity by considering the available bandwidth and the available power [25]. Given a 200-nm total available bandwidth (25,000 GHz at 1.5 μ), and assuming either a time-division or a wavelength-division modulation format that allows the channels to be crowded together with a spacing of about 2.5 Hz per bit per second, the number of nodes that can transmit simultaneously at 1 Gb/s is about 10^4. The other way of looking at it starts by assuming that each node transmits 1 milliwatt of laser power (7.5×10^{15} photons per second for light at 1.5 μ), and that each receiver requires 750 photons per bit, a number that will assure a bit error rate of about 10^{-9} with the better commercial-grade photodetectors. If each node transmits at 1 Gb/s, then clearly there are not enough photons to go around if the power must be split among more than 10^4 of them.

We can take this practical limit of 10^4 nodes at 1 Gb/s and trade off bitrate against number of stations, as in the far right diagonal line of Figure 1-8. The figure also plots various other limits on achieving a given number of nodes at a given bitrate per node. As will be seen in Chapter 6, commercially available doped fiber photonic amplifiers have a bandwidth of about 35 nm, considerably smaller than the 200 nm of the fiber. This constraint is plotted as the next diagonal line, and the 9-nm tunability range of the commercially available three-section tunable laser diodes to be discussed in Chapter 5 as the line next to that. The promise of these lasers is not that they tune over a wide range, but that they can tune in nanoseconds.

On the other hand, the Fabry-Perot tunable filter, or *etalon* (to be discussed in Chapter 4), can be made to have essentially unlimited tuning range, but a relatively slow tuning time, of the order of a millisecond. The channel selectivity of these filters is expressed by a selectivity-related figure of merit called the *finesse*. The nature of the off-resonance skirts of the filter's frequency response is such that the number of channels is usually limited by the crosstalk from adjacent channels, and not by the total bandwidth available. Thus, the limits for systems based on etalons are horizontal lines in Figure 1-8. This limit on crosstalk occurs if the number of channels exceeds about two-thirds the finesse [26].

Also shown in the figure is a line representing probably the most advanced second-generation system, FDDI, having a bitrate of 100 Mb/s totalled over all the stations on the ring.

It is seen from the figure that the throughput capabilities of fiber optic networks are on a level truly different from anything that has come before. Other advantages will be discussed later, such as protocol transparency, potentially higher reliability (because so much of the network is passive and unpowered), and potentially low cost (because of the structural simplicity). But the main point is that the capacity numbers shown in Figure 1-8 promise a way of doing networking that exploits in an imaginative way the bandwidth potential of the fiber optic medium and offers a physical base on which to build the applications of the future.

1.7 Preview of Later Chapters

The next chapter discusses applications of third-generation networks. Part II then introduces each important type of photonic device, treating each as a "building block" or "black box," whose external properties define how it can be employed to construct the total system discussed later in Part III. Part II begins with Chapter 3, which analyzes fiber transmission, couplers, and splitters (including such issues as attenuation, dispersion, nonlinearities, and polarization considerations). Tunable filters are next presented (Chapter 4), both as key elements in wavelength-division systems and as the preliminary material on resonant structures necessary for the understanding of fixed-tuned and tunable semiconductor lasers, the subject of Chapter 5. The treatment of stimulated emission introduced in that chapter leads naturally to a discussion in Chapter 6 of photonic amplifiers, which can be thought of as lasers that have been modified to amplify rather than to lase.

Chapter 7 describes how the binary information stream to be transmitted can be modulated onto the optical transmitted signal. In Chapter 8, we switch from the lightwave transmitter to the receiver, which is described from both the component and the detection-theory points of view.

Chapter 9 discusses systems in which each laser is driven not by one bitstream, but by an additive frequency-division sum of microwave subcarriers, each modulated with its own bitstream. Chapter 10 deals with problems of stabilizing the wavelength of various components.

Part III is devoted to networking issues, which tend to manifest themselves in issues of topology and protocols. Chapter 11 deals with the many topological issues in system design. Ways of deploying couplers, taps, and amplifiers are introduced; then the discussion turns to the important differences in the various topologies of choice. Topological aspects of various hybrid schemes, such as wavelength-space or wavelength-time hybrids, are seen to offer ways of overcoming component technology shortcomings. Chapter 12 takes the basic idea of an access path, representing the logical topology of connection between users, and develops it into the layered architecture model of computer communication and also the network control sequence model. The layer model is then reexamined, layer by layer, in the light of the large impact caused by introducing third-generation optical technology at the lowest (physical) protocol level. In Chapter 13 various multiaccess protocols by which the access path is set up are discussed and quantitatively compared. These protocols reside in the Media Access Control (MAC) portion of the physical protocol layer.

Part IV presents the synthesis of all the earlier ideas into total systems. By this time the reader has developed a sufficient background that it is only necessary to give, as case histories, each of several third-generation lightwave systems that have been built. Chapter 14 discusses single long-haul terrestrial and undersea links, Chapter 15 developments in fiber-to-the-home or office, and Chapter 16 recent computer network implementations.

The various symbols used are listed in the Appendix.

1.8 Problems

1.1 DS-0 is the standard *pulse-code-modulation* (PCM) format for voice. It has a sampling rate of 8000 per second, with 8 bits per sample. What was the first year that fiber technology became capable of supporting 2000 DS-0 voice-grade channels with the repeaters being spaced more than 100 kilometers apart?

1.2 A common optical component is the equal-power splitter, which splits the incoming optical power evenly among M outputs. By reversing this component, we can make a combiner, which can be made to deliver to a single output the sum of the input powers if multimode fibers are used, but which splits the power incoming to each port by a factor of M if single-mode fiber is used. (A) Compare the loss in decibels between the worst-case pair of nodes for the three topologies (star, bus, tree) shown in Figure 1-5 if the total number of nodes is $N = 128$ and multimode fiber is used. Neglect attenuation on the links. Assume that, for the tree, there are 32 nodes in each of the top two clusters and 64 nodes in the bottom one. (B) What would these numbers become if single-mode fiber were used? (C) How would you go about reducing the very large accumulated splitting loss for the bus?

1.3 A lightwave communication link, operating at a wavelength of 1.5 μ and a bitrate of 1 Gb/s, has a receiver consisting of a cascaded optical amplifier, narrow optical filter, and photodetector. It ideally takes at least 130 photons per bit to achieve 10^{-15} bit error rate. (A) How many photons per bit would it take to achieve the same error rate at 10 Gb/s? (B) At this wavelength, 1 milliwatt of power is carried by 7.5×10^{15} photons per second. What is the received power level for 10^{-15} bit error rate at 1 Gbs? Express the answer in dBm (decibels below 1.0 milliwatt). (C) At 10 Gb/s?

1.4 Consider a switch of the type shown in Figure 1-7, having 4096 input ports and 4096 output ports. (A) How many crosspoints are required to achieve nonblocking performance? (B) How many connecting links?

1.5 Assume a transmitter power of 1 milliwatt, a detector sensitivity of 750 photons per bit, and no loss during propagation. (A) What is the maximum number of 10 Gb/s wavelength-division nodes one could have in a star-connected all-optical network for tunable-filter finesses of 100, 1000, and 10,000? (B) What are these numbers for a time-division such network, assuming the 1 milliwatt to be a limit on the average power? (C) What would these numbers be if the 1 milliwatt were a peak power limitation?

1.9 References

1. S. E. Miller, I. P. Kaminow, ed., *Optical Fiber Communications − II*, Academic Press, 1988.
2. C. Lin, ed., *Optical Technology and Lightwave Communication Systems*, Van Nostrand Reinhold, 1989.
3. F. Tosco, ed., *CSELT Fiber Optic Communications Handbook - Second Edition*, TAB Professional and Reference Books, Blue Ridge Summit, PA, 1990.
4. C. K. Kao and G. A. Hockham, "Dielectric-fiber surface waveguides for optical frequencies," *Proc. IEE*, vol. 113, no. 7, pp. 1151-1158, 1966.
5. R. Dupuis, N. Holonyak Jr., M. I. Nathan, and R. H. Rediker, "Papers on early history of the semiconductor laser diode," *IEEE Jour. Quan. Elect.*, vol. 23, no. 6, pp. 651-695, 1987.
6. S. E. Miller and I. P. Kaminow, eds., *Optical Fiber Telecommunications−II*, Academic Press, chap. 21, pp. 781-831, 1988.
7. E. Desurvire, "Lightwave communications: The fifth generation," *Sci. American*, vol. 266, no. 1, pp. 114-121, 1992.
8. N. A. Olsson, J. Hegarty, R. A. Logan, L. F. Johnson, K. L. Walker, L. G. Cohen, B. L. Kasper, and J. C. Campbell, "68.3 km. transmission with 1.37 Tbitkm/s capacity using wavelength division multiplexing of ten single-frequency lasers at 1.5 micrometers," *Elect. Ltrs.*, vol. 21, pp. 105-106, January, 1985.

9.   N. K. Cheung, K. Nosu, and G. Winzer, eds., "Special Issue on Dense Wavelength Division Multiplexing for High Capacity and Multiple Access Communications Systems," *IEEE Jour. Sel. Areas in Commun.*, vol. 8, no. 6, August, 1990.

10. C. A. Brackett, ed., "Special Issue on Lightwave Systems and Components," *IEEE Commun. Mag.*, vol. 27, no. 10, October, 1989.

11. High-performance parallel interface – Mechanical, electrical, and signalling protocol specification, American National Standards Institute X3.183, 1991.

12. F. E. Ross, Fiber Digital Data Interface - Token Ring Media Access Control, American National Standards Institute X3.139, 1987.

13. "Draft IEEE Standard 802.6, Distributed Queue Dual Bus (DQDB) Metropolitan Area Network (MAN)," *IEEE Computer Society*, June, 1988.

14. P. Zafiropulo, "Gigabit local area network," *IEEE Network Magazine*, vol. 6, no. 3, 1992.

15. I. Cidon and Y. Ofek, "Metaring, a full-duplex ring with fairness and spatial reuse," *Conf. Record, IEEE INFOCOM*, pp. 969-981, 1990.

16. "Gigabit network testbeds," *IEEE Computer Magazine*, vol. 23, no. 9, pp. 77-80, 1990.

17. A. S. Acampora, "A multi-channel multihop local lightwave network," *Conf. Record, IEEE Globecom*, pp. 37.5.1 - 37.5.9, November, 1987.

18. M. G. Hluchyj and M. J. Karol, "ShuffleNet: An application of generalized perfect shuffles to multihop lightwave networks," *Conf. Record, IEEE INFOCOM*, pp. 4B4.1 - 4B4.12, 1988.

19. J. Baldini, ed., "Special Issue on Photonic Switching," *IEEE Commun. Mag.*, vol. 25, no. 5, May, 1987.

20. J. E. Midwinter and P. W. Smith, eds., "Special Issue on Photonic Switching," *IEEE Jour. on Sel. Areas in Commun.*, vol. 6, no. 6, August, 1988.

21. CXT-1000 Integrated Communications System, Ancor Communications, Inc., 1991.

22. H. Ahmadi and W. E. Denzel, "A high-performance switch fabric for integrated circuit and packet switching," *Conf. Record, IEEE Infocom*, pp. 1A2.1 - 1A2.9, 1988.

23. H. Ahmadi and W. E. Denzel, "A survey of modern high-performance switching techniques," *IEEE Jour. Selected Areas in Commun.*, vol. 7, no. 6, pp. 1091-1103, 1989.

24. J. Y. Hui, *Switching and Traffic Theory for Integrated Broadband Networks*, Kluwer Academic Publishers, 1990.

25. P. S. Henry, "High-capacity lightwave local area networks," *IEEE Commun. Mag.*, vol. 27, no. 10, pp. 20-26, October, 1989.

26. P. A. Humblet and W. M. Hamdy, "Crosstalk analysis and filter optimization of single- and double-cavity Fabry-Perot filters," *IEEE Jour. Sel. Areas in Comm.*, vol. 8, no. 6, August, 1990.

CHAPTER 2

Applications
and Their Requirements

2.1 Overview

In this chapter we continue to distinguish between links, multipoints, and networks, and also continue to emphasize networks. We shall attempt here to define the technical requirements on third-generation systems by inference from several applications that already require third-generation network architectures in order to be adequately supported.

When an entire technology class has not completely emerged, it is risky trying to be too specific about all possible uses to which it will be put. When that technology offers such aggressive improvement in performance parameters as third-generation lightwave communication systems offer, we are on particularly shaky ground. All we can say is that it will probably be those applications that we can only dimly imagine today that will be enabled, take off, and ultimately dominate the system usage of the future. This has always been the case in the past. The originators of the ARPAnet [1] thought its principal application would be "resource sharing," the ability of everyone to be in touch with some supercomputer or data file. Instead, the remarkable success of the ARPAnet was that it allowed everyone to be in touch with everyone else. E-mail turned out to be the dominant application, and the connectivity it provided, so different in kind from anything connected with the conventional thinking of the time, created a social revolution.

In spite of our inability to make correct or even half-correct predictions, we have to start somewhere. In this chapter we shall try to converge on what the technical requirements are, as inferred from the gigabit applications that are real today.

2.2 Gigabit Applications

The most successful networks, it has been wisely pointed out [2], have not been those that deliver the most bandwidth to a few users but those that provide connectivity in an easy-to-use way to the widest possible community of users. "Easy-to-use" means affordable, among other things. We can think of several examples. In technical terms, Ethernet is a bandwidth-starved technology, with only 10 Mb/s to share across all the attached users, and with a multiaccess protocol having serious performance limitations. But its ease of attachment, low cost, and the flexible access it provided to a wide user group made it a historic success story of the first order. Conversely, the idea of Picturephone, basically a high-bandwidth concept, offered too much of the wrong thing. It offered TV-level personal interaction of a constrained sort; the application was very narrowly focussed on images of persons.

As will be seen in the remainder of this book, third-generation optical-system technology and architecture is, at this writing, at a stage of evolution where the costs are still high. Yet there is plenty of bandwidth in the fiber medium, fibers are being laid down everywhere, and the simplicity of the third-generation technology offers the promise of great cost improvability. All-optical solutions will thus be able to support modes of communication having an unprecedented degree of connectivity and ease of use.

However, this will take time. Before pervasive access to the 25,000 GHz of fiber bandwidth is widely available, along with the very low-cost photonics technology for which the compact audio disc offers an existence proof, things must evolve and markets must develop.

Therefore, the prospect is that all-optical solutions will first be deployed for the high-end "gigabit applications" that are already appearing, and that these will spread outward in breadth of user community and downward in cost.

2.3 Requirements Parameters

One of the frustrating things about communication-system innovation is that we are usually unsure what to optimize. Other branches of engineering often have an easier job of it. The semiconductor-device person knows that smaller lithographic scale increases speed, reduces power needs, and improves cost. He or she can concentrate on linewidth, knowing that almost all other performance parameters come back to this one variable.

In communications, it is not so clear. Performance is a vector, not a scalar. In different circumstances, different components of the vector dominate. Even the statement made before that pervasiveness is a more important parameter than bitrate will be inapplicable in some situations. Nevertheless, this vector is of finite dimension, and we shall have to be content with making a short list of its components and discussing their relative importance.

Pervasiveness

Ideally, it should be possible to connect to another user anywhere. Every day, about 800 miles of new fiber is being installed in the U.S. alone (counting each strand in a bundle as a separate fiber) [3]. By the end of 1990, over five million miles of fiber had been installed, 50 percent of this installed base being "dark," meaning that no terminal equipment had been attached. While 2 million miles is in intercity trunking of the major carriers, 3 million miles is in the local exchange carriers and in the so-called "alternate-access" carriers. This latter group is putting fiber inside subway tunnels, inside the hollow uppermost (ground) conductor on high-tension power lines (the so-called *optical ground*), and inside unused oil and gas pipelines.

Another major provider is the cable television industry. Where the telephone companies have been investigating *fiber to the home*, and have, for the moment decided that *fiber to the curb* is more attractive economically [4], the CATV (community antenna television) industry has decided similarly [5]. Fiber is being run from the *head end* (often a satellite ground station serving tens of thousands of homes) out to a "fiber node" on a utility pole, from which the distribution to homes is by coax. The battle over who will provide ubiquitous access by optical fiber all the way out to homes and office buildings is being waged between the telephone industry and the cable industry. It promises to be an epic one, with the future prosperity of both industries very much at stake.

Whether we are talking about telephony's fiber out to the curb or CATV's fiber out to "fiber node," the fact remains that fiber, much of it of the single-mode type, will be in place in the next several years reaching out to within a kilometer of most homes and offices in medium and large cities. Fiber is also being installed at a rapid rate contiguously to the end user, for example in most new office skyscrapers, industrial parks, and university campuses. Many existing such structures are being retrofitted with fiber.

On this increasingly pervasive base it is possible to build a huge infrastructure of third-generation multipoints and networks, provided the entire optical spectrum can be accessed. As matters now stand, the owners of the city-wide and intercity fiber, whether telephone companies or cable companies, are using one or two Ghz of the roughly 25,000 to provide highly structured digital formats aimed not at developing the full potential of the medium, but at developing the maximum amount of revenue. As mentioned in Chapter 1, as the technology of all-optical networks becomes more cost-effective, as the applications build up that can only be satisfied by all-optical solutions, and as the provision of dark fiber becomes identified as a question of the common interest, we may safely predict that the present reluctance to consider dark fiber a desirable standard tariffed offering will gradually erode. As MIT's Michael Dertouzos wrote recently [6], "The telephone companies will play a major role in laying down the physical fibers for the network and so bear the responsibility for providing flexible information transport services.... But neither the telephone companies nor any other centralized body should decide what common communication conventions should be offered."

Thus, there is the requirement of pervasiveness of connections, and this pervasiveness is being provided rapidly with fiber optic technology. In addition to the fiber infrastructure, a second infrastructure is being installed, one affording connectivity to *untethered* points (cars, boats, individuals) using technologies such as cellular phones, pagers, and so forth, that employ radio-frequency and infrared free-space transmission. Thus, what is clearly emerging is a two-tier world with communication between fixed points using fiber and beyond that to almost anywhere using radio and IR.

Reachability

It does little good to have the physical transmission medium installed pervasively if means are not provided for users to reach each other with minimum impediments. The ultimate in any-to-any reachability is, of course, the telephone system, which is really an interconnection of a great many individual telephone systems. Any subscriber has connectivity with any other subscriber in the world.

Similarly, with third-generation networks, means will have to be found to connect individual LAN or MAN systems together seamlessly, and suitable directory function will have to be provided for any user who has the requirement to connect to any one of a large number of other users. Not all applications require a high degree of any-to-any reachability, but a significant number do, as we shall see.

Another aspect of reachability concerns how many other parties it is desired to reach at the same time. As pointed out earlier, the importance of packet switching is not in providing line economies by exploiting the burstiness of user traffic by "time-slicing" the capacity so as to share it across users. Its true importance is in using time-slicing to allow one node to be in "simultaneous" session with many others, all these conversations being carried by one data stream entering and one leaving the node. This ability to support "many logical ports per physical port" is required by most computer applications. Today, even small desktop computers have multitasking operating systems (or soon will); the communication network certainly must provide the same capability of deriving *virtual* concurrency from *real* sequentiality.

Bitrate

In the first chapter the unprecedented carrying capacity of a single fiber passband was mentioned, roughly 25,000 GHz. This is large enough for a single third-generation LAN or MAN network to support very large numbers of nodes at bitrates up to 1 Gb/s, as Figure 1-8 showed, something that is impossible with first- or second-generation networks. However, this is not a large enough base on which to build a total WAN communication fabric, either as a public network or as a more limited private one, for example a corporate network. In order to do that, it will be necessary to partition the system into separate subnetworks and reuse the same bandwidth in the many subnetworks.

The bitrate requirement is thus closely coupled to the reachability requirement. The higher the bitrate per user, the smaller the number of users that can be supported within one network.

Latency

Many applications have strong time sensitivities, and since there is not much we can do about the propagation latency, these applications will require minimizing the number of round-trip protocol handshakes, minimizing accesses to rotating storage (where milliseconds can be expended), and maximizing speed of execution of the communication code and minimizing software pathlengths in this code.

Error rate

Some applications place strong limits on the maximum tolerable error rate; with others this is not the dominant concern. Fortunately, as will be seen in Chapter 8, with fiber optic transmission it is possible to achieve arbitrarily small error rate, without special coding, by improving the link budget. This is in contrast to the more familiar telephone-line or coax situation, in which electromagnetic interference of various sorts placed a floor on the error rate that can be achieved without extensive coding. For lightwave networks the errors will usually be manmade, being caused by such things as buffer overflow, bit-pattern dependencies of clock sync and other such artifacts. In other words, errors are caused by randomness in usage of the network rather than randomness due to natural phenomena.

Availability

Nonstop operation is an imperative for some applications, for example certain on-line medical applications. In the last few years, as more and more enterprises "bet their business" on the continued availability of processing power and data records, nonstop operation has changed from a niche requirement to a majority requirement. Fortunately, all-optical networks have an inherent availability advantage, because the network proper is entirely passive and unpowered (as can be seen from Figure 1-5).

Security and privacy

This requirement is one for which the all-optical systems of Figure 1-5 do not provide a good solution. Even the bus architecture shown operates on the broadcast principle, producing a serious security exposure. Thus, for the all-optical networks in the figure, the need for encryption at higher protocol levels is much more urgent than it is for those first- and second-generation systems that are typified by mesh connections in which one node can freely receive only a small fraction of all the traffic in the network.

2.4 Requirement Differences: Links, Multipoints and Networks

From the requirements point of view, it seems clear that one profound difference between links, multipoints and networks is that the class of applications that directly influence the system design becomes broader in progressing through this sequence.

Links

For links, typified by the undersea cable [7] or the intercity trunk, essentially the only requirement is to pass bits at the maximum possible rate over the largest unrepeatered distance, and with the minimum possible error rate. Whether the bits are voice bits, data bits, TV bits, or any other kind of bits has been lost in the very large traffic aggregation that has taken place.

As pointed out in Chapter 1, most of the brief history of the fiber optic communication era has been one of improving the bitrate and repeater-distance parameters. Therefore, the literature on the link requirements and link technical solutions is considerably more copious than that for multipoints and networks. References such as [8-10] and also [7] provide a complete background for the interested reader.

Multipoints

The dominant multipoint application is one-to-many multimedia distribution. This primarily involves delivery of entertainment to the home, rather than services to businesses. There are certain niche exceptions, such as the distribution of stock quote information, but they are very small niches compared to the home TV, music, and eventually multimedia entertainment and education applications. Thus, while the link serves a totally unfocussed application set, the multipoint serves a well-focussed set.

The bitrate requirements are considerable. Uncompressed digital NTSC television requires 100 Mb/s, compressed NTSC 1.5 Mb/s, uncompressed HDTV 1.2 Gb/s, and compressed HDTV at least 150 Mb/s [11]. Compact-disk-quality stereo requires 4.32 Mb/s uncompressed. With multipoints pervasiveness of the service is quite important, latency is not an issue, and rapid switching speed (such as would be required for packet switching) is not an issue.

Networks

It is with networks that the entire array of gigabit and subgigabit applications come into view, along with the specific technical requirements posed by each.

A small handful of applications are widely thought of as today's genuine *gigabit applications*, meaning that each user requires a bitrate of the order of 1 Gb/s. Table 2-1 lists the four most frequently mentioned, along with a gross statement about most of the requirements parameters listed in Section 2.3.

Table 2-1.   Requirements dictated by gigabit applications.

Application	Bitrate required (Gb/s)	Pervasiveness and reachability	Packet switch required?	Low latency required?	Error rate required
Visualization	>1	Low	No	Yes	10^{-9}
Medical image	<1	Low	No	No	10^{-15}
Conferencing	1	High	Yes	No	10^{-9}
CPU interconnect	1	Low	Yes	Yes	10^{-15}

We now briefly discuss each of these four network applications, provide references for further reading about each one, and try to get at the technical requirements for third-generation networks.

2.5 Supercomputer Visualization

As supercomputers came to be more widely used by chemists, physicists, meteorologists, and other investigators, at least three things became clear.

First, the models became so accurate that it was usually easier and cheaper to do the experiment in the computer rather than build the real experimental structure and take data on it. Thus the computer became the laboratory.

Second, real laboratory experiments have always involved an interactive process between experimenter and experiment, with measurement equipment and conditions of the experiment being actively manipulated in real time as the experiment was run. Therefore, it became the norm that there would be real-time interaction between the investigator and his executing supercomputer program.

Third, it was found that the ability of the creative mind to understand the data and to get new inspirations about its meaning was tremendously enhanced if the results were output as images (preferably in color and as motion video) rather than as numbers [12].

All three of these add up to a requirement for very small latency and gigabit rates. (At 2000 × 2000 12-bit color pixels per frame, and 24 frames per second, this adds up to a little over 1 Gb/s.) The error-rate requirement is not extreme, and the system need be only pervasive enough to support several dozen concurrent users per supercomputer. Today, the communication requirement is usually served by the 1 Gb/s High Speed Parallel Interface (HIPPI) [13], a four-byte-parallel point-to-point copper connection.

The supercomputer visualization requirement is not just a matter of one supercomputer supporting a fixed set of workstations. Not only may several supercomputers be involved in one unit of work at the workstation, but there may also be the need to support pre- and postprocessors with the network. For example, it is common to provide a back-end processor for the supercomputer in the form of a commercial mainframe to do data archiving and data management support functions, while a front-end machine, such as the IBM Power Visualization System [14], converts numbers into pictures. These interconnections are typically at 1 Gb/s.

2.6 Medical Image Access and Distribution

The main problems in making use of medical images are database size, speed of access, and the lack of remote connections [15]. Hospitals of any size are drowning in X-ray and other images that cannot be thrown away and that need to be accessed, sometimes at very rapid speeds. In a life-threatening emergency, it is common to convene several physicians to look at a rapidly sequenced set of, say, a half-dozen $2000{\times}2000$-pixel images in color or very fine gray-scale in order to make decisions in a few minutes.

Consultant advice is accessed remotely only by voice today, and there is a pressing need to be able to do this consultant function in a way that is supported by remote image access.

The prescription for relief from this present condition is widely held to be the production of the original images directly in digital form, and the storage and networking technology to support it. The communication bitrate requirement is not as extreme as with supercomputer visualization, because there are not as many frames per second. This is somewhat offset, however, by the fact that, whereas video image compression is an option for visualization, the medical profession is insistent that no transformations on the original image take place, lest some minor detail in the picture be lost.

2.7 Multimedia Conferencing

Research and prototyping of multimedia teleconferencing systems is at least 25 years old. Underlying the dozens of active and imaginative conferencing projects has been the dream of "exchanging communication for transportation" by creating a natural electronic environment in which cooperative work can be accomplished easily between physically separated parties. Progress has been glacial, and the only real success seems to have been the telephone conference call, and even it has proved unsatisfactory unless the participants are few, the agenda fairly fixed, the voice quality high, and the balance of voice signal levels identical within a small number of dB. Voice-quality impairments that would be accepted easily in one-on-one conversations effectively ruin the effectiveness of voice conferencing.

Video conferencing systems vary all the way from physically separated, very expensive, specially equipped conference rooms to "desktop conferencing" between multitasking multi-windowed personal computers fitted with special voice and live-motion video plug-in cards [16]. To do truly effective multimedia conferencing seems to be partly a question of communication bandwidth and error rate, but only partly so. Suppose we had a small group of people in one conference room communicating with another such group elsewhere, with

- One full-motion NTSC video local camera and remote monitor for each person (100 Mb/s per person),

- Compact-disk-quality quadraphonic audio (5 Mb/s), lip-synced to the video,

- For presentation material, 24-bit color graphics such as is available on today's high-performance workstations, with one new foil or real-time sketch per minute per person (1 Mb/s),

- Display of textual material at full-page resolution, with browse capability (1 Mb/s).

This all adds up to close to 1 Gb/s per conference site, obviously a strain on today's economical long-haul communications. And yet, a number of experimental conferencing systems have approached this degree of coverage closely enough to show that bandwidth isn't everything. Probably it is safest to say that when successful multi-media conferencing systems are eventually developed, they will not only need a gigabit communication infrastructure, but will need at least as urgently a better understanding and mastery of human-factors considerations than exists today.

2.8 CPU Interconnection

Most companies of any size have one or more mainframes, which, together with many tens of direct-access storage devices (DASD) and printers, are herded into a computer center or "glass house," often in scarce office building space that might better be used for offices. Sometimes such machines are smaller and more dispersed as individual *servers*. Traditional means of interconnecting these machines to each other and to the peripherals have involved parallel electrical connections having a limited distance capability, usually several tens of meters.

The protocols and technology are becoming available to allow these resources to be physically separated by many kilometers using serial optical fiber links, today at bitrates like the 200 Mb/s of IBM's ESCON (Enterprise System Connection) system [17]. In the future, there will be the Fiber Channel Standard of the American National Standards Institute (ANSI) [18] running at 1 Gb/s. Both these architectures are not just serial links, but involve a central space-division switch that works in packet-switch mode, the switching time being of the order of a microsecond or two. This fast switching speed is an unavoidable requirement, since the messages (packets) going between computers or between a computer and its peripherals are short (hundreds of bytes) and their destination varies rapidly from packet to packet.

The future availability of all-optical networking to support this application will allow several important flexibilities that are not now available with the old parallel, copper-based, limited-distance connections. These come in two flavors, moving function away and moving it in.

Under the former comes the ability to remove most of the space-consuming peripherals, particularly the "DASD farm," to remote sites. (A DASD is a direct-access storage device, typically a large disk drive). Another has to do with "remote-site recovery," assuring that at most one day's worth of work will be lost in case of catastrophic loss of the glass-house site. This is usually done by making a daily journal backup-to-tape of all files on DASD, after which the tapes are removed to a distant safe-storage location. It is obviously most desirable to do this by communicating directly to tape drives at the off-site safe-storage location. The third requirement for moving function away is to shed load, which is often done today to another machine in the same glass house, but ought to be possible to machines many kilometers away.

As for moving function in, when the terminal users attached to one machine have the ability to move large databases and programs rapidly from a remote machine to their own, it is as though that remote machine were really local. On a finer granularity of time scale, the ability to run individual procedures within a local program on a remote machine (by the use of *remote procedure calls*) as though they were local, requires that the access time be pared down to the very minimum. With gigabit connections the limitation ideally becomes one of propagation latency alone.

2.9 Conclusions

From this brief consideration of gigabit applications that are visible today, we have set the stage for consideration in the remainder of the book of technologies and architectures for satisfying the key requirements of the future.

It is seen that the bitrate per user is seldom likely to exceed a gigabit per second, but that the number of users who will want more or less sustained access to gigabit connectivity can eventually become extremely large. Numbers at least in the hundreds and sometimes in the thousands or tens of thousands or even millions can be expected. Past experience has often taught that providing a user access to many other users is more important than sheer bandwidth. We shall see that the technology and architecture is essentially already in place to support hundreds or perhaps a thousand one-gigabit sustained connections, but that to go beyond that point, interconnection of individual third-generation LANs and MANs to form WANs is required.

It has also been learned that some of the important applications absolutely require a packet-switched mode of operation. At one gigabit per second, this means that the photonic technology must permit a change of connection (perhaps a wavelength retuning) in submicrosecond times.

As for bit error probability, those applications that require extremely low such probability can be supported by suitably engineering the link budget.

Security is a higher-protocol-level problem, since it cannot be easily solved by controlling access by any photonic means.

Once these capabilities are provided, low latency is exposed as the one requirement that can be only incompletely satisfied. The approach will have to be to reduce the end-to-end latency for a communication to a bare-bones minimum excess over the one-way propagation time.

2.10 References

1. A. S. Tanenbaum, *Computer Networks - Second Edition*, Prentice Hall, 1988.
2. D. D. Clark, "Abstraction and sharing," *Conf. Record, IEEE LEOS Topical Meeting on Optical Multiaccess Networks, Monterey, CA*, July, 1990.
3. J. M. Kraushaar, "Fiber deployment update—End of year, 1990," *Federal Commun. Commission, Common Carrier Bureau, Washington, DC 20554, March*, 1991.
4. P. W. Shumate and R. K. Snelling, "Evolution of fiber in the residential loop plant," *IEEE Commun. Magazine*, vol. 29, no. 3, pp. 68-74, 1991.
5. J. A. Chiddix, "Fiber backbone trunking in cable television networks: An evolutionary adoption of a new technology," *IEEE LCS Magazine*, vol. 1, no. 1, pp. 32-37, February, 1990.
6. M. L. Dertouzos, "Communications, computers and networks," *Sci. Amer.*, vol. 265, no. 3, pp. 62-69, September, 1991.
7. P. K. Runge and P. R. Trischitta, eds., *Undersea Lightwave Communication*, IEEE Press/IEEE Commun. Soc., New York, 1986.
8. S. E. Miller and I. P. Kaminow, eds., *Optical Fiber Communications—II*, Academic Press, 1988.
9. C. Lin, ed., *Optical Technology and Lightwave Communication Systems*, Van Nostrand Reinhold, 1989.
10. F. Tosco, ed., *CSELT Fiber Optic Communications Handbook - Second Edition*, TAB Professional and Reference Books, Blue Ridge Summit, PA, 1990.
11. R. Kishimoto and I. Yamashita, "HDTV communication systems in broadband communication networks," *IEEE Commun. Mag.*, vol. 29, no. 8, pp. 28-35, August, 1991.
12. G. M. Nielson, ed., "Special Issue on Visualization in Scientific Computing," *IEEE Computer Magazine*, vol. 22, no. 8, 1989.
13. High-performance parallel interface—Mechanical, electrical, and signalling protocol specification, American National Standards Institute X3.183, 1991.
14. Power Visualization System, IBM Corporation, number G225-4402-00., Hawthorne, NY 10532, July, 1991.
15. W. J. Dallas, "A digital prescription for X-ray overload," *IEEE Spectrum*, vol. 27, no. 4, pp. 33-36, 1990.
16. I. Brackenbury and H. Sachar, "Desktop conferencing—You see what I see," *Conf. Record, IEE Colloquium on Multimedia: The Future of User Interfaces?*, vol. 1990/64, November, 1990.
17. J. C. Elliott and M. W. Sachs, "The IBM Enterprise Systems Connection Architecture," *IBM Jour. of Res. and Dev.*, vol. 36, no. 1, 1992.
18. D. C. Hanson, "Progress in fiber optic LAN and MAN standards," *IEEE LCS Magazine*, vol. 1, no. 2, pp. 17-25, May, 1990.

Part II

BUILDING BLOCKS

CHAPTER 3

Fibers, Couplers, and Taps

3.1 Overview

Figure 3-1 shows the construction of typical optical fiber, both single-mode and multi-mode [1, 2]. There is a central *core*, within which the propagating field is to be confined, surrounded by a *cladding* layer, which in turn is surrounded by a thin protective *jacket*, and also perhaps one or more layers of protective sheathing, or *buffer*. The core and the cladding are made of glass, while the jacket and buffer are made of plastic. Perhaps the reader has seen yellow or orange fiber cable being run within buildings; if the outside buffer is colored orange it is *multimode* fiber, and if yellow it is *single-mode* fiber. A *mode* can be thought of as a pattern of standing waves that the field pattern forms on the cross-section of the fiber. If there are many half-cycles of such a standing waves across a fiber core diameter, the fiber is operating as a multi-mode fiber; if there is only one, it is operating in single-mode fashion.

One should be careful to distinguish fiber modes from laser modes, the latter being a topic that will be discussed in Chapter 5. Some lasers radiate many "modes" at the same time, meaning that they are emitting on many wavelengths simultaneously. These are different from the "modes" in a fiber; the latter usage means that *at the same wavelength* the fiber can support simultaneously more than one spatial distribution of energy across the cross-section of the core.

It is really quite astonishing, when you think about it, that this physical medium that has stirred up such a communication revolution should be so physically unprepossessing. Essentially the fiber itself is made out of sand (silicon dioxide, silica) and is almost invisible. (Of course, it is pretty exotic sand, with a great deal of

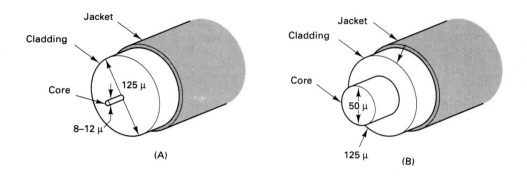

Figure 3-1. Geometry of typical single-mode and multimode optical fibers.

attention given to reducing certain impurities to extremely low levels).   As the
numbers in the figure indicate, a single-mode fiber has a core diameter of 8-10 μ,
which compares with about 50 μ for the diameter of a human hair. This latter figure
is also typical of the core in a multimode fiber.

   As we shall see, the confinement of the propagating field within the core takes
place by making the refractive index of the core about one percent greater than that of
the cladding, and this serves to turn back into the core the light that would otherwise
escape. As you might imagine, fiber of larger core diameter is easier to manufacture
than that of such small diameter as 8-10 μ, but for the wavelengths at which the attenuation is low enough to be useful, this increase in size allows more than one propagating mode to be supported simultaneously. Since these modes propagate at slightly
different velocities from one another, the longer the fiber the greater the time smearing
that will be suffered by any signal waveform. In many applications where the distances and bitrates are modest, this time smearing may not matter.

   Because of its much larger core diameter, it is is much easier to splice multimode fiber, to couple segments together with low loss and to do other mechanical
operations. Techniques to do these things with both kinds of fiber are improving gradually [3], but the fact still remains that single-mode fiber is relatively unpleasant to
deal with. As we shall see, however, its low dispersion properties make it an imperative for networks extending more than a kilometer or so.

   Three properties of fibers give them their unique position as a communication
technology: their large bandwidth, their low attenuation, and their small size. Probably
a fourth should be added: cost. As production volumes increase, costs should continue
to drop, since optical fibers are made from one of the most plentiful materials on earth.
Unlike copper, silica will be in plentiful supply permanently, and the development of
raw-material sources will be of low cost and negligible impact on the environment.

As the author of *Future Shock* pointed out some years ago, "The same ton of coal required to produce ninety miles of copper wire can turn out eighty thousand miles of fiber"[4] .

Figure 3-2 diagrams the progress made through recorded history in reducing the attenuation of various kinds of glass [5]. The glass made by the ancient Egyptian civilization was sufficiently transparent at visual wavelengths that presence or absence of light could be perceived on the other side (say, 20 dB attenuation) only if the glass were thinner than 1 millimeter, while today that number is 100 kilometers for silica fiber at 1.5 μ in the near infrared. Figure 3-3 shows in spectral detail the history of attempts to reduce silica fiber attenuation since the beginning of the fiber optic era. In addition to work on silica materials, there is ongoing laboratory research on exotic materials that could have attenuations less than 0.01 dB/km. If these ever become practical, one would be able to look into one fiber end on one side of the Atlantic and see someone blinking a flashlight into the other end on the other shore.

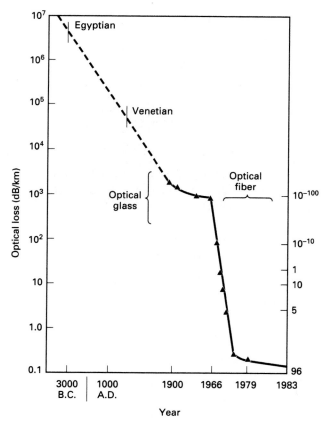

Figure 3-2. Attenuation history of glass materials (from [5], © 1988 IEEE).

Figure 3-3 indicates the wide bandwidth available. In the vicinity of 1.5 μ, the attenuation is about 0.2 dB per kilometer, and there is a window about $\Delta\lambda = 200$ nm wide between wavelengths having double that number of dB per kilometer. From $\lambda = c/f$, where c is the free-space light velocity, 3.0×10^8 meters/sec, f is the frequency in cycles per second (Hertz), and λ is the free-space wavelength in meters per cycle, the bandwidth expressed in frequency is

$$\Delta f = \frac{c\Delta\lambda}{\lambda^2} = 25,000 \quad \text{GHz.} \tag{3.1}$$

It is handy to remember that at 1.5 μ wavelength, one nanometer in wavelength is 133 GHz in frequency.

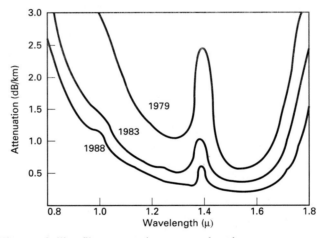

Figure 3-3. History of silica fiber attenuation vs. wavelength.

At 1.3 μ there is another low attenuation window, also about 25,000 GHz. wide. Both these operating regions can be used with laser diodes based on the indium phosphide materials system (Chapter 5) and photodetectors using either germanium or such alloys as InGaAsP (Chapter 8). At 0.85 μ there is another interesting region whose attenuation is about 2 dB per km, which, while less attractive from the attenuation standpoint, has the advantages that the lasers and high-speed transmit electronics can be made from the same materials system (gallium arsenide) and the detectors from silicon, the most common material for the electronics. In the 0.85 μ band there is a useful bandwidth of some 25,000 GHz, just as with the 1.3 and 1.5 μ bands. This band is defined not by a minimum in fiber attenuation but by the range over which gallium arsenide emitting components can be easily made.

Dispersion is one of the parameters that limits the bitrate on a link. Dispersion is a difference in velocities of propagation across the signal spectrum. Even if the fiber attenuation is sufficiently low that enough photons per bit arrive at the receiver to

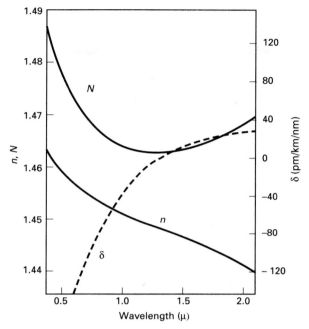

Figure 3-4. Dispersion δ vs. wavelength. Also shown are the refractive index n, and the "group index" N. (From [6], by permission).

normally give adequately low error rate (Chapter 8), when dispersion is present, the arriving pulses will suffer a certain amount of time smearing or *intersymbol interference*, which then produces random data-dependent bit errors of its own. Figure 3-4 shows the dispersion of silica fiber as a function of wavelength, neglecting the "modal dispersion," the difference in arrival times of different *spatial* modes, but including some other effects that we shall discuss shortly. The quantity $\delta(\lambda)$ is the dispersion that would be suffered in using a single-mode fiber, given in picoseconds per kilometer per nanometer of bandwidth $\Delta\lambda$, meaning that two signals one nm apart in wavelength would suffer δ picoseconds relative time skew for every kilometer of travel.

Thus, given the dispersion δ, the wavelength λ, and the link length L, we can determine the maximum bitrate for the time-delay difference between band edges to remain below, say, one-tenth of a bit time. An exact expression will be derived later in this chapter, but meantime it is useful to show how to make a "back-of-the-envelope" estimate based on the assumption that a digital signal of bitrate B can be represented as two sinusoids spaced by $\Delta f = B$ Hz. Equating the total delay in picoseconds between these two sinusoids to one-tenth of the bit time gives (using Equation 3.1)

$$L\,(\Delta f)^2 \;=\; \frac{10^2\,c}{\lambda^2\,\delta} \tag{3.2}$$

where L is in kilometers, Δf is in Hertz, c is 3×10^8 km/sec, λ is in meters and δ is in picoseconds per kilometer per nanometer.

The small size of optical fibers means that we may bundle many fibers together in a small cable. In Figure 3-5 is shown a typical AT&T 144-fiber cable used for intercity transmission. High bundling density offers some favorable options for architecting fiber optic networks, but is even more useful in applications involving very short distances of millimeters up to several meters. Such an application area is *optical interconnect*, the interconnection of chips to chips, boards to boards, chips to boards, and so forth within the covers of a single switching or processing unit. Composite bundles have been made that embody parallel single-mode fiber paths packed to a density of 10^6 per cm^2. Such bundles are available in lengths up to 15 feet long [7].

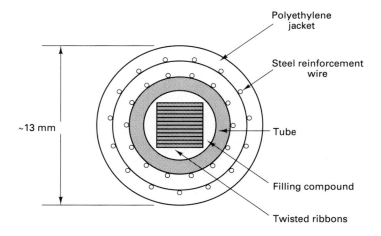

Figure   3-5. Standard AT&T 144-fiber cable. (From [8], Academic Press, by permission).

In this chapter we shall first introduce the simple terminology and concepts of propagation of a plane wave in a homogeneous medium, starting with Maxwell's equations and from them defining the ray picture of *geometric optics*. Using this ray model we can deduce what happens when a light wave meets an interface at which it is partially reflected and partially transmitted. This picture of a propagating wave at an interface is then used to discuss how energy is coupled into an optical fiber. Propagation down the fiber takes place within one or more possible *propagation modes*, but the methodology of geometric optics is insufficient for any realistic discussion of them. Tedious solutions of the wave equation in cylindrical coordinates are required instead that express how light in the core is guided down the fiber by radial dependence of the refractive index, especially by the contrast in refractive index between core and cladding. Highlights of this analysis are given. The emphasis in this book is on single-mode technology, because of the high bitrates that third-generation lightwave networks must support. Nevertheless, multiple-mode propagation is included in the discussion.

We then go on to the problem of how light can be tapped off or coupled sideways between two fibers very close together. This form of coupling will be seen to be the basis for the taps, splitting and merging couplers, and star couplers that will figure so importantly in the later network architecture discussions.

It is possible to have unwanted random fluctuations of received signal strength just due to unwanted effects associated with the fiber itself. Also, certain types of receivers can be adversely affected by slow random drifts in the polarization state, and this must be taken into account. Finally, if the power level gets sufficiently large, it is possible that the tiny glass fiber can be driven into slightly nonlinear behavior, which can either produce harmful intermodulation effects or be played off against the dispersion to permit extremely high-bitrate *soliton pulse* propagation. All these matters are discussed in sequence in the remainder of this chapter.

These discussions will all be at quite an elementary level, accurate enough quantitatively to account with adequate precision for what is going on in real systems, but not physically complete. A number of pointers are given to particularly understandable detailed treatments elsewhere in the literature.

3.2 **Propagation in an Infinite Uniform Medium**

As the first step in building up to a discussion of how light travels down a fiber or is coupled in and out of the fiber, it is necessary to understand the simple case of propagation in a medium that extends infinitely in all directions, and is homogeneous, isotropic, nonconducting, devoid of all free charges, lossless, dispersionless and linear [9].

The reader will see that the flow of ideas in this chapter amounts to discussing propagation in such an idealized medium, and then introducing, one by one, the exceptions to this idealized picture that describe the real situation. Real fibers have some attenuation and some dispersion. They cannot be considered to be of infinite extent, except along the axis, and the finite size leads to transverse standing-wave patterns constituting the permitted propagation modes. If these modes are not handled correctly, harmful dispersion and signal-interference effects can occur. There are also small departures from linearity that must be considered.

A simple form of Maxwell's equations will be sufficient to handle the situations to be met in this chapter on fiber propagation. However, when we discuss lasers in Chapter 5, we shall be dealing with material that is conducting, has free charges, is lossy, has dispersion and is nonlinear. There, instead of trying to treat the device as one within which waves propagate, we shall not use the Maxwell equations, but different equations, called the *rate equations*, that express the rate at which different quantities increase and decrease within a single small volume.

Maxwell's equations give, for any point in the medium at any instant of time, the interrelationships between the electric field vector **E** and the magnetic field vector **H**.

$$\nabla \times \mathbf{E} = -\mu \dot{\mathbf{H}} \qquad (3.3)$$

$$\nabla \times \mathbf{H} = \epsilon \dot{\mathbf{E}} \qquad (3.4)$$

$$\nabla \cdot \mathbf{E} = 0 \qquad (3.5)$$

$$\nabla \cdot \mathbf{H} = 0 \qquad (3.6)$$

where μ and ϵ are the *magnetic and electric permittivities* of the medium, respectively, and the dot over a quantity indicates its time derivative. We shall use boldface to denote a vector. quantity and the equivalent non-boldface character to represent the magnitude of that vector.

In anisotropic media, such as some we shall meet in Chapters 4, 5 and 6, μ and ϵ are *tensors*, expressing their directional dependence. It will turn out that the component and technology ideas to be discussed in this book can be characterized with sufficient quantitative and pedagogical accuracy by treating the variables as scalars, and that is the way they will be dealt with. This simplification is particularly valid in fiber material, whose microscopic properties can always be considered to be independent of direction.

The unification that Maxwell effected with his equations in 1865 amounted to combining four phenomenological ideas [10, 11] going back into the previous century, one for each equation, respectively: Faraday's Law that changing magnetic fields produce electric fields, Ampère's Law that moving charges generate magnetic fields, Gauss' Law about electric field lines emanating from charges, and the magnetostatic law that says that magnetic field lines are closed paths.

It is easy to be confused about the difference between *electric permittivity*, *dielectric constant* and *index of refraction*. The electric permittivity is more conveniently written as

$$\epsilon = K_e \, \epsilon_o \qquad (3.7)$$

where ϵ_o is the electric permittivity of free space and K_e is the dielectric constant. The same goes for

$$\mu = K_m \, \mu_o \qquad (3.8)$$

As we shall see shortly, the *phase velocity* of propagation is

$$v_p = 1/\sqrt{\epsilon \mu} \qquad (3.9)$$

Writing the free-space version of this as

$$c = 1/\sqrt{\epsilon_o \mu_o} \tag{3.10}$$

we have

$$v_p = c/n \tag{3.11}$$

where

$$n = \sqrt{K_e K_m} \tag{3.12}$$

is the index of refraction, a dimensionless quantity, so called because it affects the refraction and reflection of waves, as will be seen in Section 3.7.

For free space, n is unity, but for silica fiber it is about 1.45, so instead of being 3.0×10^8 m/sec, the phase velocity of light in fiber material is around 2.1×10^8 m/sec.

For those who, like the author, easily confuse dielectric constant with index, it is helpful to think of the dielectric constant as something the designer of a capacitor would use to relate the physical geometry of the device to the capacitance, but not involving the magnetic properties. The refractive index, on the other hand, is some-thing useful in optics and involves the quantity μ, since it has to do with a propagating wave.

We shall now use the Maxwell equations to get the *wave equation* describing how the wave looks as a function of position and time and also the *Poynting vector* **P**, the vector that defines the *intensity* (power density) of the light and its direction. At any given location in the homogeneous medium, the Poynting vector **P** poynts along the *ray path*.

The use of rays rather than a complete vector description of the electromagnetic field [10] constitutes the approach called *geometric optics*, and is a handy way to see what is going on that is strictly valid only when the wavelength of the light is much smaller than the scale of the discontinuities of the region being analyzed. Given this restriction, it is clear that geometric optics is not really valid for treating propagation in real fibers where the wavelength can be, say, 1.5 μ and the core diameter 8-10 μ. Nevertheless, tracing rays using the apparatus of geometric optics will permit consider-able insight to be gained and will often be quite effective in giving an understanding that is quantitatively adequate for dealing with lightwave networks.

Taking the curl of Equation 3.3 and applying the identity

$$\nabla \times (\nabla \times \mathbf{A}) = \nabla(\nabla \cdot \mathbf{A}) - \nabla^2 \mathbf{A} \tag{3.13}$$

to Equation 3.3, substituting 3.4 and noting from Equation 3.5 that the first term on the right of 3.13 is zero, we obtain the wave equation for the electric field

$$\nabla^2 \mathbf{E} = \mu\epsilon\ddot{\mathbf{E}} \tag{3.14}$$

and using (3.4), (3.3) and (3.6) similarly,

$$\nabla^2 \mathbf{H} = \mu\epsilon\ddot{\mathbf{H}} \tag{3.15}$$

for the magnetic field.

The Poynting vector is determined as follows. Consider the volume dV in space, as shown in Figure 3-6. The electric and magnetic fields throughout that volume are **E** and **H**, respectively. The electrical and magnetic energies stored in the volume are

$$\mathcal{E}_e = \iiint \frac{\epsilon \mathbf{E}\cdot\mathbf{E}}{2}\, dV \quad \text{and} \quad \mathcal{E}_m = \iiint \frac{\mu \mathbf{H}\cdot\mathbf{H}}{2}\, dV \tag{3.16}$$

respectively. As long as no energy is being generated or dissipated in the volume, the (scalar) rate at which energy leaves the volume per unit time is

$$W = -\dot{\mathcal{E}} = -(\dot{\mathcal{E}}_e + \dot{\mathcal{E}}_m) = -\iiint (\epsilon\mathbf{E}\cdot\dot{\mathbf{E}} + \mu\mathbf{H}\cdot\dot{\mathbf{H}})\, dV \tag{3.17}$$

Let us consider the power leaving the volume to have a direction, so that it can be represented as a vector **P** over the surface **S** enclosing the volume V. Using Equations 3.3 and 3.4 on the right side of Equation 3.17, we have

$$W = \iint \mathbf{P}\cdot d\mathbf{S} = -\iiint (\mathbf{E} \bullet \nabla \times \mathbf{H} - \mathbf{H}\cdot\nabla \times \mathbf{E})\, dV \tag{3.18}$$

From the vector identity

$$\nabla \bullet (\mathbf{A} \times \mathbf{B}) = \mathbf{B}\cdot\nabla \times \mathbf{A} - \mathbf{A} \bullet \nabla \times \mathbf{B} \tag{3.19}$$

we have

$$\iint \mathbf{P}\cdot d\mathbf{S} = \iiint \nabla\bullet(\mathbf{E} \times \mathbf{H})\, dV \tag{3.20}$$

which, from the divergence theorem that the volume integral of $\nabla \bullet \mathbf{A}$ equals the surface integral of **A** over **S**, the surface of the volume, is

$$\iint \mathbf{P}\cdot d\mathbf{S} = \iint (\mathbf{E} \times \mathbf{H})\cdot d\mathbf{S} \tag{3.21}$$

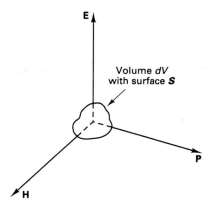

Figure 3-6. A unit volume, showing the **E**, **H** field vectors and **P**, the Poynting vector of light intensity.

Shrinking the volume to the point where **E** and **H** are constant within it, we have

$$\mathbf{P} = \mathbf{E} \times \mathbf{H} \tag{3.22}$$

The Poynting vector **P** represents the direction and magnitude of the energy flow per unit time per unit total cross-sectional surface area from the differential volume.

3.3 Complex Envelope and Phasor Convention

So far in this chapter, **E** and **H** were physical vectors whose waveforms as a function of time could be anything. They were always real (not complex) quantities. Similarly, E and H were the magnitudes of these physical vector quantities and they too were real (not complex) quantities.

From this point onward, when such variables are sinusoids, unless otherwise stated, we shall replace these real variables with the equivalent complex envelope quantities.

The reason is to gain convenience in representing and manipulating the particular form of waves and signals that are sinusoids or can be decomposed by Fourier analysis into sinusoids. The *complex envelope* or *complex amplitude* notation suppresses a lot of sines and cosines, and much bookkeeping with respect to phase angles.

Consider a sinusoidally time-varying electric field vector

$$\mathbf{E}'(t) = \mathbf{E}'' \cos (\omega t + \phi) \tag{3.23}$$

(where ω is the angular frequency in radians per second and ϕ is the phase in radians). Both $\mathbf{E}'(t)$ and $\mathbf{E}''$ are real physical vectors and $\mathbf{E}'$ is also a function of time.

The complex envelope representation of this same equation is gotten by saying that $\mathbf{E}'(t)$ is *the real part of a fictitious complex quantity* $\mathbf{E} \exp(-j\omega t)$.

$$\mathbf{E}'(t) = Real \ \{ \mathbf{E} \exp(-j\omega t) \} \tag{3.24}$$

where $j = \sqrt{-1}$ and

$$\mathbf{E} = \mathbf{E}'' \exp(-j\phi) \tag{3.25}$$

The quantity $\mathbf{E}$ is called the *complex envelope* or *phasor* of $\mathbf{E}'$. The original phase angle ϕ has been absorbed into the complex envelope, and the frequency separated out into the explicitly understood exponential factor, which replaces the trigonometric cosine function.

A representation that shows a sinusoid as a vector in a two-space of in-phase component and quadrature component is called the *phasor* representation.

To repeat, for the rest of this book, unless otherwise stated explicitly, when dealing with sinusoids, a given electric, magnetic, acoustic or other signal will be considered to be complex, and to get the right answer, **it is understood that** the real part of the final result is to be taken.

Handled with care, the complex envelope convention saves much notational encrustation, but there are circumstances in which it can give the wrong answer, namely when taking the *products* or *powers* of sinusoidal functions of time. In those cases the clumsier real trigonometric form of the sinusoids must be used.

If the light is all at one frequency, it is said to be *monochromatic*, and then the in-phase and quadrature components of the phasor $\mathbf{E}$ are invariant with time, and so is the phase ϕ. We shall limit most of the discussion in this chapter to this monochromatic situation. *Incoherent light* not only has a finite nonzero bandwidth, but acts mathematically like filtered noise, which has randomly time-varying phase and has in-phase and quadrature components that are usually statistically independent of one another and of the phase. We shall meet an exception to this independence in the case of laser phase noise (Section 5.5).

3.4 The Propagation Vector

Returning to the wave equations 3.14 and 3.15, and considering only sinusoidal time functions, it can be shown that for, say, the electric field at point $\mathbf{r}$ from the origin, and at time t, there are two possible solutions, travelling sinusoids of the form

$$\mathbf{E}(t, \mathbf{r}) = \mathbf{E}_o \exp[\pm j(\omega t - \mathbf{k} \cdot \mathbf{r})] \tag{3.26}$$

where ω in (radians per second) is the temporal angular frequency of the sinusoid and $\mathbf{k}$ is the spatial angular frequency, or *wave number* in radians per meter. We shall arbitrarily pick the one with the minus sign. The magnitude of the vector **k** is k. To visualize what is going on, if one were to pick a certain place on this travelling wave and somehow mark it (with red paint, for example), then at a fixed physical position as a function of time this red dot would arrive at a certain instant as a part of a sinusoidal function of time having a certain phase. Similarly, if we took a snapshot at one instant of time of the entire function of space and looked for the red dot, it would be seen to be a particular point at a certain phase on the sinusoid as a function of space.

The (scalar) velocity at which the marked spot travels is called the *phase velocity*. From reckoning distance travelled per unit time from the argument in the Equation 3.26, this is seen to be

$$v_p = \frac{\omega}{k} \tag{3.27}$$

which gives for non-free space conditions, using (3-9) through (3-12),

$$k = \frac{\omega n}{c} = \frac{2\pi n}{\lambda} \tag{3.28}$$

where λ is the wavelength the wave would have in free space.

The vector **k** is called the "k-vector," "wave number vector," "propagation vector," or "wave vector." We shall employ the terms *propagation vector* for **k** and *propagation constant* for k.

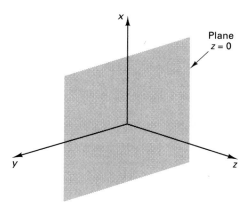

Figure   3-7. Coordinate system for a wave propagating in the z-direction.

It remains to show that, for the uniform medium we have been considering, not only are **E** and **H** orthogonal, but also the propagation direction is in exactly the same direction as the Poynting vector. Consider the rectangular coordinate system of Figure 3-7, in which there is a time-varying **E** field in the x-direction that is constant over the infinite plane $z = 0$, i.e.,

$$\mathbf{E} = \mathbf{u}_x E_x \, (z = 0, \; t) \; \text{ with } \; E_y = E_z = 0 \qquad (3.29)$$

($\mathbf{u}_x$ being the unit vector in the x-direction). Since $\mathbf{E}$ does not vary with x or y, the propagation vector $\mathbf{k}$ can only lie in the z-direction. Expanding (3.3),

$$\nabla \times \mathbf{E} = \mathbf{u}_y \frac{\partial E_x}{\partial z} = - \mu \dot{\mathbf{H}} = - \mathbf{u}_y \, \mu \dot{H}_y \qquad (3.30)$$

so the $\mathbf{H}$ field lies along the y-axis, which is the same result as with the Poynting vector equation 3.22, which says that if $\mathbf{E}$ lies in the x-direction and $\mathbf{H}$ lies in the y-direction, then $\mathbf{P}$ must lie in the z-direction.

In this case, where the sinusoidal wave propagates along the z-axis, (3.26) becomes

$$\mathbf{E}(t, z) = \mathbf{u}_x E_x \exp[\, -j \, (\omega t - kz)] \qquad (3.31)$$

where $\mathbf{k} = \mathbf{u}_z \, k$

3.5 Attenuation in Fibers

If there is any attenuation in the medium, the field strength of the wave defined in Equation 3.31 will suffer an exponential decay with z, which can be accounted for by writing the propagation constant k as a complex quantity, where the real part represents propagation and the imaginary part represents attenuation

$$k = \beta + j\alpha \qquad (3.32)$$

whereupon (3.31) becomes

$$\mathbf{E}(t, z) = \mathbf{u}_x E_x \exp(\, -j\omega t) \exp[\, - (\alpha + j\beta)z] \qquad (3.33)$$

or

$$\mathbf{E}(z) = \mathbf{u}_x E_x \exp[\, - (\alpha + j\beta)z] \qquad (3.34)$$

The shape of the curves of attenuation versus wavelength given in Figure 3-3 can be explained as representing the combined effect of several different physical phenomena [12] that act to absorb the energy of the light entering the fiber or to scatter it out through the cladding, where it is lost. The important ones are four in number, and we shall now discuss them. Their effect as a function of wavelength is shown in Figure 3-8.

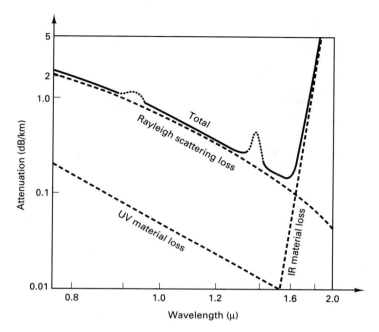

Figure   3-8. Spectra of three kinds of fiber attenuation in silica fibers. In addition, the loss due to OH⁻ molecular resonances is given by the dotted lines. (From [12], p. 84, by courtesy Marcel Dekker, Inc.).

At the long-wavelength end, there is *infrared material loss*, the tail of an absorption-loss spectrum that consists of peaks out beyond 9 μ having an attenuation of over 10^{10} dB/km! These fundamentals and their harmonics are due to excitation of molecular vibrations by the incident photons.

The incident photons can also stimulate electron transitions if the photons are in the ultraviolet, and therefore the long wavelength tail of this *ultraviolet material loss* also contributes to the overall loss spectrum.

These two effects are the principal unfixable sources of loss, but there are two other forms of loss that can be controlled to some extent.   These have been subject to steady improvement through the years.   It is quite conceivable that small further improvements can be made, particularly on the second one, but today this exercise is probably already past the point of diminishing returns.   This can be seen from the fact that the curve in Figure 3-2 has become almost flat in recent years.

The first is the *Rayleigh scattering loss*, the same kind of λ^{-4} loss that so skews toward shorter wavelengths the spectrum of sunlight scattered from microscopic scattering centers in the earth's atmosphere that the daylight sky looks blue. In silica fiber, the scattering centers (each of whose size is much smaller than λ) are density and compositional inhomogeneities that are frozen into the glass during manufacture.   Of these two, the compositional fluctuations are the easier to fix.   Scattering wastes some of the energy by misdirecting it away from the desired propagation direction.

Perhaps the area where progress has been most spectacular has been in reducing *OH⁻ molecular absorption*, another case in which the incident photons excite vibrations of an entire molecule. This loss component takes the form of peaks near harmonics of the basic absorption peak at 2.8 μ, namely 1.4, 0.9, and 0.7 μ. The principal physical source of the OH^- ion is water, and the history of fiber-attenuation improvement given in Figure 3-3 has been in large part a history of progress in reducing the water that is left within the glass during fabrication.

From Figures 3-3 and 3-8 it is seen that Rayleigh scattering, infrared absorption and OH^- absorption are the main contributors to silica fiber attenuation, with ultraviolet absorption playing only a minor role. There are several other sources of loss [1, 13], of which the most important is the class of losses referred to as *waveguide attenuation*. This includes leakage into the cladding, loss at bends, losses at connectors and splices, and waveguide imperfections that cause scattering of the light.

An exhaustive discussion of connectors and various forms of splices is given in [3]. For single-mode fibers, we may expect losses averaging around 0.5 dB at most of today's connectors, and about 0.2 dB for *butt splices* in which two fiber ends are cleaved to present smooth surfaces perpendicular to the longitudinal axes and then held mechanically together, perhaps with an *index matching fluid* applied between the surfaces to minimize the amount of index change as the light hits the fiber end. As we shall see in Section 3.7, the amount of light reflected from an interface depends on the difference in index across the two sides of the interface. *Fusion splices*, in which electric arcs are used to melt the two butt ends together, can routinely be made in the field with less than 0.05 dB of loss. Most field-usable semiautomatic fusion splicers now use TV imaging and pattern-recognition algorithms to microposition the two cores correctly before firing the arc.

Poor splices and connectors not only can lose signal energy in a fiber system, but can produce reflections that are capable of upsetting the functioning of lasers and photonic amplifiers by forming an undesired *external cavity* between the device and the reflecting point. Even if these reflections were eliminated, there is always the small background of Rayleigh *backscattering*, that portion of the isotropic directivity pattern of Rayleigh scattering that is captured by the fiber and sent back toward the source.

3.6 Group Velocity and Dispersion in Fibers

Recall that in a homogeneous nondispersive medium, the light wave propagates at a velocity v_p, the phase velocity. Even if the medium is dispersive (velocity a function of wavelength), a monochromatic single sinusoid will propagate at velocity v_p. But in real life, in order to send information over the fiber, the signal must have a nonzero bandwidth, whereupon dispersion enters the picture.

We can see what happens by considering the simple case of launching at the origin ($z = 0$) the sum of two sinusoidal waves at slightly different angular frequencies ($\omega + \Delta\omega$) and ($\omega - \Delta\omega$).

$$E(t, 0) = \exp[-j(\omega - \Delta\omega)t] + \exp[-j(\omega + \Delta\omega)]t \tag{3.35}$$

We shall neglect attenuation and deal with β, the real part of k in Equation 3.32. At the point z, we have

$$\begin{aligned} E(t, z) = &\exp\{-j[(\omega - \Delta\omega)t - (\beta - \Delta\beta)z]\} \\ &+ \exp\{-j[(\omega + \Delta\omega)t - (\beta + \Delta\beta)z]\} \end{aligned} \tag{3.36}$$

in which β is to be considered a linear function of ω such that at ($\omega \pm \Delta\omega$) the propagation constant is ($\beta \pm \Delta\beta$). Equation 3.36 can then be rewritten as

$$E(t, z) = 2 \exp[-j(\omega t - \beta z)] \cos(\Delta\omega t - \Delta\beta z) \tag{3.37}$$

which is a high-frequency wave (at angular frequency ω) whose amplitude varies at a low frequency rate as $\cos(\Delta\omega t - \Delta\beta z)$, a variation in both time and distance analogous to the single-frequency wave of Equation 3.26. If we now mark a particular spot on this low-frequency *envelope*, for example with blue paint, the marked spot will be seen to move at an envelope velocity or *group velocity*

$$v_g = \frac{d\omega}{d\beta} \tag{3.38}$$

From this, we infer that any waveform having a nonzero bandwidth will propagate not at v_p, but at v_g, as given by Equation 3.38. So the actual information will be transmitted at velocity v_g (blue dot), while the phase front is advancing at velocity v_p (the red dot of Equation 3.26).

A more general expression for the waveshape that a pulse will take after being propagated through a dispersive medium can be derived as follows [8]. At distance z from the origin, a z-propagating pulse whose spectrum is $A(\omega)$ at the origin will arrive as the superposition of infinitely many plane waves

$$a(z, t) = \int_{-\infty}^{\infty} A(\omega) \exp[-j(\omega t + \beta z)] \, d\omega \tag{3.39}$$

Suppose the medium exhibits first-order dispersion – that is, the group velocity is a linear function of frequency. Let ω_o be the center frequency of the pulse spectrum. Then β can be represented as

$$\beta(\omega) = \beta_o + \beta'(\omega - \omega_o) + \frac{1}{2}\beta''(\omega - \omega_o)^2 \qquad (3.40)$$

where β' is the first derivative $d\beta(\omega)/d\omega$, and β'' is the second derivative $d^2\beta(\omega)/d\omega^2$. β'' is the proportionality constant by which the group velocity is a linear coefficient of frequency. Changing variables by $u = (\omega - \omega_o)$,

$$a(z, t) = \exp[-j(\omega_o t + \beta_o z)]$$
$$\int_{-\infty}^{\infty} A(u + \omega_o) \exp\left[-j\left[(t + \beta'z)u + \frac{1}{2}\beta''zu^2\right]\right] du \qquad (3.41)$$

$$a(z,t) = A'(\omega) \exp[-j(\omega_0 t + \beta_o z)] \qquad (3.42)$$

In other words, the arriving pulse is smeared out to the form $A'(\omega)$, specified by the integral factor in Equation 3.41. This is the more exact form of the "back-of-the-envelope" Equation 3.2.

There are several physical causes of dispersion in fibers [1, 2, 13], of which the most significant are *material dispersion, waveguide dispersion*, and *modal dispersion*.

For single-mode fibers, the dominant form of dispersion is material dispersion, a consequence of the refractive index in Equation 3.11 being a function of wavelength, as illustrated in Figure 3-4. This has also been known for centuries as *chromatic dispersion*. Equivalently to Equation 3.11, instead of saying that the phase velocity is c divided by the actual refractive index, n, we can say that group velocity is c divided by a fictitious quantity called the *group index N*, obtained as follows:

$$N = \frac{c}{v_g} = c\frac{d\beta}{d\omega} = c\frac{d}{d\omega}\left(\frac{\omega n}{c}\right) \qquad (3.43)$$

Assuming no attenuation, we have $k = \beta$. Now, with dispersion, $n = n(\lambda)$, and using the fact that $\omega\lambda/2\pi = c$,

$$N = n + \omega\frac{dn}{d\omega} = n + \omega\frac{dn}{d\lambda}\frac{d\lambda}{d\omega} = n + \frac{2\pi c}{\lambda}\frac{dn}{d\lambda}\left(-\frac{2\pi c}{\omega^2}\right) \qquad (3.44)$$

$$N = n - \lambda\frac{dn}{d\lambda} \qquad (3.45)$$

Both the actual index $n(\lambda)$ and the group index $N(\lambda)$ for silica material are shown in Figure 3-4, as well as $\delta(\lambda)$, the dispersion expressed in picoseconds per kilometer of length per nanometer of bandwidth. The dispersion δ was introduced earlier in connection with Equation 3.2, and now Figure 3-4 has given us a more complete picture of it.

Since the vertical axis is greatly expanded, it is seen that the dispersion is actually quite a small effect; it is the large fiber lengths and high bitrates that make fiber transmission so sensitive to dispersion that these small values actually become quite influential in the design of the system.

The dispersion is seen to be zero at a wavelength around 1.3 μ. At this point, $dN/d\lambda = 0$, which, from Equation 3.45, means that the second derivative of the index $d^2n/d\lambda^2 = 0$.

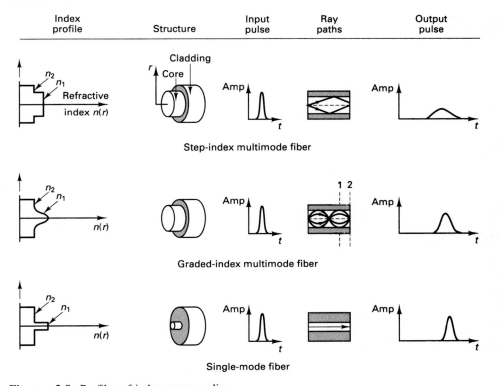

Figure   3-9. Profiles of index versus radius.

One can actually make *dispersion-shifted fiber* [6], in which the dispersion minimum, normally at 1.3 μ, is shifted to around 1.5 μ, the attenuation minimum. Or we can make *dispersion-flattened fiber*, in which δ changes less slowly with wavelength than with a normal fiber. Both these modifications are effected by introducing a controlled amount of *waveguide dispersion*, of the right sign. Waveguide dispersion is defined as the wavelength dependence of a given mode due to the shape and index profile along a radius of the waveguide region. We shall discuss waveguide dispersion further when we get to Figure 3-21. Since it is a property of any mode, it is present in both single-mode and multimode fiber.   By controlling both the core-cladding boundary and the index profile within the core, as in Figure 3-9, various waveguide dispersion characteristics can be obtained.

For multimode fibers, index profile shaping is not just a minor issue, it is **the** issue, because of the third kind of dispersion, *modal dispersion* or *multipath dispersion*, present only in multimode fibers. This form of dispersion occurs when there are multiple modes because these modes propagate at different phase and group velocities from one another. We shall discuss these modes, their velocities, and the resulting modal dispersion in Section 3.9.

The history of multimode fibers has essentially amounted to trying different ways of shaping the index profile within a multimode fiber so as to guide the energy down the center, in other words to try to achieve single-mode characteristics as closely as possible. These efforts have been successful only to the extent of providing multimode fiber that will suffice for short distances and low bitrates, and the only way to get true single-mode performance is to go to the extreme in profile shaping, make the core so small that only one mode will propagate. Figure 3-9 previews the next two sections of this chapter by showing two of the approaches used to minimize modal dispersion in multimode fibers, *graded-index* profiling and *step-index* profiling.

Modal dispersion, unlike material and waveguide dispersion, is present even when the signal bandwidth is very small, so we can represent it as δ in picoseconds per kilometer, essentially a function of distance only, unlike material dispersion. It is so large in any multimode fiber that it dominates the material dispersion. Its calculation is quite tedious, so it will be sufficient for present purposes to simply note that commercial multimode fiber has values of modal dispersion of 0.3 - 1.0 nsec. per km for graded-index fiber and some 50 nsec per km for the step-index case.

3.7 Reflection and Refraction at an Interface

Understanding what happens when a plane electromagnetic wave encounters a plane boundary between two media of different dielectric constants [14, 15] is the key to understanding how light is confined within a fiber, how lenses act to couple light into the fiber, how light is coupled between closely adjacent fibers in a fiber tap or splitter, and also how lasers and certain tunable optical filters work. The basis of this understanding is contained in the two classic relations, *Snell's Law*, and the *Fresnel Equations*.

The situation is shown in Figure 3-10, which depicts a ray of light incident on an infinite plane interface between two ideal media having refractive indices n_1 and n_2. The angles θ_i, θ_r, and θ_t are the departures from normal of the incident, reflected, and transmitted rays, respectively. All three of these rays, plus the normal to the plane, lie in the same plane, called the *plane of incidence*. Two cases are shown, where the electric field vector is into the page (left diagram) and in the plane of the page in the arrow directions (right diagram).

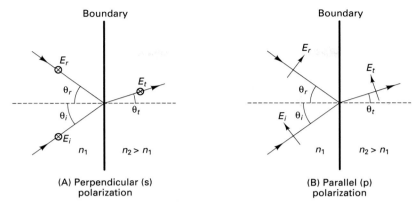

Figure 3-10. Reflection at a boundary for the two cases in which the **E**-vector is perpendicular to the plane of the paper (plane of incidence), and in that plane.

The reflected ray departs at an angle $\theta_r = \theta_i$ by mirror reflection. Snell's law gives the direction of the transmitted ray as

$$\frac{\sin \theta_t}{\sin \theta_i} = \frac{n_1}{n_2} \qquad (3.46)$$

While Snell's law tells us what the angles are, it says nothing about the field strength, relative phase, or polarization of the light represented by the rays. The set of two *Fresnel Equations* provide this information. They define the *reflection coefficients* (reflected field strength as a fraction of the incident field strength) as [15]

$$\rho_p = \frac{-n_2{}^2 \cos \theta_i + n_1 \sqrt{(n_2{}^2 - n_1{}^2 \sin^2 \theta_i)}}{n_2{}^2 \cos \theta_i + n_1 \sqrt{(n_2{}^2 - n_1{}^2 \sin^2 \theta_i)}} \qquad (3.47)$$

and

$$\rho_s = \frac{n_1 \cos \theta_i - \sqrt{(n_2{}^2 - n_1{}^2 \sin^2 \theta_i)}}{n_1 \cos \theta_i + \sqrt{(n_2{}^2 - n_1{}^2 \sin^2 \theta_i)}} \qquad (3.48)$$

These reflection coefficients ρ_p and ρ_s apply separately to that part of the incident light that is *polarized* in the plane of incidence or perpendicular to it, respectively.

By convention, if the light wave is of the simple sort discussed earlier in connection with Figure 3-7, with the **E**-vector in the *x*-direction, the wave is said to be *plane polarized* or *linearly polarized* along the *x* direction. The state of polarization

(SOP) of a light wave matters a lot with various devices used in optical networking, and we shall discuss it at length in Section 3.13.

The *reflectances* are the ratios of reflected to incident *intensities*

$$R_p = |\rho_p|^2 \quad \text{and} \quad R_s = |\rho_s|^2 \tag{3.49}$$

and the *transmittance* is the ratio of transmitted to incident intensities

$$T_p = 1 - R_p \quad \text{and} \quad T_s = 1 - R_s \tag{3.50}$$

We now discuss, one by one, the relevant consequences of the Snell and Fresnel equations.

Normal incidence

When $\theta_i = 0$, the two Fresnel equations become identical,

$$\rho = \frac{n_1 - n_2}{n_1 + n_2} \tag{3.51}$$

and

$$R = \left(\frac{n_1 - n_2}{n_1 + n_2} \right)^2 \tag{3.52}$$

The concept of a plane of incidence becomes inapplicable, since all rays are collinear. As it leaves the interface, the transmitted wave has the same polarization and phase as the incident light, but the reflected wave has its phase shifted by π if $n_2 > n_1$, but unshifted otherwise. As can be seen clearly from the right diagram in Figure 3-10, in the limit as $\theta_i = 0$, the reflected **E**-vector is directed oppositely from the incident one. (In the diagram, the approaching ray has the **E** vector pointing to the left in the plane of incidence; the receding ray has this vector pointing to the right).

Critical angle

When the angle of incidence θ_i is greater than zero, things rapidly get much more complicated [9, 15]. Figure 3-11 shows some of what happens. If $n_1 < n_2$, some light is reflected for all values of θ_i, as indicated in Figure 3-11(A). This is what happens when you are looking through a glass window. The case $n_1 > n_2$ (Figure 3-11(B)) is much more interesting to us, because this is the situation within an optical fiber. As θ_i (which for $n_2 < n_1$, is smaller than θ_t) increases, as in Figure 3-12, eventually the latter becomes 90°. At this point, the transmitted ray is pointed along the boundary.

The value of θ_i at which this happens is θ_c, the so-called *critical angle*,

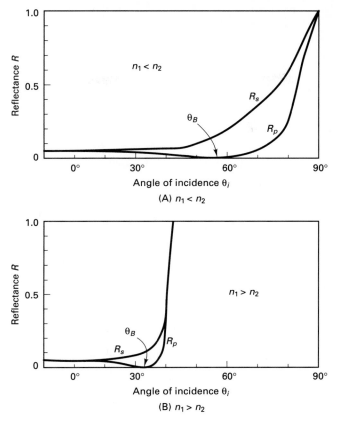

Figure 3-11. Reflectances R_p and R_s versus angle of incidence θ_i for (A) The case of $n_1 < n_2$ and (B) Vice versa. The two indices are 1.0 and 1.5. (From [14], p. 65).

$$\sin \theta_c = \frac{n_2}{n_1} \tag{3.53}$$

For $\theta_i \geq \theta_c$, as in Figure 3-12(C), the radical terms in *both* the Fresnel equations become imaginary, and for both of them the magnitude of ρ takes the form

$$|\rho| = \frac{|A - jB|}{|A + jB|} \tag{3.54}$$

where both A and B are real numbers, so that $R = |\rho|^2$ is unity for all such angles $\theta_i \geq \theta_c$. Inside a fiber, then, for all incidence angles greater than the critical angle, we have *total internal reflection*. For example, if $n_1/n_2 = 1.01$ (a typical value for the core-cladding interface), the critical angle θ_c is 88.3°, only 1.7° from a grazing angle to the core-cladding interface. Above the critical angle where the magnitude of the

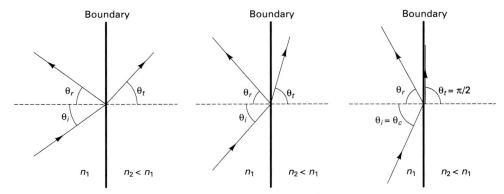

Figure   3-12. Showing how, for the $n_1 > n_2$ case, as the angle of incidence θ_i increases, eventually $\theta_t = 90°$ and the transmitted wave becomes evanescent.

reflection coefficient is unity, the phase is neither exactly 0 nor 180°, but varies between these values (Figure 3-13), as specified by the Fresnel equations.

Brewster angle

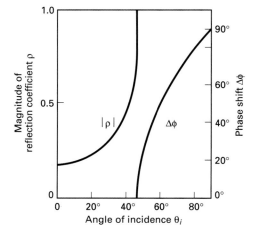

Figure   3-13. Magnitude of the reflection coefficient and phase shift on reflection for transverse electric waves (the "perpendicular" case in Figure 3-10) versus angle of incidence θ_i for a glass-air interface ($n_1 = 1.5$ and $n_2 = 1.0$). (From [16], John Wiley & Sons, Inc., by permission).

At this angle, θ_B (not to be confused with the critical angle), the component whose **E**-vector is parallel to the plane of incidence (Figure 3-10(A)) is not reflected at all. The Brewster angle θ_B appears in Figure 3-11(A) as the values of θ_i where ρ_p vanishes. Setting the numerator of (3.47) to zero, we get

$$\tan \theta_B = \frac{n_2}{n_1} \tag{3.55}$$

Evanescent wave

At angles $\theta_i \geq \theta_c$, the boundary conditions on **E** and **H** require that, even though there is complete reflection, some field on the right side of the boundary must be present in order to satisfy the conditions on continuity of fields at the interface given by the last two Maxwell equations (3.5 and 3.6). The incident light being an infinite plane wave, to the left of the boundary of Figure 3-10 there is a *standing-wave pattern* set up by the two waves that are counterpropagating to each other with respect to the normal to the boundary. Figure 3-14 shows the amplitude of this standing wave to the left of the boundary and the *evanescent wave* to the right. The evanescent wave, in which the **E** − and **H**-fields are 90° out of phase (so that no power is transmitted), is something you might dismiss as a weird phenomenon of no practical interest, except that the construction of *fused biconical couplers*, so widely used in lightwave networks, is based on coupling between mostly evanescent fields in closely adjacent fibers. We shall discuss such couplers in Section 3.10.

The evanescent field has a purely imaginary Poynting vector (because the electric and magnetic fields are in phase quadrature) that is aimed parallel to the boundary in the plane of incidence.   The field strength decays exponentially with distance z away from the boundary on the n_2 side as $(\exp - \alpha z)$, where the attenuation constant α is

$$\alpha = \frac{2\pi}{\lambda} \sqrt{n_1{}^2 \sin^2 \theta_i - n_2{}^2} \tag{3.56}$$

This can be seen by considering the field strength, which must be proportional to some $\exp(-j k_z z)$. To the n_2 side of the boundary,

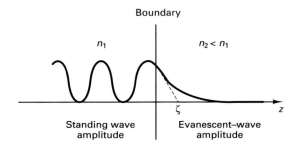

Boundary

n_1 $n_2 < n_1$

ζ z

Standing wave   Evanescent–wave
amplitude   amplitude

Figure   3-14. Amplitude of standing and evanescent waves as a function of z, the distance into the material normal to the boundary.

$$k_z = \frac{2\pi n_2}{\lambda} \cos \theta_t \qquad (3.57)$$

which, from Snell's Law, Equation 3.46, is

$$k_z = \frac{2\pi n_2}{\lambda} \sqrt{1 - \left(\frac{n_1}{n_2}\right)^2 \sin^2 \theta_i} \qquad (3.58)$$

which becomes imaginary for $\theta_i > \theta_c$ or $k_z = j\alpha$, where α is given by Equation 3.56.

From this it is seen that deeply penetrating evanescent waves occur for θ_i just beyond the critical angle, but that as θ_i increases, the evanescent wave is confined closer and closer to the boundary. It is absent altogether for $\theta_i < \theta_c$. We shall see just this behavior for modes propagating in the cladding of an optical fiber.

3.8 Coupling Light Into and Out of the Fiber; Diffraction

Numerical aperture

As indicated in Figure 3-15, there is a certain *acceptance angle*, shown as θ_a, within which light entering a fiber must fall in order to be totally reflected inside the fiber. If light comes in along a ray path at an angle larger than θ_a, then when the ray path hits the core-cladding boundary at Point B, it does so at an angle of incidence smaller than θ_c and the light will be passed into the cladding and lost. There are three indices involved in determining the acceptance angle θ_a, those of the air outside the fiber, the core, and the cladding, n_0, n_1, and n_2, respectively.

By Snell's law (3.46)

$$n_o \sin \theta_a = n_1 \sin \theta_1 \qquad (3.59)$$

which, considering the right triangle ABC, is

$$n_o \sin \theta_a = n_1 \cos \theta_c = n_1 \sqrt{1 - \sin^2 \theta_c} \qquad (3.60)$$

and, using (3.53),

$$n_o \sin \theta_a = \sqrt{n_1^2 - n_2^2} \qquad (3.61)$$

The fiber, being circular in cross-section, will actually have a *cone of acceptance* with a vertex angle of θ_a, and any lens system to couple light into the fiber must deliver its light within this cone; light arriving at wider angles will not propagate down the fiber.

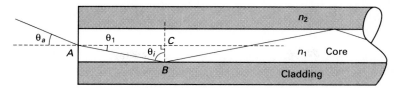

Figure 3-15. Side view of the end of an ideal fiber showing angle of acceptance θ_a when the ray path in the fiber has θ_i exactly at θ_c. For any greater θ_1, the light will escape from the core.

The quantity $n_0 \sin \theta_a$ is called NA, the *numerical aperture* of the fiber. We shall discuss it at greater length in the next section. Since we are usually speaking of the end of a fiber in air, we may replace n_o by unity, and since the difference in core and cladding indices is usually much smaller than unity, we may replace $\sin \theta_a$ by θ_a, so that

$$NA = \theta_a = \sqrt{n_1{}^2 - n_2{}^2} \approx n_1\sqrt{2\Delta} \tag{3.62}$$

where Δ is the fractional index difference

$$\Delta = \frac{n_1 - n_2}{n_1} \tag{3.63}$$

For example, for fiber having $n_1 = 1.45$ and a one percent fractional difference Δ between indices, $NA = 0.21$ radian, or $12°$.

Even though it was derived using geometric ray optics, Equations 3.61 and 3.62 turn out to be sufficiently accurate for engineering purposes, even for single-mode fibers whose radii are only two or three times λ. Incidentally, implicit in the above calculation was the assumption that the rays entering the fiber were *meridional rays*, those whose plane of incidence contained a diameter of the fiber (as distinguished from *skew rays*, whose planes of incidence do not pass through the center of the fiber core).

Coupling with lenses

Light is usually coupled into the fiber using either the conventional lens of Figure 3-16 or a graded-index lens (*GRIN lens*) of Figure 3-17.

The GRIN lens is like a short section of graded-index multimode fiber, but used in a novel and clever way. Notice in Figure 3-9(A) that the light rays passing through the two planes 1 and 2 have a very interesting relationship. When the rays at 1 are all parallel to the z-axis, down the line at 2 they are all convergent onto one focus. If the angle of convergence can be made less than the NA of an immediately adjacent butted end of a fiber core at 2, then this short section of graded-index material can be used as a lens to couple light into the fiber very efficiently. The nice thing about GRIN lenses

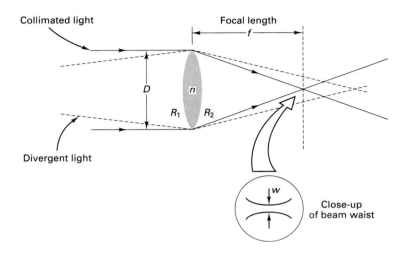

Figure   3-16. Parallel light impinging on a lens having radii R_1 and R_2, index n and thus focal length f. The inset is a magnified view of the light bundle at the focal point.

is that they can be made quite small. Typical commercial GRIN lenses for service with single-mode fibers are about 2 mm long and 2 mm in diameter.

With the conventional lens of *focal length* f, the relation between distance d_o from the lens to an object on one side of the lens to d_i the distance from the lens to the plane of an image of the object on the other side is [14]

$$\frac{1}{f} = \frac{1}{d_o} + \frac{1}{d_i} \tag{3.64}$$

The focal length is seen to be that distance at which an entering parallel (*collimated*) bundle of light will be focussed to a point. A collimated beam might, for example, be considered to be one that originated at an infinite distance to the left of the lens in the figure. As the source at the left moves closer, the ray paths act like levers, and the point at the right where the rays focus moves out to the right. Equation 3.64 is useful for designing systems to couple uncollimated light into the fiber. The lens will have a finite diameter D, and the ratio f/D is called the *f-number* of the lens (measured in *diopters*).

In real systems, where the lens diameter is not an infinite multiple of the wavelength (the ratio D/λ is not infinite), *diffraction* enters the picture. This form of diffraction is the same situation an engineer faces in designing an antenna which requires that, in order to get a narrower and narrower antenna beam, the antenna size broadside to the axis must be a larger and larger multiple of the wavelength.

There is a most useful Fourier-transform relationship [11] between the distribution of field strength of a plane wave across a plane aperture S and that observed at a single point at a distance r in the *far field* (roughly, many wavelengths away from the aperture). The plane-wave assumption means that the phase is constant across the

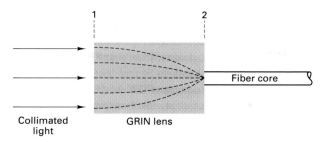

Figure   3-17. A graded index (GRIN) lens. Note the similarity of the ray paths to those in the graded-index multiimode case of Figure 3-9.

aperture. Let $\psi(x, y)$ be either E or H across the plane aperture, and $\psi'(x', y')$ be the observed value at the point (x', y') in a plane perpendicular to the axis and many wavelengths from the (x,y) plane. If the distance of the point (x', y') from the axis is a small fraction of r, the distance along the axis, so that cosines can be approximated by unity and sines by zero, then

$$\psi'(x', y') = \frac{j \exp(-jkr)}{\lambda r} \iint \psi(x, y) \exp[jk(xx' + yy')/r] \, dx \, dy \qquad (3.65)$$

This relationship is very useful generally, for example in deducing the directivity pattern of light radiated from the plane aperture of a semiconductor laser, but is also useful in talking about lenses. In the lens case, the wave across the aperture is no longer plane in the sense that the phases are all equal; there is a steady decrease of phase retardation from the axis outward toward the thinner parts of the lens. However, it can be shown [17] that the Fourier-transform relationship still holds.

Thus, thinking of the distribution $\psi(x, y)$ as being that across the plane of a lens, this lens acts as a Fourier transformer of collimated light. If the distribution of field strength across the aperture is gaussian, then that across the far-field plane will be gaussian. If the distribution across a rectangular aperture is uniform, the illumination across the far-field plane will be a $\sin^2 x / x^2$ function of x' and y'. If the aperture is a uniformly illuminated circle we get the Bessel function equivalent, giving the well-known circular *Fresnel interference fringes*. The finite nonzero width of the beam at the focal plane at the right in Figure 3-16 is called the *beam waist, w*.

One practical consequence of diffraction is that in coupling light into a fiber, the diameter of the lens must be large enough that the beam waist will be small enough to illuminate only the core of the fiber. Furthermore, the acceptance angle of the fiber, θ_a, given by Equation 3.61, must be larger than the cone angle of of the lens, which from Figure 3-16 is $D/2f$ for small θ,

$$\theta_a \geq \theta \approx \frac{D}{2f} \qquad (3.66)$$

Since, in addition to this, the lens diameter D must be large enough for the beam waist to be smaller than the fiber diameter, this implies a large focal length.

3.9 Modes in Fibers

Figure 3-18 shows a view of a step-index optical fiber. If the wavelength were much smaller than the fiber radius ($\lambda \ll a$), we could easily calculate several things. Since the core-cladding interface could be considered plane, Equation 3.53 would predict the value of θ_c, the value of θ_i below which the rays would pass into the cladding and be lost. For larger values of θ_i, the penetration depth into the cladding could be computed from Equation 3.56, in order to design the cladding to be thick enough so that a negligible amount of light would be lost. Modes of propagation would correspond to standing waves across the diameter $2a$, each mode having an integral number of half waves across this distance.

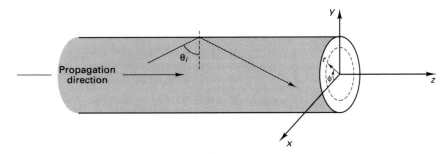

Figure 3-18. View of the end of a fiber, showing the cylindrical coordinate system and an idealized ray path.

Of course, all this would give quantitatively wrong answers, because geometric optics is in fact insufficient to deal with the real situation where the wavelength and diameter are of the same order of magnitude. What is required is to express the wave equation in cylindrical coordinates, and solve it subject to boundary conditions that represent the core-cladding boundary. In addition, if the core has a *graded index*, as in the middle diagram of Figure 3-9, the entire index profile $n(r)$ is part of the boundary conditions. For simplicity's sake, we shall confine ourselves to the step index case.

It will turn out that one mode can be supported for all wavelengths. This one mode will have two states that are *degenerate* with respect to each other (transverse component of the **E**-vector in orthogonal directions). As the ratio of core radius a to free-space wavelength λ increases, more and more modes become supportable (and each has a number of degenerate states that can be equal to or greater than two), and for any value of the ratio of radius to wavelength, the number of higher-order modes permitted is finite. By operating the fiber at a wavelength longer than that which will support two modes, the single-mode situation results, (with the single mode being composed of the two degenerate states or submodes). As we have seen, single-mode propagation is highly desirable because modal dispersion is then absent.

If the fact that even a "single" mode consists of two submodes is confusing, all this should become much clearer in Section 3.13. It will not be very important in the discussion before that point.

Particularly lucid explanations of the complexities of the complete cylindrical coordinate analysis are given in Section 8.1B of [18] and in [19]. We shall first present some pictures of mode patterns and then show most of the steps in the analysis that leads to them.

The wave we have been talking about in previous sections, a plane wave propagating in an infinite uniform medium, is called a *TEM* wave, to express the fact that both **E** and **H** are exactly transverse to the propagation vector **k**, i.e. the z-axis of the fiber. The fiber modes that emerge from the analysis are be called $TE_{\ell,m}$ modes when the **E** field is exactly perpendicular to z, and $TM_{\ell,m}$ when it is the **H** vector that has this perpendicularity. If *both* have nonzero z-components, the nomenclature $EH_{\ell,m}$ is used if the E_z-component has the larger effect, and $HE_{\ell,m}$ if the H_z-component predominates. Figure 3-19 shows the **E** vector across the fiber cross-section for several of these modes.

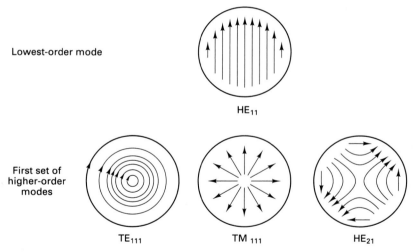

Figure 3-19. Distribution across the fiber cross-section of the **E**-vector for the first two sets of modes. (From [20], © McGraw-Hill, Inc.).

The parameters ℓ and m are integers. The subscript ℓ is the number of multiples of 2π that the **E** field seems to rotate to a non-rotating observer in making one round trip in azimuthal angle ϕ around a circumference within the core. The other integer m represents the number of quasi-sinusoidal "half-cycles" going radially in r from core center to core-cladding boundary at $r = a$. One can see this from Figure 3-19.

The solution of the wave equation to get this modal structure proceeds as follows. Assuming a lossless medium and a sinusoidal form for **E** and **H**, the wave equations (3.14) and (3.15) become

$$\nabla^2 \mathbf{E} + \left(\frac{n\omega}{c} \right)^2 \mathbf{E} = 0 \tag{3.67}$$

(where **E** is the vector complex envelope) and similarly for **H**. This equation, when rewritten in cylindrical coordinates, becomes

$$\nabla^2 \mathbf{E} = \mathbf{u}_r \left(\nabla^2 E_r - \frac{2}{r^2} \frac{\partial E_\phi}{\partial \phi} - \frac{E_r}{r^2} \right) + \mathbf{u}_\phi \left(\nabla^2 E_\phi - \frac{2}{r^2} \frac{\partial E_r}{\partial \phi} - \frac{E_\phi}{r^2} \right)$$
$$+ \mathbf{u}_z \nabla^2 E_z = - \left(\frac{n\omega}{c} \right)^2 (\mathbf{u}_r E_r + \mathbf{u}_\phi E_\phi + \mathbf{u}_z E_z) \tag{3.68}$$

Notice that the equation for E_z (the z-component of the complex envelope of the **E** field) is in a particularly simple form. The analysis exploits this fact. Writing out the expression for just this component

$$\frac{\partial^2 E_z}{\partial r^2} + \frac{1}{r} \frac{\partial E_z}{\partial r} + \frac{1}{r^2} \frac{\partial^2 E_z}{\partial \phi^2} + \frac{\partial^2 E_z}{\partial z^2} + \left(\frac{n\omega}{c} \right)^2 E_z = 0 \tag{3.69}$$

Given the solutions for E_z and H_z, we can obtain the equations for E_ϕ, E_r H_ϕ, and H_r.

We are interested in solutions that constitute waves travelling in the z-direction with some propagation constant β, and thus the z dependence of E_z is of the form $\exp(-j\beta z)$. Also, E_z must be periodic in ϕ, so the ϕ dependence will be $\exp(-j\ell\phi)$, where ℓ is the integer referred to several paragraphs ago. The thing we don't know yet is what the r dependence might look like. Substituting the form of the assumed z and ϕ dependences,

$$E_z(r, \phi, z) = E_z(r)\exp(-j\ell\phi)\exp(-j\beta z), \quad \ell = 0, \pm 1, \pm 2, \ldots \tag{3.70}$$

into Equation 3.69 gives

$$\frac{d^2 E_z(r)}{dr^2} + \frac{1}{r} \frac{dE_z(r)}{dr} + \left(\frac{n^2 \omega^2}{c^2} - \beta^2 - \frac{\ell^2}{r^2} \right) E_z(r) = 0 \tag{3.71}$$

which is the Bessel differential equation.

Now to impose the boundary conditions. Suppose the only change in index radially is at the core-cladding boundary $r = a$, where n changes from n_1 to n_2. This is the situation with single-mode fibers and step-index multimode fibers. Let us define

$\beta_1 = n_1\omega/c$ for the core and $\beta_2 = n_2\omega/c$ for the cladding. These quantities β_1 and β_2 are the propagation constants one would have had for *TEM* propagation in the core and cladding materials, respectively.

The condition that the wave will be guided is that it be propagating power in the core and be an evanescent wave in the cladding. This requires that the actual propagation constants be some $\beta_{\ell m} < \beta_1$ in the core and $\beta_{\ell m} > \beta_2$ in the cladding.

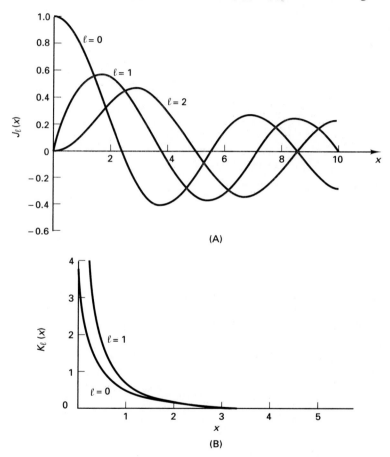

Figure 3-20. Bessel functions (A) Of the first kind (J_ℓ) and (B) Modified Bessel function of the first kind (K_m).

It is convenient to define

$$u_{\ell m}{}^2 = \beta_1{}^2 - \beta_{\ell m}{}^2 \tag{3.72}$$

and

$$w_{\ell m}^2 = \beta_{\ell m}^2 - \beta_2^2 \qquad (3.73)$$

for the core and cladding, respectively. Since the right sides are positive, both $u_{\ell m}$ and $w_{\ell m}$ are real. Equation 3.71 can then be written as two Bessel differential equations, one for the core

$$\frac{d^2 E_z(r)}{dr^2} + \frac{1}{r}\frac{dE_z(r)}{dr} + \left(u_{\ell m}^2 - \frac{\ell^2}{r^2}\right) E_z(r) = 0 \qquad (r < a) \qquad (3.74)$$

and one for the cladding

$$\frac{d^2 E_z(r)}{dr^2} + \frac{1}{r}\frac{dE_z(r)}{dr} - \left(w_{\ell m}^2 + \frac{\ell^2}{r^2}\right) E_z(r) = 0 \qquad (r > a) \qquad (3.75)$$

Notice that the big difference in these two equations is the sign before the last term. Excluding functions that approach infinity at $r = 0$, we get, finally for the solutions to Equations 3.74 and 3.75

$$E_z(r, \phi) \propto \begin{cases} J_\ell(u_{\ell m}r)\cos\ell\phi & r < a \quad \text{(core)} \\ K_\ell(w_{\ell m}r)\cos\ell\phi & r > a \quad \text{(cladding)} \end{cases} \qquad (3.76)$$

where $J_\ell(x)$ is the Bessel function of the first kind and order ℓ, and $K_\ell(x)$ is the modified Bessel function of the first kind and order ℓ. Sure enough, as Figure 3-20 shows, the J_ℓ functions look somewhat like decaying sinusoids, while the K_ℓ functions look like decaying exponentials, which is exactly what we would expect from our earlier discussion of evanescent waves in the plane-wave case.

Now that we have determined the z-components E_z and H_z, they can be substituted in the Maxwell equations 3.3 and 3.4 to get (tediously) $E_r(r, \phi)$, $H_r(r, \phi)$, $E_\phi(r, \phi)$, and $H_\phi(r, \phi)$.

Figure 3-21 summarizes most of the important information about the various modes. The abscissa is the very useful dimensionless *V-number* parameter

$$V = a\sqrt{u_{\ell m}^2 + w_{\ell m}^2} = \frac{2\pi a}{\lambda}\sqrt{n_1^2 - n_2^2} = \frac{2\pi a}{\lambda}\,\text{NA} \qquad (3.77)$$

The ordinate is also dimensionless and expresses the actual propagation constant β of each mode. It is seen to lie between the two *TEM* propagation constants $\beta_2 = 2\pi n_2/\lambda$ and $\beta_1 = 2\pi n_1/\lambda$, just as we would expect intuitively from imagining that λ is so small that the core dominates or so large that the cladding dominates, respectively.

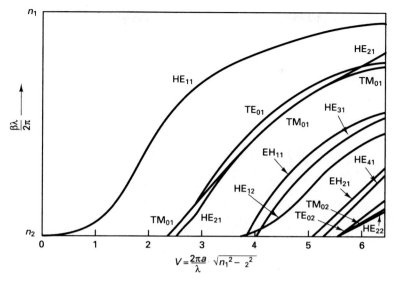

Figure   3-21. Normalized propagation constant versus "V number" for the first few orders of fiber modes. (From [21], © 1971 Optical Society of America).

Thus, to make a single-mode fiber for operation at a given λ, a small radius is used as well as a small contrast in the indices of core and cladding. Incidentally, the fact that fiber has lower attenuation at the longer near-infrared wavelengths is a piece of good luck, since it is obviously easier to fabricate single-mode fiber the longer the wavelength, since the core diameter can be larger.

By doing a numerical analysis of the modes, we can show that if the V-number is less than the magic number 2.405, then only single-mode propagation can exist. All the modes except HE_{11} (see Figure 3-19) have *cutoff wavelengths* above which no energy will propagate.   Below $V = 2.405$ the optical frequencies of all modes except HE_{11} are "below cutoff."   The HE_{11} mode has no cutoff and ceases to exist only when the core diameter goes to zero.

Figure 3-21 is also useful in connection with two more rules of thumb [20]. The first says that the number of modes that can be excited in a multimode fiber is approximately

$$M = \frac{V^2}{2} \qquad (3.78)$$

for large V, if we count the two degenerate states of the same mode as separate. The second says that the ratio of power carried by the cladding to that carried by both core and cladding is

$$\left(\frac{P_{clad}}{P_{total}} \right) = \frac{2\sqrt{2}}{3V} \tag{3.79}$$

Figure 3-21 is also useful in determining the *waveguide dispersion* for all modes above their cutoff by reading off $d\beta/d\lambda$. (Both the dependent and independent variables contain λ).

3.10 Taps, Splitters and Couplers

Physics of fused biconical tapered couplers

Figure 3-22 shows what happens if we place two pieces of fiber side by side in a flame and draw them into a *fused biconical tapered coupler*. Within each fiber, there is a long tapered section, then a uniform section of length Z where they are fused together, and then another positive taper back to the original cross-sectional configuration of two separate fibers.

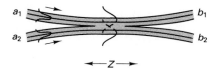

Figure   3-22. A fused biconical coupler, showing the coupling region of length Z and the squeezing of the core field out into the "air cladding."

The tapers are very gradual ("adiabatic"), so that a negligible fraction of the energy incident from either of the left-hand ports is reflected back to either left-hand port. For this reason, these devices are sometimes called *directional couplers*.

A variety of useful optical components can be made [22] that exploit the fact that the power coupled from one fiber to the other can be varied by changing Z, the length of the *coupling region* over which the fields from the two fibers interact, a the core radius in the coupling region, and Δa, the difference in core radii in the coupling region. The equation that governs the power coupling coefficient from one fiber to the other is

$$\alpha^2 = F^2 \sin^2 \left(\frac{CZ}{F} \right) \tag{3.80}$$

where F expresses the effect of the core-diameter difference,

$$F^2 = [1 + (234 \, a^3 / \lambda^3)(\Delta a / a)^2]^{-1} \tag{3.81}$$

and

$$C = 21\lambda^{5/2}/a^{7/2} \tag{3.82}$$

expresses the coupling between the fields of the two fibers. Since α^2 is a raised sinusoid, there are many tricks that can be played by trading off Z, λ, a, and Δa against one another. For example, a coupler could be made that delivers almost all the energy around 1.5 μ to one output and almost all the energy around 1.3 μ to the other. This is done by making a peak of α^2 in Equation 3.80 lie at 1.3 μ and a trough at 1.5 μ. By conservation of energy, this situation corresponds to a trough at 1.3 and a peak at 1.5 for the other output.

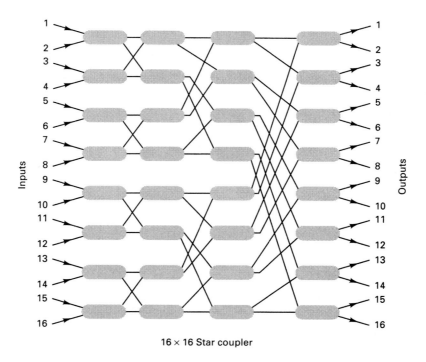

16 × 16 Star coupler

Figure   3-23. One way of using couplers to form an $N = 16$ network. A single 16 × 16 star coupler is made up of 32 fused biconical couplers.

To understand the physics, consider just one member of the pair of fused fibers and imagine what happens to its core and cladding fields as the core radius a is gradually reduced as the light propagates along the taper and into the coupling region. From examining Equation 3.77, it is clear that there is a significant decrease in V number, because of the reduction in the ratio a/λ. As Equation 3.79 shows, this decrease in V means that much of the field is squeezed out of the fiber (which now

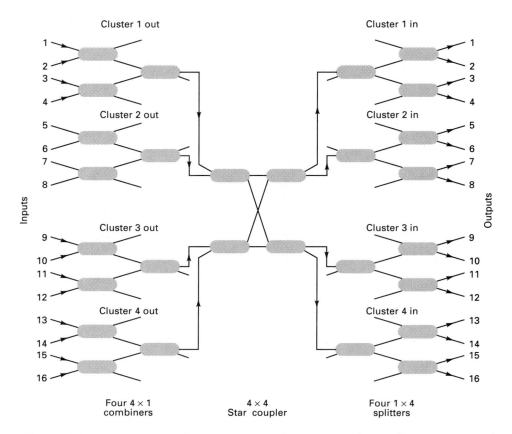

Figure   3-24. A second way of using couplers to form an $N = 16$ network, an arrangement in clusters of 4 nodes.

consists effectively of the glass core and glass cladding combined) into the air "cladding," as shown by the little curves in Figure 3-22.

It is this squeezing of the light out of the core that forms the basis of the coupler. Clearly, at a fixed λ, the coupling ratio α^2 can be made anything between zero and unity by varying Z and by adjusting the two core diameters in the coupling region. The coupling ratio into the other output tap will then be $(1 - \alpha^2)$. This is the basis of *taps and splitters*.

The parameters can be adjusted so that the power division is exactly fifty-fifty (that is, half the light at either input port at the left will appear in each output port at the right), and a number of these *3-dB couplers* can be connected in the *perfect shuffle* [23] topology of Figure 3-23 to form a *star coupler*. It is seen that the number of couplers required is

$$N_c = \frac{N}{2} \log_2 N \tag{3.83}$$

Figure   3-25. A commercial 32×32 single mode star coupler made from 2×2 fused biconical couplers. (Courtesy of Gould, Inc.).

and by following the light paths through the device, we can satisfy ourselves that indeed $1/N$th of the power at each input port will appear at all output ports.

It is customary to refer to couplers and splitters in terms of the number of input and output ports provided. For example, the device of Figure 3-23 would be called a "16×16" star coupler. An $N = 16$ network can be built by attaching each transmitter on one side and each receiver on the other. By populating only the diagonal of the matrix and the upper right hand quadrant, we might have a 1×16 *splitter*, or using the diagonal and the lower left quadrant, a 16×1 *combiner*. Figure 3.24 shows another way of building an $N = 16$ network using only 4×4, 1×4, and 4×1 devices.

The control over manufacturing processes is today so precise that 32×32 single-mode star couplers having only about 1.0 dB excess loss (over and above the splitting loss) and 0.5 dB rms scatter of port-to-port overall loss are available commercially. A photograph of such a unit is shown in Figure 3-25. The state of the art has now advanced to the point where each cell of which a large star coupler is made no longer must be a 2×2 unit, but can be at least a 4×4 unit. This again involves drawing the fiber so that the field in the core is now squeezed out, but now it must couple evenly to three other nearby other such fiber cores.

Planar couplers

As a final note on the physics of taps, couplers and splitters, it should be pointed out that they are particular lightwave network components that are immediate beneficiaries of the emerging *integrated optics* or *planar waveguide* technology. (Another component is the Mach-Zehnder interferometer form of tunable filter to be discussed in the next chapter). Planar waveguides are light paths that have been carefully fabricated as a region of increased refractive index either within some low-loss transparent substrate material (such as silica or certain polymers) or upon such a substrate. Comprehensive treatments of this field are given in [24, 25], and it can be seen from such discussions, that most of the ideas in this chapter apply essentially verbatim to integrated optic light paths, except that the waveguide cross-sections are usually rectangular, not circular.

The great appeal of integrated optic components is economic, the prospect of taking "lumped circuit" devices, such as the 32×32 star coupler of Figure 3-25, and implementing them lithographically. The key problem with integrated optical components is their attenuation, partly a matter of the bulk properties of the material and partly that of keeping the edges of the waveguide so smooth that scattering is negligible. Silica-on-silicon techniques have achieved attenuations of less than 0.1 dB/cm, and polymers a figure of 1.0 dB/cm. Figure 3-26 shows experimental 32×32 star couplers fabricated in silica on silicon.

Another way of building star couplers with planar components is shown in the 19×19 *planar lens* coupler of Figure 3-27 [26]. By shaping the curved region and the aiming of the different ports appropriately, the excess loss, including the port-to-port variation of field strength, was kept to within 3 dB.

3.11 The Coupler as a Network Component

We have just seen that, by adjusting the length of the coupling region, the fraction of the power presented to one input that appeared at one of the two outputs could be varied between zero and unity. The remainder of the power appears at the other output port. We shall now analyze 2×2 couplers (Figure 3.22) more thoroughly as a four-terminal black box, one having two inputs and two outputs [27]. The results will be seen to set some important constraints on network topology.

In the figure, the upper and lower input complex electric field strengths are represented as a_1 and a_2, respectively, and the output field strengths as b_1 and b_2. The relationship between these can be represented in terms of the *scattering matrix S* of the device [11]. S defines the relationship between the column matrix a of input complex envelopes and the column matrix b of output complex envelopes thus:

$$b = Sa \quad \text{where} \quad b = \begin{bmatrix} b_1 \\ b_2 \end{bmatrix}, \quad a = \begin{bmatrix} a_1 \\ a_2 \end{bmatrix}, \quad \text{and} \quad S = \begin{bmatrix} s_{11} & s_{12} \\ s_{21} & s_{22} \end{bmatrix} \quad (3.84)$$

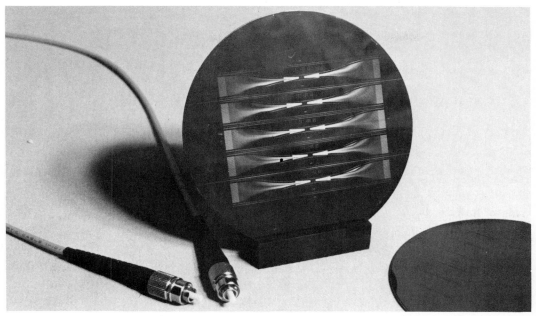

Figure   3-26. Four 32×32 single-mode splitters lithographed in silica on silicon by flame hydrolysis deposition and reactive-ion etching. (Courtesy of Photonic Integration Research, Inc.).

In general S is a 4×4 matrix because there can be an input and an output at each of the four ports. However, we shall analyze here the simpler 2×2 case, where only two of the ports receive inputs and only two deliver outputs, as shown in the figure. For fused biconical tapered couplers, as used in the systems we shall be discussing, this is a reasonable assumption, for two reasons. First, in the topologies to be dealt with, a distinction is always made between an input port and an output port. We shall see that the device of Figure 3-22 is never used in such a way that an input is applied to one of the two right-hand ports, nor is an output ever taken from one of the two left-hand ports. Second, the taper of fused couplers is so gradual that negligible energy is reflected back from the device toward either input. Therefore the two ports at the left of the diagram will be considered to be strictly input ports and the two at the right output ports.

There are two restrictions on any S characterizing a physically realizable passive device. First, as a consequence of the fact that Maxwell's equations have two solutions in counterpropagating directions, it can be shown [28] that the *reciprocity condition* that follows from this requires that

$$S^t = S \qquad (3.85)$$

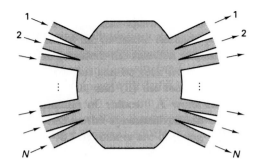

Figure   **3-27.** Geometry of an integrated planar lens star coupler (From [26], © 1989 IEEE).

(where t indicates the transpose). For example, the matrix element s_{12} must equal s_{21}. Second, if the device is lossless, the sum of the output intensities I_o must equal the sum of the input intensities I_i

$$I_o = b_1{}^*b_1 + b_2{}^*b_2 \; = \; I_i = a_1{}^*a_1 + a_2{}^*a_2 \quad \text{or} \quad b^+b = a^+a \qquad (3.86)$$

where the superscript * indicates complex conjugate and the superscript + indicates conjugate of the transpose. Expanding the left side b^+b by use of Equation 3.84 and the identity $(XY)^t = Y^tX^t$ gives

$$b^+b = (Sa)^+Sa = a^+S^+Sa = a^+a = a^+Ia \qquad (3.87)$$

in which I, the diagonal matrix of unity values, has been introduced in the last step. This requires that

$$S^+S = I \qquad \text{or} \qquad S^*S = I \qquad (3.88)$$

from Equation 3.85.

Writing out this last matrix equation, element by element, gives the following four equations, one for each element:

$$|s_{11}|^2 + s_{12}{}^*s_{21} = 1 \qquad (3.89)$$

$$s_{11}{}^*s_{12} + s_{12}{}^*s_{22} = 0 \qquad (3.90)$$

$$s_{21}{}^*s_{11} + s_{22}{}^*s_{21} = 0 \qquad (3.91)$$

$$s_{21}{}^*s_{12} + |s_{22}|^2 = 1 \qquad (3.92)$$

From these equations we can obtain most of the interesting properties of the 2×2 coupler as a system element.   Assume that the splitting ratio has been set so that $s_{11} = +\sqrt{1 - \alpha}$, a real number between 0 and 1. In other words, a fraction $(1 - \alpha)$ of the optical power at input 1 appears at output 1, the rest being tapped out to output 2. Without loss of generality, the electric field at output 1 is assumed to have a zero phase shift relative to the input.   The questions to be answered include the following. What fraction of the signal appears as output b_2?   What is its phase?   What constraints, if any, are there on the combined use of inputs 1 and 2 simultaneously? (Perhaps surprisingly, there is one, an important one).

From Equations 3.89 through 3.92, respectively, and using the fact that $s_{12} = s_{21}$, we have

$$s_{12}{}^* s_{21} = |s_{12}|^2 = \alpha \tag{3.93}$$

$$\sqrt{1 - \alpha}\ s_{12} + s_{12}{}^* s_{22} = 0 \tag{3.94}$$

$$\sqrt{1 - \alpha}\ s_{12}{}^* + s_{22}{}^* s_{12} = 0 \tag{3.95}$$

and

$$|s_{12}|^2 + |s_{22}|^2 = 1 \qquad \text{or} \qquad |s_{22}|^2 = (1 - \alpha) \tag{3.96}$$

from which

$$s_{22} = \sqrt{1 - \alpha}\ \exp(j\phi_{22}) \tag{3.97}$$

The phase angle ϕ_{22} is for the moment unknown.   Let $s_{12} = |s_{12}| \exp(j\phi_{12})$. Then Equation 3.90 gives

$$\sqrt{1 - \alpha}\ |s_{12}| \left[\exp(j\phi_{12}) + |s_{22}| \exp(j\phi_{22}) \exp(-j\phi_{12}) \right] = 0 \tag{3.98}$$

which is only possible for

$$\phi_{12} = \frac{(\pi + \phi_{22})}{2}, \frac{3(\pi + \phi_{22})}{2}, \frac{5(\pi + \phi_{22})}{2}, \dots \tag{3.99}$$

If we impose the condition of zero phase shift going from input 2 to output 2, then $\phi_{22} = 0$, and

$$\phi_{12} = \frac{\pi}{2}, \frac{3\pi}{2}, \frac{5\pi}{2} \ldots \tag{3.100}$$

from which

$$S = \begin{bmatrix} \sqrt{1-\alpha} & j\sqrt{\alpha} \\ j\sqrt{\alpha} & \sqrt{1-\alpha} \end{bmatrix} \tag{3.101}$$

which for a 3-dB coupler ($\alpha = 0.5$) is

$$S = \frac{1}{\sqrt{2}} \begin{bmatrix} 1 & j \\ j & 1 \end{bmatrix} \tag{3.102}$$

So we see that, independently of α, there is a 90° phase shift from input 1 to output 2 relative to that of output 1.

It is also clear that to deliver power from input 1 to output 1 we want α small, but this in turn diminishes the amount of power reaching output 1 from input 2. Thus there is no way that at one particular wavelength all the power from both inputs 1 and 2 can be delivered to the same output simultaneously. A 2×2 coupler cannot be used as an equal power combiner of light at the same wavelength without each input appearing at the output reduced by a factor of two in power.

The impossibility of getting, at one wavelength, full power from both inputs into one output has important consequences for real networks. Look again at the two toy examples of Figures 3-23 and 3-24. The arrangement of Figure 3-23 supports a 16-node network using one 16×16 star coupler. The splitting-loss component of the link budget is $10 \log_{10} 16 = 12$ dB. Real networks usually come in clusters of nodes, so it will often be desirable to connect nodes as shown in Figure 3-24, in which there are four clusters of four nodes each. A 4×1 combiner is formed by using output 1 of a 4×4 star coupler and leaving out those 2×2 elements in the upper right quadrant. Similarly, on receiving, a 1×4 splitter is formed from a 4×4 structure by feeding input 1 and leaving out those elements in the lower left quadrant. Now the total loss is $10 \log_{10} 4 + 10 \log_{10} 4 + 10 \log_{10} 4 = 18$ dB. We might have hoped that the combining of lightwave signals done in the leftmost combining couplers of Figure 3-24 could have been done losslessly, but alas, such is not the case.

In general, then, one must pay an extra price of $10 \log_{10} M$ dB for every M-fold combining operation just as with every M-fold splitting operation. The latter is to be expected, but it is an unpleasant surprise that the former is also the case.

This entire discussion holds true for the single-mode situation at one wavelength. In a single-mode 2×2 coupler, as long as α is not strongly wavelength dependent, if it has been adjusted to deliver power α at one wavelength to output 1, then a unit with power only $(1 - \alpha)$; the combined effect of the two conditions of reciprocity and power conservation is to force this condition on the elements of S.

For multimode couplers, the situation is much more complex, since energy entering at one wavelength in one mode may exit at the same wavelength in a completely different mode. As long as the four fibers entering the coupling region have the same diameter, the number of modes each will support is the same, by Equation 3.78. If we were to repeat the above analysis one mode at a time, we would find that there was a different α for every mode, but that by the time all the powers were added up, the same restrictions apply; we lose power by $10 \log_{10} M$ in combining M signals in a coupler. If the output-fiber diameter is made larger than the input-fiber diameter in order that the output fiber support many more modes than the input fiber, then this loss can be decreased.

Interconnections between widely separated taps, star couplers, splitters and combiners constitute the "glass network," the most basic functional level of the third-generation network that this book is all about, constituting the Physical Protocol Level mentioned in Chapter 12. If the reader goes all the way back to Figure 1-5, these components are plainly in evidence. They are the key physical component involved in organizing the system topologically, the subject of Chapter 11.

3.12 Sources of Multiplicative Noise in Fibers

As suggested earlier, fiber optic links have many orders of magnitude less severe a noise problems than electrical or satellite links, because even though there are noise sources associated with the lightwave detection process (Chapter 8), there is no noise being generated inside the fiber, and "since moving photons do not interact, whereas moving electrons do," there are no optical equivalents to electrical pickup, ground-loop voltages, and so forth. However, when dealing with fiber systems there are several sources of *multiplicative* noise. That is, there are random fluctuations of the overall end-to-end transmission loss producing effects that disappear when the transmitted signal is turned off. *Additive noise* components, such as thermal receiver noise, are present even when no signal is being sent from the transmitter.

The two forms of multiplicative noise in a fiber link, *modal noise* and *mode partition noise* [13], must be carefully controlled to obtain the performance that the fiber is capable of providing.

Modal noise can occur when there are many possible *propagating modes*, as with multimode fibers. With mode partition noise, it is the various *radiating modes* of the optical source that are referred to, specifically those radiated by lasers that emit more than one optical frequency simultaneously or in rapid succession.

Parenthetically, modal noise is sometimes called *speckle noise*, since it is associated with the familiar speckle pattern that one sees when many laser constructive and destructive interferences are visible across an aperture many wavelengths wide. In the multimode fiber case, the fiber cross section is sufficiently large that many modes are supported, as discussed earlier. If such a fiber is excited by a narrowband (high-coherence) source, such as a narrowband laser, any slight changes in a physical parameter such as bend curvature, pressure, temperature, reflectivity from a splice or connector, or the frequency of the laser (e.g., *laser chirp*, a frequency shift under modulation) will cause a shift of the phase and amplitude at various points in the speckle pattern across the fiber cross-section. That is, there will be a redistribution of energy between the multiple modes that the fiber supports. When you realize that you are dealing with frequencies of 10^{15} Hz and long distances, it is easy to see that tiny physical changes with time can produce large time-varying phase changes. In a real highly monochromatic multimode system it is virtually impossible to collect up into the detector all the energy from all these fluctuating modes, and so anything along the path that is at all mode selective (for example, a splice or connector with a slightly misalignment between fiber cores) will introduce a fluctuation in the signal delivered to the receiver.

The antidote to modal noise is to make sure that there is a great deal of averaging over these fluctuations. We may get rid of modal noise by using an optical source that is wideband, either because it is a laser with many modes (at many frequencies) or because it is a totally incoherent source such as a *light-emitting diode (LED)*. In such cases, there is so much fluctuation between modes during one bit time that there is little likelihood that during the duration of any bit the arriving signal will be in a null. Stated more quantitatively, the length of the link must be many times the source's *coherence length L_c*, the length in space corresponding to the bandwidth W_s of the source's spectrum

$$L_c \geq v_g / W_s \qquad (3.103)$$

where v_g is the group velocity.

Mode partition noise is one of the dominant sources of received signal fluctuation in single-mode fiber systems using ordinary Fabry-Perot lasers (which radiate at many frequencies) as distinguished from *single-frequency lasers*. These differences in forms of laser will constitute much of the material in Chapter 5. Mode partition noise is due to the interaction of the nonzero bandwidth of the laser source and the dispersion of the fiber. (An LED would not be used with single-mode fiber over anything but the shortest distances and lowest bitrates because its wide optical bandwidth of 20-30 nm would invite excessive time smearing due to dispersion). Simpler forms of laser diode radiate at several discrete frequencies corresponding to the resonant cavity modes of the active region where lasing is taking place (Chapter 5). Even when the total laser output power is constant, the actual radiation can jump around from one of these resonances to the other. When such an output is transmitted down a long fiber having nonzero dispersion, the various pieces of the waveform launched at different frequencies as a function of time will arrive time-overlapped and therefore be subject to randomly varying constructive and destructive interference.

To sum up, modal noise is a problem only with multimode fiber and narrowband sources, and mode partition noise is a problem only with single-mode fiber and wide-band (multifrequency) laser or LED sources. If one is interested in building a WDM system, single-mode fibers are required to avoid modal noise and then narrowband sources are required, not only to be able to crowd the channels closely together but to avoid mode partition noise.

These two sources of noise will not enter into our discussion of detection theory for optical systems in Chapter 8, because they can be engineered out of the system, unlike other sources of noise to be discussed. Detectability analyses exist that do treat modal noise [29] and mode partition noise [30] as statistical problems that form a part of an overall system model.

3.13 Polarization Effects in Fibers

State of polarization (SOP); birefringence

The state of polarization of the light travelling in a fiber optic network must be taken into account, since semiconductor laser diodes emit light having a specific polarization direction, and even more importantly, the functioning of many of the components of which the system is built depend on the state of polarization of the light at their input. These include various forms of tunable filters, amplifiers, modulators and receivers (Chapters 4, 6, 7, and 8, respectively). Fortunately, some components, such as direct photodetectors and most Fabry-Perot optical filters, are not polarization sensitive.

As discussed in Section 3.9, a wave with the $\mathbf{E}$ vector in a particular direction orthogonal to the propagation direction, for example the HE_{11} mode in Figure 3-19, is said to be polarized in the vertical direction. (We ignore the axial component in making this distinction). There is implicitly a possible second wave at 90°, a horizontally polarized component that is *degenerate* with the one shown. That is, any plane-polarized wave at some in-between angle can be composed of a linear combination of these two, one with $\mathbf{E} = \mathbf{u}_x E_x$, and the other with $\mathbf{E} = \mathbf{u}_y E_y$, where each $\mathbf{u}$ is a unit vector ($\mathbf{u}_z$ is assumed normal to the page in the figure). Therefore, what is nominally a "single-mode fiber" really has a split of this single HE_{11} mode into two degenerate modes, constituting the x- and y-components of the $\mathbf{E}$ vector.

It turns out [10] that if the fiber is not completely circularly symmetric, then E_x and E_y have phase velocities that are very slightly different, i.e., the medium is slightly *birefringent*. (By convention in discussing birefringence, the slower of the two is called the *ordinary* submode and the faster the *extraordinary* submode).

Moreover, as the light travels along a long fiber it inevitably encounters small imperfections such as bends and inhomogeneities that are not at all circularly symmetric and therefore affect light of two polarizations differently. Thus, light can get coupled from one polarization direction into the other. For short distances, such effects would not be noticed, but over many meters or kilometers of fiber that is never completely circular in cross section and never without small inhomogeneities and bends, they become important. The combined effect of the birefringence due to noni- deal circular symmetry and the small discontinuities is to produce a situation in which light launched with a particular state of polarization (SOP) will, in general, change its SOP gradually along the path.

For the example at the top of Figure 3-19, the SOP of the two degenerate states of HE_{11} would be said to be "pure vertical" or "pure horizontal," respectively. The specification of the SOP includes any phase offset between degenerate states. For example, if vertical and horizontal components are exactly in phase, some angle of *linear polarization* results, the angle depending on their relative strengths. If they are in phase quadrature, a pure *circular* SOP results, provided $E_x = E_y$, or *elliptical* polarization if they are unequal. Figure 3-28 shows various elliptical SOPs in terms of the time trajectory (arrowhead) of the instantaneous **E** vector during one cycle. In other words, the **E** vector makes one round trip around the ellipse in the direction shown by the arrowhead once every $2\pi/\omega$ seconds.

The figure illustrates the trajectory of the instantaneous **E** vector for various values of the ϕ, the phase difference between E_x and E_y. The figure describes the case

$$\mathbf{E} = \mathbf{u}_x |E_x| + \mathbf{u}_y 0.7 |E_x| \exp(-j\phi) \qquad (3.104)$$

The Poincaré sphere

We now introduce the *Poincaré sphere* of Figure 3-29, a handy graphical way of representing not only the SOP of a lightwave signal, but also the interaction of this signal with a polarization-sensitive device upon which the signal might impinge.

The *Stokes vector* [15] is not a real vector in space, but a four-dimensional vector in the matrix sense. It has the components

$$
\begin{aligned}
S_0 &= |E_x|^2 + |E_y|^2 \\
S_1 &= |E_x|^2 - |E_y|^2 \\
S_2 &= 2|E_x||E_y| \cos \phi \\
S_3 &= 2|E_x||E_y| \sin \phi
\end{aligned}
\qquad (3.105)
$$

The Stokes vector has a practical significance, since the components can be measured with suitable instruments (*polarimeters*). The physical interpretation is that S_0 represents the total power, S_1 represents the excess of horizontally polarized over vertically polarized power, S_2 represents the amount of 45° polarized power, and S_3 the amount of right-circular polarized power.

The Poincaré sphere is a graphical way of representing the SOP of the light, as expressed by the Stokes vector, in a cartesian coordinate system (X, Y, Z) of S_1, S_2, and S_3, respectively. The radius is the total power S_0. The reason that three Poincaré-

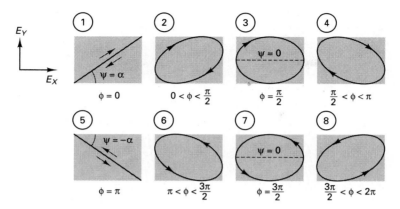

Figure 3-28. Elliptical polarization with various values of the phase difference ϕ between vertical and horizontal components, for a particular choice of the ratio of $|E_y|$ to $|E_x|$. The angle ψ is the tilt of the polarization ellipse.

sphere coordinates are sufficient, while there are four Stokes-vector components, is that one of the four is redundant $(S_1^2 + S_2^2 + S_3^2 = S_o^2)$. Completely right-circular polarization $(S_1 = S_2 = 0, \ S_3 = S_0)$ is represented by the north pole, and left circular ($S_1 = S_2 = 0, \ S_3 = -S_0$) by the south pole.

All purely linear polarization states ($\phi = 0$ or π, i.e. $S_3 = 0$) are represented by points on the equator. Of these, linear horizontal ($S_0 = S_1, \ S_2 = 0$) and linear vertical ($S_1 = -S_0, \ S_2 = 0$) are at the opposite ends of the X-axis.

The three-dimensional coordinate system of the Poincaré sphere must not be confused with either (1) the physical coordinate system of the travelling radiation, (2) the physical two-space of the ellipse in Figure 3-28, or (3) the electrical engineer's representation of a sinusoid as a phasor, a vector in a two-space whose unit vectors are a unit in-phase signal and a unit quadrature-phase, respectively. Note that orthogonality of two signals in the latter space (e.g., a 90° phase shift) is represented by two phasors being at 90° from one another, whereas in the Poincaré sphere, orthogonality is represented by the two SOPs being *antipodal* to each other. For example, right-circular light (north pole) is 50 percent correlated with linear (equator), but is orthogonal to (i.e., has zero correlation with) left-circular light (south pole), and so forth for other pairs of SOP points.

As will be discussed in connection with signal detection (Chapter 8) and multiaccess protocols (Chapter 13), the correlation between two functions of time is a normalized time average of their product. In general, if the SOPs of two monochromatic light waves are separated by an angle y on the sphere, the normalized correlation is [31]

$$\rho = \cos \frac{y}{2} \qquad\qquad (0 \leq \rho \leq 1) \qquad\qquad (3.106)$$

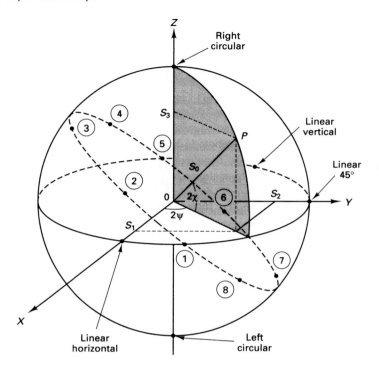

Figure 3-29. Poincaré sphere representation of the state of polarization of completely polarized light. The numbered dashed line indicates the evolution of polarization state depicted in the example of Figure 3-28.

Phase difference ϕ can also be visualized on the sphere. The Stokes-parameter representation in the spherical coordinate system (S_0, 2χ, 2ψ) of the Poincaré sphere is, given, respectively, by the total power S_0, by

$$
\begin{aligned}
S_1 &= S_0 \cos 2\chi \cos 2\psi && \text{by} \\
S_2 &= S_0 \cos 2\chi \sin 2\psi && \text{and by} \\
S_3 &= S_0 \sin 2\chi
\end{aligned}
\tag{3.107}
$$

where ψ is the spatial tilt of the ellipse of Figure 3-28 and χ is related to the phase difference ϕ by

$$
\sin 2\chi = \sin 2\alpha \, \sin \phi \quad \text{where} \quad \alpha = \arctan \left(|E_y| / |E_x| \right)
\tag{3.108}
$$

Thus, if the horizontal and vertical components are equal, so that $\alpha = \pi/4$, then lines of equal phase shift ϕ between the two components will be lines of equal latitude on the sphere, with $\phi = 0$ corresponding to the equator ($\chi = 0$). If the two are 90° out of phase, such that $\phi = \pi/2$, then $2\chi = \phi = +90°$ (north pole) or $-90°$ (south pole), depending on the sign of the phase shift. If the two components are unequal, as in Figure 3-28, the equi-phase shift lines are still lines of equal latitude, but they do not go all the way to the poles; they go only as far as $\sin 2\alpha$. The trajectory travelled on the Poincaré sphere by the sequence of examples in Figure 3-28 is depicted as the dotted line in Figure 3-29.

Birefringence and beat length

Suppose a wave is launched having a given SOP, as given by a particular point on the sphere. As the wave propagates down the fiber, one of the degenerate submodes (the extraordinary one) will have a slightly faster velocity than the other (the ordinary one). This difference in propagation velocities, birefringence, will figure importantly not only in our discussions of propagation in fibers, but also in certain forms of tunable filters (Chapter 4) and modulators (Chapter 7).

In a long fiber, the slight birefringence, together with the mode coupling due to inhomogeneities, causes a progressive increase of the relative phase ϕ and eventually a gradual change of the relative strengths of E_x and E_y. The SOP point begins to change (Figure 3-30) and move over the surface of the sphere. The physical distance the wave travels before the phase shift ϕ completes a change of 2π radians is called the *beat length*, and is given by the condition

$$L_{beat}\,\Delta\beta = 1 \qquad \text{i.e.,} \qquad L_{beat} = \frac{1}{\beta_o - \beta_e} \qquad (3.109)$$

where the subscripts "o" and "e" refer to the *ordinary* and the *extraordinary* wave, respectively. In the example of Figure 3-30, the light is launched at 45° polarization.

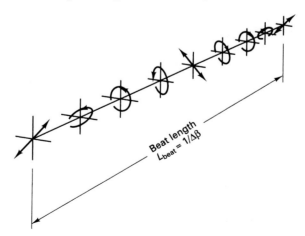

Figure 3-30. Showing one evolution of the polarization state along a fiber. The light traverses a distance L_{beat} before returning to the original polarization state.

Typical values for the beat length of commercial single-mode fiber are one-half to several meters. Lest this strike the reader as a short distance, think of how many wavelengths this is. By the time the wave has propagated over network distances, there is no usable relationship left between the transmitted SOP and the received SOP. It is perfectly possible to launch an optical signal as right-circular polarized and have it arrive left-circular. Fortunately, the received SOP does not vary rapidly, the time constant ranging from measured values of a few seconds while the fiber is being installed [32] to a few tens of hours for fiber already installed in ducts [33].

Because of the slightly different propagation velocities of the the two polarizations, we might also worry that the resultant pulse smearing (*polarization dispersion*) could lengthen the received pulse enough to limit the achievable data rate. This effect has been found to be too small to cause any significant problem for even long-distance single links, much less LANs and MANs. Polarization dispersion has been measured at about 0.8 picoseconds per 100 km of fiber [34].

We see that some unpleasant things happen to the polarization of light during propagation, in particular the unpredictable and slowly time-varying change of the received SOP. But several even more unpleasant things do not happen. The fluctuation of polarization is very slow, many orders of magnitude too slow for the bandwidth of monochromatic light to increase to the point where the two components could become decorrelated and thus the light could become *depolarized*.

If the light is non-monochromatic (has a nonzero spectral width) **and** random, so that the two components E_x and E_y have zero correlation, the light is said to be *completely depolarized*, and is represented in the Poincaré sphere coordinate system as a point at the origin. If the two components are not 100 percent correlated, the light is said to be *partially polarized*, and the point is at a fractional distance between the origin and the outer surface of the sphere given by the percent correlation. While a partially polarized or depolarized light wave has finite bandwidth, the converse is not necessarily true; if the light has nonzero bandwidth, it is still possible for the two components to be completely correlated wideband functions of time.

Another useful thing to know about polarization states is that if we send two light signals into the fiber and they have two orthogonal SOPs (antipodal on the sphere), they will arrive orthogonally polarized [35]. This will prove to be important for the form of modulation called *polarization shift keying*, where two orthogonal states represent binary 0 and 1.

Optical isolators and Faraday rotation

Certain optical components, notably laser diodes (Chapter 5) and amplifiers (Chapter 6), are very sensitive to light reflected back into their outputs from such things as splices, connectors, or filters. Therefore it is often necessary to place an *isolator* [36] immediately downstream from the device.

Isolators for use in fiber optic communication systems all use the nonreciprocal nature of *Faraday rotation*, a change of SOP that occurs in some materials in the presence of a magnetic field. If a plane-polarized wave is incident on an appropriate piece

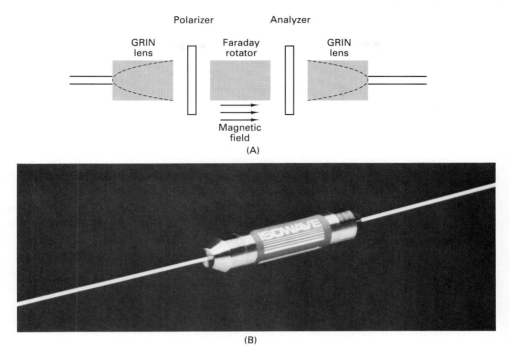

Figure   3-31. Isolators using Faraday rotation (A) Principle, (B) A commercial pigtailed unit. (Courtesy Isowave, Inc.).

of transparent material exhibiting Faraday rotation, and if there is a magnetic field parallel to the propagation vector, then the piece of material will rotate the plane of polarization by an angle proportional to the product of magnetic field strength and interaction length.

Now what is interesting is that if the output light encounters a reflection downstream from the device and passes backward through it, the device does not undo the rotation, it rotates the polarization plane by the same amount again.

Figure 3-31(A) shows how an isolator works. The *polarizer* passes only the vertically polarized component. The Faraday rotator has the magnetic field and interaction length adjusted so as to produce a 45° rotation, and this light is passed by the polarization *analyzer*, a component identically with the polarizer, but rotated by 45°. Light returning from some undesired reflection, being polarized at 45° from having passed through the device once, will be rotated by another 45° in the Faraday device and will thus be orthogonal to the incoming plane-polarized light.

Isolators have been made that have over 30 dB of isolation while introducing only a dB or two of loss. Such a unit is shown in Figure 3-31(B). It is clear that an isolator, in addition to being complex and therefore expensive, will introduce a significant loss if the incoming light is not properly plane-polarized. This is not as bad as it sounds, since isolators are usually used at the transmitting end, where the isolator can be adjusted to match the laser diode, which supplies highly linearly polarized light, as we shall see in Chapter 5. When isolators are used nearer the receiver, there is the loss to be concerned about, but this can be overcome either by more amplifier gain or by *polarization diversity*, involving two isolators operating in polarization quadrature, as will be described in Section 8.12.

3.14 Nonlinearities in Fibers

Because of the tiny cross-sectional area of the fiber core, it should not be surprising that, for system designs having high per-node power levels, many nodes, and long runs of fiber, nonlinear effects must be thought about in the system design. In the next section, we shall describe one way that has been found to turn the small nonlinearity of refractive index to advantage. Here we discuss the upper limits on power levels imposed by the possibility for such effects as interchannel crosstalk in dense wavelength-division systems. Crosstalk is the contamination of the signal in one channel by the modulation in another.

The system architect has to work between two levels of power, a maximum given by the nonlinear effects or safety considerations or both, and a minimum set by the weak signal-detectability thresholds to be dealt with in Chapter 8. This is a great deal of latitude with present systems, but it is still worth reviewing here briefly where the upper power limits lie beyond which nonlinearities will produce either a significant loss of power per channel or an unacceptably high level of crosstalk. In this section we shall briefly review the four nonlinear effects that have been judged most important for lightwave systems, giving essentially a summary of a single paper, the tutorial treatment of [37].

Figure 3-32 summarizes the upper limits on power per channel versus number of channels before each of the four nonlinearities causes the overall system penalty to exceed 1 dB. This penalty could be due to a typical signal being depleted by 1 dB or due to the crosstalk of unwanted modulation from other channels effectively degrading the received signal to noise ratio by 1 dB. The calculation assumes a wavelength-division system working along a 22-km fiber of 8-μ core diameter in the 1.5-μ band, the channels being 10 GHz (0.08 nm) apart. This set of numbers is representative of a third-generation MAN with up to 1000 channels (which would span 80 nm of the low-attenuation band). The limits in the figure may appear somewhat alarming until you realize that they contain some fairly pessimistic assumptions. For example, if we consider a network of N nodes, each feeding a power level P into a star coupler, such as the one shown in Figure 3-23, and then follow the course of these signals through the the star coupler, we find that there is no place in the network, not even inside the coupler, where the power level is greater than P. Problems with nonlinear effects would be likely to occur in network geometries where many power sources are combined at full power over a single long run of fiber, for example the tree topology of Figure 1-5(C).

Two of the nonlinear effects are forms of scattering, defined as the combined effect of many random interactions of light with solid, liquid or gaseous material. We

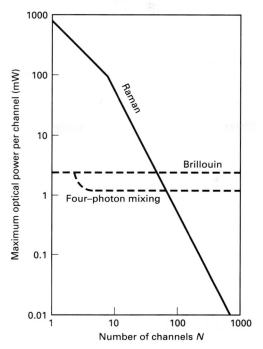

Figure 3-32. Maximum power limits per channel set by fiber nonlinearities for wavelength division systems as a function of number of channels. (From [37], © 1990 IEEE).

have already met Rayleigh scattering, in which light at one wavelength had no effect on light at other wavelengths but did have a wavelength dependence of scattered field strength. The two other important forms of scattering are *stimulated Raman scattering (SRS)*, and *stimulated Brillouin scattering (SBS)*. Both of them involve light at one wavelength affecting what is going on at another wavelength. We introduce them here and discuss them further in connection with amplifiers in Chapter 6.

Stimulated Raman scattering (SRS)

SRS is an effect in which the energy from a photon incident on a molecule delivers part of its energy to mechanical vibration of the molecule and part into reradiated light (*Stokes light*) of longer wavelength than the incident light (because energy has been lost). The process can be *stimulated* by some second photon that happens along at this longer wavelength, so that one such incident photon emerges as two of them, the process called *stimulated Raman amplification*. For a long time, the Raman effect was considered the most promising direction for lightwave amplifiers made out of fiber, but recently the doped-fiber approach that will be discussed in Chapter 6 has proved much more successful.

The efficiency of the stimulated emission process (Raman gain) increases more or less linearly with the wavelength difference between the two photons (the one supplying the energy and the one receiving it) up to a wavelength difference of about 120 nm. This means that there will be crosstalk between signals at two wavelengths. The curve in Figure 3-32 is somewhat pessimistic, having been derived by assuming that all channels are transmitting continuously, rather than doing a statistical average over all possible bit patterns. Also, the analysis did not include the dependence of group velocity with wavelength (dispersion), which would tend to desynchronize the bit pulses at different wavelengths.

SRS is more or less isotropic in its directionality, so a propagating lightwave signal can be affected by other light that is propagating in the same or in the opposite direction.

Stimulated Brillouin scattering (SBS)

SBS, on the other hand, is highly directional, with a null in the forward direction, so that in a fiber essentially all of the scattered energy is counterpropagating with respect to the signal. SBS is due to interaction between the travelling light wave, composed of photons, and a travelling sound wave that it induces, which can be considered as composed of quantum sound particles, *phonons*. The phonon model of sound will prove very useful in the next chapter when we discuss optical filters based on acoustooptic interactions.

For present purposes it is sufficient to know that an incident photon of a certain frequency (and travelling at light velocity) can produce a phonon travelling at sound velocity in the other direction and a Stokes photon that is down-shifted in frequency roughly by the ratio of sound velocity to light velocity. For 1.5-μ light this shift is some 11 GHz. Any light travelling in the reverse direction at just the right 11-GHz offset will stimulate the Brillouin scattering process and thus be amplified. The frequency selectivity of this effect is very narrow, some 20 MHz. In other works, approximately 11 GHz won't do; it has to be exactly some particular frequency within a few MHz. Each wavelength channel acts independently of all the others; hence the horizontal line in the figure for SBS, which assumes that all channels are sending continuous (CW) sinusoids.

SBS, which may be useful for extremely narrowband light amplification, is not expected to pose a significant system problem unless there are many transmitters all using external on-off modulation of extremely narrowband laser sources. Only then does very much of the transmitted energy fit into the narrow acceptance band. With external modulation (Section 7.3) the on-off bit pattern interrupts a continuous sinusoid rather than starting off with a new value of time-origin phase ϕ in Equation 3.23 with each new bit.

The two remaining nonlinear effects are a consequence of the same phenomenon, the fact that the dielectric constant of the silica medium is very slightly nonlinear, the index becoming slightly greater as very high power density levels are reached.

Four-photon mixing

This is also known as "three-wave mixing." These are just fancy terms for third-harmonic distortion, well known to radio engineers and hi-fi buffs. When a medium is slightly nonlinear, the first few terms in a series expansion are the important ones. Second-harmonic distortion simply produces sum and difference terms that usually lie outside the band of interest (see Section 9.3), but third-harmonic distortion can have an interfering effect in frequency-division multiplex schemes where the various frequencies are exactly equally spaced. Call the indices of three of these frequencies i, j, and k. The third-order cross-product of these three would be indexed i, j, k, from which we have intermodulation cross-product terms like

$$f_i - f_j + f_k \qquad (3.110)$$

which, for the equally spaced situation, is none other than f_j for all cases in which i, j, and k are consecutive integers. Thus, the light signal at f_j can be contaminated by crosstalk involving f_i, f_j, and f_k. The term four-photon mixing comes from the fact that when there are three interfering photons of the three frequencies there is a fourth cross-product resulting. For each signal channel there are a number of such cross-terms, and the net result is that in an ideal nondispersive medium the cross-products are in phase and severe contamination of any one channel by its neighbors can occur.

In actuality, two effects lessen the amount of four-photon mixing that will occur. There is dispersion, which disrupts the phase coherence among cross-products. Moreover, the frequencies in real networks will not usually be exactly equally spaced. An exception could occur when one synthesizes all the frequencies in a wavelength-division network from one source and distributes them perfectly to all nodes, a possibility that is discussed in Section 10.4.

Carrier-induced phase modulation

This is the effect that occurs when the phase of a signal arising in a given channel is shifted randomly due to strong fluctuating signals in all other channels causing changes of index. CIPM is relevant only in systems that use phase-sensitive forms of detection, such as phase-shift keying or differential phase-shift keying. We shall see in Chapter 8 that this restricts the discussion to coherent receivers and that these have other problems that must be overcome that are more serious than CIPM.

3.15 Solitons

Some years ago, mathematicians playing with nonlinear differential equations discovered that there were certain singular conditions under which very stable "solitary" solutions were obtained that seemed to defy the normal rules. When applied to wave propagation, these results said that, under such conditions, it was possible to launch a single pulse of a very special shape and have it propagate indefinitely without being smeared out by dispersion, even though the medium did have a dispersive character to any other kind of signal. Such a pulse is called a *soliton*, and the surprising thing is that optical fibers can support soliton transmission. In principle, solitons could eventually open the way to extremely high-speed long-distance lightwave systems using very short time-division pulses.

The bad news is that although the soliton idea will work (and work for essentially arbitrary pulse duration), to propagate as a soliton the pulses must be of not only exactly the right shape, but of the right amplitude too, which means that they must be amplified periodically along the path. The good news includes the fact that, since pulse distortion due to dispersion does not occur, the time duration of a bit can be very short (tens of picoseconds). Also, the several-hundred-milliwatt power level required for such bitrates is achievable by lasers and photonic amplifiers. Furthermore, if the launched pulse is not quite the right shape, as it propagates it will converge to the right shape and maintain that shape.

The soliton effect depends on playing off the slight dispersiveness of the fiber against the slight nonlinearity of index with field strength. Both effects lead to a "chirping" of the pulse, a change of optical frequency with time. With the index nonlinearity, this will obviously occur at the part of the pulse having the highest amplitude. With dispersion, it occurs where the pulse is richest in harmonic content.

Now for wavelengths greater than the zero point at 1.3 μ, the two effects are of opposite sign. Therefore it seems plausible that we ought to be able to discover some pulse shape that would propagate as a stable soliton. Such a pulse ought to be one that is richest in harmonics in the vicinity of the amplitude peak, and very smoothly varying elsewhere. But what exactly should that pulse shape be? One could try different pulse shapes for years looking for the right one before thinking of trying the reciprocal hyperbolic cosine, or one could just give up and go to Reference [38], where there is particularly clear line of argument that shows that the reciprocal hyberbolic cosine is, in fact, the right answer. This function is shown in Figure 3-33.

The analysis goes as follows. Assuming that the index dependence on field strength is quadratic ($n = n_o + n_2 |E|^2$), a nonlinear differential equation can be obtained describing the pulse $a(z, t)$ received when the pulse launched at the origin has a complex spectrum $A(\omega)$. To show the effect of dispersion, a second such equation can be obtained from Equation 3.42, which likewise gives the pulse shape $a(z, t)$ at some distance z from the origin. Adding the two, a new nonlinear differential equation is obtained which has the solution

$$a'(z, t) = A_o[\exp(-jaz)] \operatorname{sech} \left(\frac{t - \beta'z}{\tau} \right) \qquad (3.111)$$

where

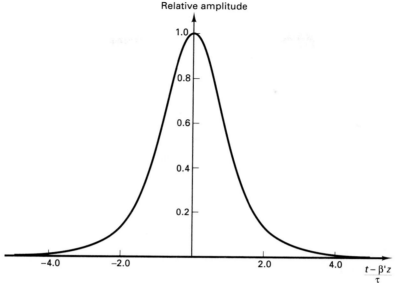

Figure   3-33. Shape of a stable soliton pulse.

$$a = \frac{\beta''}{2\tau^2} \qquad (3.112)$$

in which β' and β'' are the first and second derivatives of the propagation constant with respect to frequency, as defined in Equation 3.40, τ is 1.75 times the 3-dB pulse width for the bitrate in use, and

$$|A_o|^2 = -\frac{n_o}{n_2}\frac{\beta''}{\beta_o\tau^2} \qquad (3.113)$$

Sure enough, the plot of Equation 3.111 in Figure 3-33 shows a pulse for which the tighter curvature (containing the higher frequency components) lies nearer the amplitude peak. Equation 3.111 indicates specifically that, as the pulse propagates, there is a slow added rotation of the phase with z, but there is no z dependence of the pulse width.

Many engineering problems will have to be solved before soliton-based systems are practical, but the idea has great potential. A number of experiments have been done that exhibit soliton transmission over distances up to 10,000 km. The solitons are quite stable and not easily destroyed by small propagation inhomogeneities, as long as the power level is regenerated frequently along the path. Most experiments (e.g., [40]) have involved sending the same soliton around a loop of fiber several tens of kilometers in length with an amplification taking place every round trip. Recently,

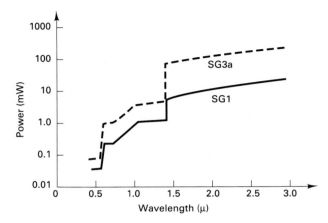

Figure 3-34. Wavelength dependence of the maximum permitted optical power to be carried in a fiber on the basis of possible eye damage. (Data from [39]).

erbium doped fiber amplifiers, to be discussed in Chapter 6, have proved particularly suitable for this function.

3.16 Fiber Power Limits Imposed by Eye Safety

The nonlinear effects that can occur at high powers are not the only phenomena that set upper limits on the optical power that can be carried in a fiber. There is also an upper limit on power beyond which it is possible to direct harmful levels of exposure into the eyes of humans that are using and maintaining the system, or, for example, are playing with the accidentally severed loose ends of fiber communication links.

These questions have been studied extensively and there are standards on the subject, e.g. Reference [39]. As that document points out, any optical fiber communication system can normally be considered quite harmless, even to service personnel, because of the concentration of light flux inside a jacketed cable, both in the link itself and at the source, which is usually a pigtailed laser diode or LED. However, the document goes on to point out, there are special high-risk situations, and these are considered important enough that surprisingly low power levels are considered to be the maximum that will be permitted in commercial equipment and systems.

One might think that the high-risk cases involve exposure to nonusers, such as the child who picks up the end of a severed fiber optic cable, and being interested in fiber optics (having perhaps read this book), starts staring at the fiber end, looking for the core. Actually the rules are aimed at service personnel. The standards document [39] says that "when an optical connector is removed during service and an optical instrument, such as an eye-loupe, hand magnifier or a microscope.... is used to view the end of an energized fiber, it is conceivable that all of the radiant energy can impinge on the eye." The standards body has also considered it conceivable that the length of time that some service person might inadvertently stare into the lit fiber can be as long as 100 seconds.

Fortunately for the communication fraternity, the longer wavelengths, 1.3 and 1.5 μ, are well absorbed in the outer layers of the lens of the eye, but as we go more toward visible wavelengths (400 to 700 nm), the lens becomes more transparent and the retinal damage for a given power level increases.

Figure 3-34 summarizes the upper limits of CW optical power. A 100-second exposure duration is assumed. Systems conforming to Service Group 1 are "exempt from all control measures and from any form of surveillance." Above the SG1 limits is the next category, Service Group 3a, and here significant network-management complexities set in. Special connectors may be required that defocus the light when the connection is disengaged. Personnel not specially trained may have to be excluded from space in which maintenance is being done, and so forth. If the power level exceeds the SG3a limits (Service Group 3b, the highest), light radiated from the fiber end is considered capable of producing eye damage to the naked eye.

Depending on the network being built, these considerations can very well impose more severe power limits on the system design than do either the unavailability of high power components or the nonlinear effects discussed in the last section.

3.17 What We Have Learned

In this chapter we used the vector Maxwell's equations as a starting point from which to develop a quick tour through all the important things that go on in the parts of the network that are "made of glass," the fiber links, couplers, and splitters. It was seen that, because the propagation distance is such a large multiple of the wavelength, and because the light is so concentrated spatially, some physical effects that are locally very tiny accumulate into large total effects and must therefore be dealt with carefully in designing the links in the system. These effects include such things as chromatic dispersion, loss due to various interactions of light with atoms or entire molecules, small changes in polarization state due to fiber eccentricity and inhomogeneity, and small changes in index at high power levels. The quantitative picture of these processes that was presented should be adequate to handle most of the relevant problems in architecting and implementing a practical optical network.

The subdivision of light travelling in a fiber into a number of possible propagation modes, and the calculation of the number of such modes, their polarization, relative power, and chromatic dispersion, has given us the information we need to decide when to use single-mode fiber and when to use the more convenient multimode fiber. It also underlies the discussion of the splitters and couplers that will be seen to exert a strong control over what forms the physical topology of the network can take.

The discussion of propagation modes will also prove relevant when we get to the light-generation processes inside a laser diode. Also relevant to the discussion of optical sources will be the treatment given in the present chapter of numerical aperture, the use of lenses to couple into fiber ends, and the Fourier-transform picture of the directivity pattern of light from a plane distribution of optical field, whether this field be that at one of the facets of a light source (laser or light emitting diode) or at the end of a fiber.

3.18 Problems.

3.1   Show that using the complex-envelope notation to get the product $a(t)b(t)$ where $a(t) = \cos\ (\omega t + \phi_a)$ and $b(t) = \cos\ (\omega t + \phi_b)$ gives the wrong answer.

3.2   (*Complex Poynting vector*)   If **E** and **H** are the physical vector complex envelopes of sinusoidal electric and magnetic fields, respectively, (A) Show that the time-averaged Poynting vector is

$$\langle\, \mathbf{P}\, \rangle\ =\ \frac{1}{2}\, \mathbf{E} \times \mathbf{H}^*$$

(B) In addition to this average, there is another term in the Poynting vector for sinusoids.   What is your interpretation of it?

3.3   (*Rayleigh backscatter*)   As Figure 3-8 demonstrates, most of the attenuation is due to Rayleigh scatter.   This form of scattering happens to be isotropic, so that some is scattered back toward the transmitter.   If you have a fiber with an NA of 0.1 for which all of its 0.5-dB/km attenuation is due to backscatter, and you send a single light pulse of duration $T = 1$ nsec into it, how many dB down will be the peak of the Rayleigh backscatter waveform?   Assume that the core index $n_1 = 1.45$.

3.4   You want to build a long 1-Gb/s link with signal-regenerator and pulse-reshaping stations spaced as infrequently as possible.   Assume a transmitter power of 1 milliwatt and a receiver that requires -30 dBm of receiver input power in order to achieve the desired bit error rate. The dispersion at 1.5 μ is 20 psec/km/nm, and the attenuation is 0.25 dB/km.   At 1.3 μ, the dispersion is zero and the attenuation is 0.5 dB/km.   Roughly how far apart would you have to place the stations if you chose (A) A wavelength of 1.5 microns and single-mode, graded-index multimode or step-index multimode fiber, respectively.   (B) Suppose you chose 1.33 μ wavelength instead?   Assume that the spectrum of the transmitted signal can be roughly approximated by a pair of sinusoids 1 GHz apart.

3.5   The cladding in a typical multimode fiber is only 35 microns thick, and yet in analyses of wave propagation in fibers it is assumed to be infinitely thick.   Assuming this is permissible only so long as the evanescent wave has attenuated 20 dB at the outer cladding boundary, how long can the wavelength of 89° incident light get before this condition is violated, assuming $n_1 = 1.5$ and n_2 is one percent lower?   Get an approximate answer by assuming a plane, not cylindrical, geometry.

3.6   A phosphorescent fish is swimming one night at a depth of 1.0 meter below a perfectly smooth surface of water ($n = 1.33$).   A bird of prey flies directly over the fish at essentially zero height and a velocity of 10 m/sec.   For how long can the bird see the light from the fish, assuming $v_{fish} = 0$?

3.7 (*Goos-Hänchen shift*) The dotted line in Figure 3-14 extrapolates the sinusoidal field to reach a zero value at some distance ζ behind the true interface. This means that a totally reflected ray appears to be reflected from a plane surface slightly behind the true interface and therefore suffers an apparent sideways displacement in the plane of incidence and in the z-direction. How big is the sideways displacement (called the *Goos-Hänchen shift*) for 1.5-μ light incident on a glass-air boundary from within the glass at $\theta_i = 60°$?

3.8 (*Lambertian sources*) A light-emitting diode (LED), for which the radiations from different parts of the emitting surface are in random phase with one another, has a Lambertian radiation pattern, in which the power radiated per unit area varies as the cosine of θ, the angle away from the axis. (A) If *brightness* of a source is defined as the radiated power per unit cross-sectional area of a bundle of rays from the source to the observer, how does the brightness of a Lambertian source vary with θ ? (B) Both an edge-emitting LED and an edge-emitting laser diode radiate light from a very narrow line source. Both will have a broad radiation pattern in the plane perpendicular to that of the line source, the LED will have a Lambertian radiation pattern in the plane of the line source, but the laser diode will have a narrow pattern in that plane, because, since the radiated light is in phase across the line source, it will act as a directive antenna in that plane. Given an LED and a laser diode of equal output power, with the laser's half-power beamwidth in the plane of the line source being 10°, how much more efficient is the laser diode in terms of power radiated in the forward direction into a nearby fiber?

3.9 (A) Given a plane wave incident on a fiber end, what fraction of the power fails to be transmitted into a fiber, assuming unity index for air and 1.45 for silica? (B) Inside a laser, the index is usually around 3.5. If the "facets" that constitute the end mirrors of the cavity have no special reflective coating, so that the interface with air constitutes the reflector, what fraction of the power is lost on each reflection?

3.10 The answer is 2.405. What is the question?

3.11 (*Wavelength demultiplexor and Dichroic coupler*) (A) Design a single mode fused biconical coupler that accepts at one input a mixture of light at 1.3 μ and 1.53 μ, and delivers 100 percent of one to one output and 100 percent of the other to the other output. (B) Repeat this process for a device that will combine light at 1.41 μ and 0.98 μ into the same output. Such a unit would be needed for optically pumping a 1.41 μ fiber amplifier with 0.98 μ energy. In both (A) and (B), assume that throughout the coupling region, both fibers have 0.2 μ diameter.

3.12 Suppose that, in order to minimize the effect of modal noise, you want to design a step-index multimode fiber that supports 100 modes or more, with a core diameter of 50 μ, using a CD-player laser of 780 nm wavelength. If the core index n_1 is to be 1.450, what should the cladding index be?

3.13 A typical 1.5-μ commercial-grade distributed feedback laser that has been well isolated from reflections will have a linewidth of less than 0.01 nm. Give a rough estimate of the coherence length in air and in fiber.

3.14 Consider light linearly polarized at 135°. (A) What are the relative electric field strengths of the horizontal and vertical components? (B) What are the four Stokes parameters and (C) The corresponding SOP point on the Poincaré-sphere representation? (D) What happens to the

SOP point as this light gets more and more depolarized? (E) What happens as the phase difference between horizontal and vertical components grows?

3.15 Single-mode fiber, operated at 1.5 μ, typically has a beat length ranging from 10 cm to 12 meters, depending on how good the circularity of the core cross-section is. Suppose the core index for the ordinary wave is 1.450. What value of the core index for the extraordinary wave would produce a beat length of 10 centimeters?

3.16 (*Properties of a typical fiber*) A typical single-mode fiber, such as those being widely installed today, has an 8.7-μ core, a core index $n_1 = 1.450$, and a cladding index n_2 that is 0.3 percent smaller. It is typically operated at 1.50 μ. Compute (A) Numerical aperture NA, (B) V-number, (C) Chromatic dispersion distance limit in km for a 10 Gb/s bitstream approximated by two sinusoids 10 GHz apart, (D) The repeater spacing, assuming that attenuation is allowed to accumulate to no more than 20 dB between repeaters, and (E) The wavelength below which fiber becomes multimoded. Assume the same attenuation and dispersion numbers as those in Problem 3-4.

3.17 Discuss the things that can happen at the system level when a fiber designed for service at 1.5 μ is used instead at 0.85 μ.

3.18 Crosstalk between channels figures importantly in multichannel communication systems. Compare qualitatively the crosstalk between two OOK (on-off keying) channels for two causes: (A) Imperfect separation by narrowband filtering: some of the power from one channel is seen in the other, and (B) Stimulated Raman scattering, where the higher-frequency signal pumps energy into the lower-frequency signal. Express your answer as pictures of Channel 1 and Channel 2 waveforms before and after the filter or Raman scattering process, assuming these to be the transmitted waveforms, with Channel 1 being at shorter wavelength than Channel 2:

3.19 References

1. J. Gowar, *Optical Communication Systems*, Prentice Hall, 1984.
2. J. M. Senior, *Optical Fiber Communications*, Prentice Hall, 1985.
3. C. M. Miller, *Optical Fiber Splices and Connectors*, Marcel Dekker, 1986.
4. A. Toffler, *The Third Wave*, Wm. Morrow, 1980.
5. S. R. Nagel, "Optical fiber − The expanding medium," *IEEE Commun. Mag.*, vol. 25, no. 4, pp. 33-43, 1987.
6. C. Lin, ed., *Optical Technology and Lightwave Communication Systems*, Van Nostrand Reinhold, 1989.
7. Fiber optics: Theory and applications − Tech. Memo, Galileo Electro-Optics Corp., Sturbridge, MA, 1990.
8. S. E. Miller, I. P. Kaminow, ed., *Optical Fiber Communications − II*, Academic Press, 1988.
9. E. Hecht, *Optics - Second Edition*, Addison-Wesley, 1987.
10. P. K. Cheo, *Fiber Optics and Optoelectronics*, Prentice Hall, 1990.
11. S. Ramo, J. R. Whinnery, and T. Van Duzer, *Fields and Waves in Communication Electronics - Second Edition*, Wiley, 1984.

12. L. B. Jeunhomme, *Single-Mode Fiber Optics – Principles and Applications*, Marcel Dekker, 1983.

13. E. E. Basch, ed., *Optical Fiber Transmission*, Sams/McMillan, 1986.

14. J. C. Palais, *Fiber Optic Communications*, Prentice Hall, 1988.

15. M. Born and E. Wolf, *Principles of Optics - Sixth Edition*, Pergamon, 1980.

16. J. E. Midwinter, *Optical Fibers for Transmission*, Wiley, 1979.

17. J. W. Goodman, *Introduction to Fourier Optics, Sec. 5.2*, McGraw-Hill, 1968.

18. B. A. Saleh and M. Teich, *Fundamentals of Photonics*, Wiley, 1991.

19. W. van Etten and J. v. Plaats, *Fundamentals of Optical Fiber Communications*, Prentice Hall, 1991.

20. G. Keiser, *Optical Fiber Communications*, McGraw-Hill, 1983.

21. D. Gloge, "Dispersion in weakly guiding fibers," *Applied Optics*, vol. 10, pp. 2252-2258, 1971.

22. V. J. Tekippe, "Passive fiber optic components made by the fused biconical taper process," *Fiber and Integrated Optics (Taylor and Francis, U.K.)*, vol. 9, pp. 97-123, 1990.

23. H. Stone, "Parallel processing with the perfect shuffle," *IEEE Trans. on Computers*, vol. 20, no. 2, pp. 153-161, 1971.

24. D. L. Lee, *Electromagnetic Principles of Integrated Optics*, Wiley, 1986.

25. J. T. Boyd, ed., *Integrated Optics – Devices and Applications*, IEEE Press, LEOS Progress in Lasers and Electro-Optics Series, New York, 1991.

26. C. Dragone, "Efficient N x N star coupler based on Fourier optics," *IEEE/OSA Jour. Lightwave Tech.*, vol. 7, no. 3, pp. 479-489, 1989.

27. R. Ramaswami, "Analysis of biconical couplers," *Private communication*, March, 1991.

28. H. A. Haus, *Waves and Fields in Optoelectronics*, Prentice Hall, 1984.

29. T. Kanada, "Evaluation of modal noise in multimode fiber-optic systems," *IEEE/OSA Jour. Lightwave Tech.*, vol. 2, no. 1, pp. 11-18, 1984.

30. G. P. Agrawal, P. J. Anthony, and T. M. Shen, "Dispersion penalty for 1.3 with multi-mode semiconductor lasers," *IEEE/OSA Jour. Lightwave Tech.*, vol. 6, no. 5, pp. 620-625, 1988.

31. T. Okoshi, "Private communication," *Symposium on Photonic Switching and Optical Networks*, Tirrenia, Italy, 1989.

32. C. D. Poole, N. S. Bergano, H. J. Schulte, R. E. Wagner, V. P. Nathu, J. M. Amon, and R. L. Rosenberg, "Polarization fluctuations in a 147-km. undersea lightwave cable during installation," *Elect. Ltrs.*, vol. 23, no. 21, pp. 1113-1115, October, 1987.

33. R. A. Harmon., "Polarization stability in long lengths of monomode fiber," *Elect. Ltrs.*, vol. 18, no. 24, pp. 1058-1060, November, 1982.

34. N. S. Bergano, C. D. Poole, and R. E. Wagner, "Investigation of polarization dispersion in long lengths of single-mode fiber using multilongitudinal mode lasers," *IEEE/OSA Jour. Lightwave Tech.*, vol. 5, no. 11, pp. 1618-1622, November, 1987.

35. L. J. Cimini, I. M. Habbab, R. K. John, and A. A. Saleh, "On the preservation of polarization orthogonality through a linear optical system," *Elect. Ltrs.*, vol. 23, pp. 1365-1366, 1987.

36. S.E. Miller and I.P. Kaminow, eds., *Optical Fiber Telecommunications – II*, Academic Press, chap. 10, pp. 369-416, 1988.

37. A. R. Chraplyvy, "Limitations on lightwave communications imposed by optical-fiber nonlinearities," *IEEE/OSA Jour. Lightwave Tech.*, vol. 8, no. 10, pp. 1548-1557, 1990.

38. S.E. Miller and I.P. Kaminow, eds., *Optical Fiber Telecommunications – II*, Academic Press, chap. 3, pp. 56-107, 1988.

39. American National Standard for the Safe Use of Optical Fiber Communication Systems Utilizing Laser Diode and LED Sources, American National Standards Institute, Z136.2, 1988.

40. L. F. Mollenauer and K. Smith, "Demonstration of soliton transmission over more than 4000 km. in fiber loss periodically compensated by Raman gain," *Optics Ltrs.*, vol. 13, pp. 675-677, 1988.

CHAPTER 4

Tunable Filters

4.1 Overview

There are two reasons to turn next to tunable filters as the second class of building blocks for third generation systems. First of all, we need to provide a foundation upon which to discuss laser structures in the next chapter. All lasers embody some kind of resonant structure to provide sufficient optical energy reinforcement just for lasing to occur at all, and furthermore, the most useful lasers from the system point of view are often those that are single-frequency and whose optical frequency can be tuned.

Second, as we saw in Chapter 1, wavelength (frequency)-division architectures are an attractive approach, and doing wavelength-division networking using incoherent detection requires optical receiver filters that are tunable. Third, fixed-tuned filters find several other uses throughout a lightwave system, for example to filter out noise generated within lightwave amplifiers.

In this chapter we first discuss the system requirements to be met by the tunable optical filter, when regarded as a black box. We then consider the current state of the art and the performance of each of a half-dozen particularly promising approaches, and conclude the chapter with an overall comparison.

The tunable filter as a black box may be represented as in Figure 4-1. Many different signals at many different optical frequencies appear at the input, but the filter is selective enough that only one appears at the output, though not so selective that the modulation sidebands on the desired signal are cut off.

Perhaps the first thing you might think of as a wavelength-selective device is a simple prism, but, besides being bulky, a prism works on a small physical effect (chromatic dispersion) and is therefore not very interesting. The filters we shall describe in this chapter all work on *interference* effects that are very wavelength-selective. We shall see that when the filter is tuned to light of a desired wavelength, there is, in one way or another, a reinforcement by constructive interference; for other wavelengths this is replaced by nonconstructive interference or even completely destructive interference (an output null). Sometimes these processes have fancy names, like *Fabry-Perot* or *Mach-Zehnder interferometry*, or *Bragg diffraction*, or *phase matching*, but the underlying process is the same, the interference between many light bundles of the same wavelength.

We shall see, over and over again, that when one filter will not do the trick, there are clever ways of cascading devices. In this way, an undesired passband peak in one filter may be killed off by placing either a high-attenuation region or an actual null of a second filter at exactly that frequency.

These two tricks of (i) playing the interference effects, and (ii) cascading, underlie all practical optical filters.

Laboratory devices that pass only a narrow band of wavelengths around some adjustable center wavelength are centuries old, going back to our first knowledge of chromatic dispersion. Emphasis on fast tuning speed did not come until the 1950s, when scanning spectrometers were introduced in astronomy and in automatic spectrometers. The emphasis on submillisecond tuning speeds is a new requirement that is now appearing for the first time, driven by the needs of fiber optic communications.

4.2 Requirements

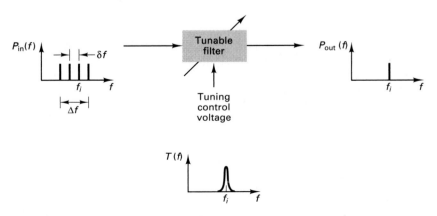

Figure   4-1. Basic function of a tunable filter in selecting one of many inputs at different frequencies.

Number of resolvable channels

This is perhaps the principal issue. There are two parts to this question, the range over which the filter can tune and the frequency-selectivity of the response, once the filter is tuned. As shown in Figure 4-1, we define Δf as the frequency difference between the lowest- and highest-frequency channels, and δf as the spacing between channels. (Sometimes it will be more convenient to deal with wavelength, and in this case we shall use $\Delta \lambda$ and $\delta \lambda$, respectively). If the tuning range is to cover a Δf equal to the entire window of low fiber attenuation at either 1.3 or 1.5 μ, then 200 nm (25,000 GHz) is probably a reasonable target for tuning range. When photonic amplifiers (Chapter 6) are part of the system, the required tuning range can often be less than this number. Presently available doped-fiber amplifiers have a maximum bandwidth of about 35 nm, but quantum-well laser diode amplifiers are capable of the of the full 200 nm bandwidth.

It is possible to deduce the maximum number of channels, starting with the tuning range $\Delta \lambda$ and the shape $T(f)$ of the filter's power transmission as a function of frequency. We calculate $N_{max} = |\Delta f / \delta f|_{max}$, the maximum number of equally spaced channels that can be crowded into the tuning range before the *crosstalk* (interference from adjacent channels) becomes high enough to exceed the desired noise-to-interference ratio at the receiver. We shall determine this quantity for each of the filter types discussed in this chapter and compare the results. We shall not be particularly interested in filter options incapable of more than several tens of channels.

Access time

The speed with which a tunable optical filter can be reset from one frequency to a new one determines the breadth of applicability of the network. If the filter requires milliseconds to retune, this could be quite fast enough for some circuit-switched applications, but much too slow for packet-switched systems. If the filter can be retuned in submicrosecond time, then it can be used for both packet- and circuit-switched applications. It should be clear from the discussion of applications in Chapter 2 that the ability to do packet switching is a very important requirement for lightwave networks. We shall not discuss tunable filter designs with access times greater than one millisecond.

Attenuation

The peak of $T(f)$, the filter's transfer function, will not in general be unity because of internal losses in the filter. If too high, the filter loss can be a serious contributor to the debit side of the link budget.

Controllability

There are several aspects to this question. First of all, the filter must be *stable*, so that once set to a particular frequency, thermal and mechanical factors will not cause the tuning to drift more than a small fraction of the bandwidth of one channel. This may mean an intrinsically stable design or it may mean the addition of some frequency-locking technique. Such techniques are the subject of Chapter 10. Second, the filter must be easily *resettable* to some designated value. Upon application of a voltage, current, or bit pattern specifying the desired frequency, the device should tune to exactly that frequency. It is an added bonus if the filter is such that its tuned frequency is set up by directly inputting a digital word. It is most desirable that the filter have a smooth characteristic curve of frequency versus applied signal; if this curve is discontinuous, this may add considerable complexity to the control circuitry.

Polarization independence

Filters that work for all possible polarization states of the arriving light are greatly to be preferred over those that are polarization-sensitive, because the latter entail the added complexity of polarization control, polarization diversity, or polarization scrambling elsewhere in the system. As will be discussed in Chapter 8, this is one of the potential advantages of many tunable-filter-based architectures over coherent-detection networks, since in the latter the polarization state of the tunable local oscillator laser must match the polarization state of the light input signal, and the latter drifts slowly in an unpredictable way, as discussed in Section 3-13.

Cost

For some designs being tested today, it is easy to see how volume production techniques can be devised that can lead to very low per-unit cost; for others it is equally clear that fabrication costs are likely to remain high. As of this writing, the ultimate cost-reduction potential of the competing designs to be discussed is unclear. Nevertheless, it is possible to speculate about two important factors that must be considered, the potential for *lithographic* realization, and the potential for reduction of *pigtailing* (fiber attachment) costs. An ideal solution would be one that followed both approaches. The lithography art that has been perfected with electronic integrated circuits can be brought to bear on several of the designs we shall discuss, but even after the filter itself has been fabricated, low-loss attachment of fibers has often remained as a problem. With others, the pigtailing problem is virtually absent, but other high-cost manufacturing steps are involved that show little potential for a high degree of automation.

Size, power consumption, and operating environment

For maximum usefulness in practical networks, tunable filters and their associated electronic circuitry should be easily accommodated in today's printed-circuit and microelectronics implementation environments. This includes the need to have the filter operate from the same supply voltages that are already available for the electronics. The device should be resistant to shock, vibration, humidity and temperature conditions likely to be met in low-cost high-usability environments.

4.3 Quantification of Crosstalk

Figure 4-2(A) shows the power transfer function $T(f)$ of a typical narrowband optical filter, and (B) shows a number N of wavelength-division channels incident upon it. These channels are assumed equally spaced in frequency by δf over a total tuning range Δf. The total number of channels is

$$N = \Delta f / \delta f \tag{4.1}$$

Assume, for the moment, that $T(f)$ is a repetitive function of frequency, as shown, so that Δf is δf less than the total spacing between repetitive peaks. The channels cannot extend over into an adjacent repetition of the periodic function $T(f)$, otherwise the filter would be unable to distinguish the repetitions from each other (*aliases*).

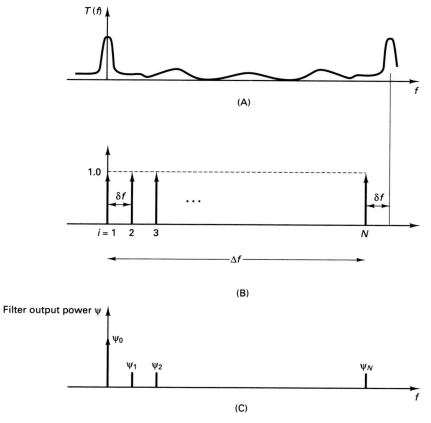

Figure 4-2. Basis of the inter-channel crosstalk calculation. (A) Power transfer function of a typical filter. (B) WDM channels appearing at the filter input. Only channel $i = 0$ is desired, since the rest produce crosstalk. (C) The desired output power ψ_o and the various crosstalk components ψ_i, $i \neq 0$.

Assuming channel number $i = 0$ to be the channel to which the filter is tuned, Figure 4-2(C) shows the output of the filter. This output is seen to consist of the power ψ_o of the desired channel, plus the sum of other ψ_i, $i = 1, \ldots, (N-1)$ that represent the undesired crosstalk.

This situation has been analyzed extensively in [1, 2] to determine, for a given $T(f)$, and a given filter tuning range Δf, how many equispaced channels can be crowded within Δf before the crosstalk power becomes excessive. In this section we repeat the essentials of this study for later use in understanding the results for the various tunable filter types that will be discussed.

The analysis is fairly straightforward if all the channels are sending at constant amplitude and constant frequency, which would be the case for *phase-shift keying*, for example, but the interesting case is the one in which all channels are carrying *amplitude-shift keying* (ASK) signals, which we shall assume are of unit power during a "1" bit and of zero power during a "0" bit. Calling the bit carried by the i th channel a_i, where i indexes the channel number and a can be either 0 or 1, the output power of the 0 th filter due to the i th signal is

$$\psi_i(a_i, N) = a_i \int_{-\infty}^{+\infty} T(f) \left| S(f - i\delta f) \right|^2 df \qquad (4.2)$$

where $S(f)$ is the complex spectrum of the transmitted waveform. References [1, 2] treat several cases in which the transmitted signal, being of finite duration, has a nonzero width of $S(f)$ due to the modulation, and they also treat the *frequency-shift keying* (FSK) case in which "0" and "1" are represented by two different frequencies. However, for present purposes of getting rough estimates only, it will be sufficient to assume ASK and to make the simplifying assumption that the modulation bandwidth is much smaller than the filter bandwidth, so that $S(f)$ may be approximated by a δ-function, whereupon

$$\psi_i(a_i, N) = a_i T(f - i\delta f) \qquad (4.3)$$

From this, the desired signal power after the filter is

$$S = \psi_o(a_o, N) = a_o T(0) \qquad (4.4)$$

and the undesired crosstalk power is

$$XT(N) = \sum_{i=1}^{N-1} a_i T(f - i\delta f) \qquad (4.5)$$

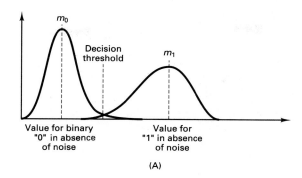

(A)

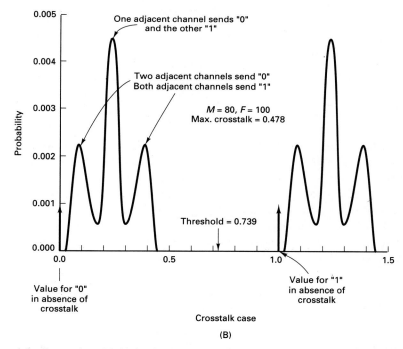

(B)

Figure   4-3. Comparing the basic thresholding operation of a binary receiver for the case of (A) Random noise, and (B) Crosstalk. The noise amplitude probability distributions have tails, but with crosstalk the probability distributions drop abruptly to zero. (From [1], © 1989 IEEE).

Adjacent-channel crosstalk is very different from random noise in an advantageous way. All the various sources of noise have amplitude probability distributions that have gradually decaying tails, as will be discussed in Chapter 8, so that when a threshold is set up in the receiver to distinguish between a "0" and a "1," as shown in Figure 4-3(A), the tails mathematically go on forever, and, in order to keep lowering the probability of bit errors, the mean values for "0" and "1," namely m_0 and m_1, have to be somehow pulled farther and farther apart. But in the crosstalk case, the undesired power $XT(N)$ that appears at the threshold decision point and messes up the decisions between "0" and "1" is bounded; the relevant amplitude probability distributions in Figure 4-3(B) have no tails. This can be seen from Equation 4.5 by noting that (i) the number of channels N is finite, and (ii) the modulation coefficients a_i are either zero or one, not some continuum of values. A worst-case result can be gotten by assuming that all the channels are bit-synchronized, so that, for example, when the desired channel is carrying a "0," all others are carrying some number of "1"s for the entire bit duration.

The result is that crosstalk, as seen at the receiver decision point, has the character shown in Figure 4-3(B), which is actually a probability density curve for the simple Fabry-Perot tunable filter we shall talk about next. It consists of many impulses close together because the number of channels N is large, namely 80 in the example, but for clarity the curves show the envelope of these many impulses). The left-hand multipeaked distribution contains all the situations for which the desired channel is carrying a "0," and is seen to go identically to zero at 0.478. The right-hand one covers all cases where the desired channel carries a "1," and it goes identically to zero at 1.015. If there were no crosstalk at all, the "0" signal would produce a zero output and the "1" signal unity output, as shown by the two impulses.

The arrows in the figure show where the different clusters of impulses are coming from, and reveals a most important point: the crosstalk is dominated by the two channels that are immediately adjacent on either side of the receiver's response peak.

The preceding paragraphs show how the crosstalk analysis is set up quantitatively. In future sections of this chapter we shall quote the results gotten using this analysis to determine N_{max}, the greatest number of nodes that can be supported with each filter technology discussed. The way N_{max} is determined is first to allow random noise to be included in what the receiver decision point sees, whereupon tails are introduced on the "0" and "1" probability distributions. Then N_{max} is calculated to be that N for which crosstalk degrades the probability of bit error by the same amount as 1.0 dB more noise would have degraded it.

4.4 The Single-Cavity Fabry-Perot Interferometer

In the rest of this chapter, we shall discuss most of the tunable filter types under active consideration today. For each of them, we first say a bit about the basic physical principles, the expected tuning range, access time, and polarization dependency of the device. We typically then define an equivalent electrical filter from which we can compute the power transmission function $T(f)$. From $T(f)$ and the acceptable level of crosstalk, we can deduce the number of channels that can be covered with each filter technology.

Basics

Figure 4-4(A) shows the basic Fabry-Perot interferometer (FPI) [3] or *etalon*, consisting of a resonant cavity formed by two parallel mirrors. Light from an input fiber is collimated (made parallel), passed through the cavity, and then refocussed onto the output fiber. We ignore the input and output fibers and lens systems and the incidental losses that may occur there, and focus attention on the cavity itself.

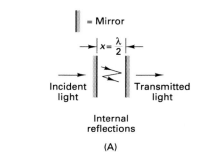

(A)

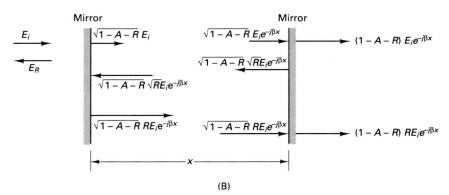

(B)

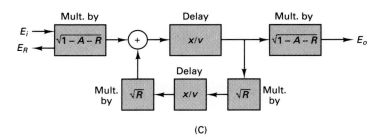

(C)

Figure 4-4. Basic Fabry-Perot interferometer (etalon). (A) Geometry of the device (i = order of the resonance), (B) Electric field strength of successive reflections, (C) Equivalent block diagram.

In the figure we show a single mirror surface at each end of the cavity without showing the other side of the piece of glass that the mirror is coated onto. In order to eliminate undesired resonances due to reflections from these outer surfaces, in practice the outer surfaces are often bevelled slightly from perpendicularity, and perhaps also given an antireflection coating. The inner mirror surfaces are maintained in precise parallelism.

The entering light reflects back and forth within the cavity, and if the cavity length x is exactly

$$x = \frac{i\lambda}{2n} = \frac{ic}{2nf} \tag{4.6}$$

(where n is the index within the cavity and i is an integer, called the *order*), then a number of reflections proportional to a quantity called the *finesse* occurs before the light intensity decays roughly to $1/\epsilon$ of its original value. Almost all the light eventually passes to the output. If the cavity length is detuned from this resonance value, completely constructive interference is no longer the case, and the light output is reduced. All the bookkeeping involved in following these reflections is given in Figure 4-4(B). This is represented in turn by the equivalent electrical block diagram of Figure 4-4(C), which shows E_i and E_o, the complex optical field strengths at input and output. The summation point is just to the right of the first mirror.

Starting from this equivalent circuit, we can derive the overall power transfer function as a function of frequency as follows. Let R be the power reflectivity (reflectance) of each of the two mirrors. That is, on each reflection, the reflected field strength is $\sqrt{R}$ times the incident field strength. Let A be the power absorption loss as the light passes through the supporting glass material and reaches the first mirror, and let the absorption passing from the second mirror to the output likewise be A.

Light of field strength E_i enters the filter, passes through the first mirror, $\sqrt{1 - A - R}\, E_i$ survives, and the rest is lost as heat or is reflected to the left from the device. The signal then proceeds into the resonant cavity (assumed lossless), where it is reflected repeatedly from one mirror to another. The light first arrives at the right-hand mirror with field strength $\sqrt{1 - A - R}\, E_i \exp(-j\beta x)$, where β is the propagation constant. A portion $(1 - A - R)\, E_i \exp(-j\beta x)$ passes to the output, and a portion $\sqrt{1 - A - R}\,\sqrt{R}\, E_i \exp(-j\beta x)$ is reflected back toward the left. This arrives at the left mirror and is reflected back to the right where it emerges as $\sqrt{1 - A - R}\, R E_i \exp(-3jkx)$.

Adding up all the successive contributions to the output E_o, and using $\beta x = 2\pi f\tau$, where τ is the one-way propagation time x/v across the cavity, the complex transfer function of the field strength is

$$H(f) = \frac{E_o(f)}{E_i(f)} = (1 - A - R) \exp(-j2\pi f\tau) \sum_{m=0}^{\infty} R^m \left[\exp(-j4\pi m f\tau)\right]$$

$$= \frac{1 - A - R}{1 - R\exp(-j4\pi f\tau)} \exp(-j2\pi\ \tau)$$

(4.7)

The power transfer function $T(f) = |H(f)|^2$ is

$$T(f) = \frac{(1 - A - R)^2}{1 + R^2 - 2R\cos 4\pi f\tau}$$

(4.8)

$$T(f) = \frac{(1 - A - R)^2}{(1 - R)^2 + R(2\sin 2\pi f\tau)^2}$$

(4.9)

$$T(f) = \left(1 - \frac{A}{1 - R}\right)^2 \left[1 + \left(\frac{2\sqrt{R}}{1 - R}\sin\frac{4\pi f\tau}{2}\right)^2\right]^{-1}$$

(4.10)

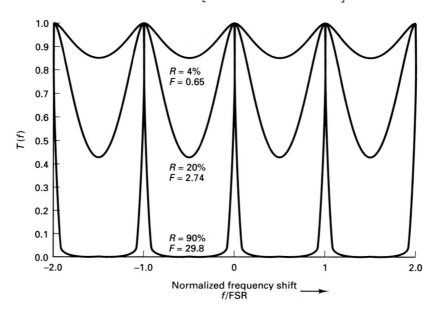

Figure 4-5. The power transfer function (Airy function) for single etalons of various finesse.

The right-hand side is known as the *Airy function*, and is shown in Figure 4-5 for three choices of R. It is seen to repeat in frequency at a regular period, which is called the *free spectral range*.

$$FSR = \frac{1}{2\tau} = \frac{c}{2nx} \tag{4.11}$$

In a wavelength-division network, we would place N equally spaced frequencies within one free spectral range, as shown. So we would set $FSR = \Delta f + \delta f$ (Figure 4-2). The *half-power bandwidth* (*HPBW*) of a peak, or its 3-dB bandwith, is often called the *FWHM* (full width at half maximum). It is given by

$$HPBW = \frac{c}{2nx} \frac{1-R}{\pi\sqrt{R}} \tag{4.12}$$

The most important performance parameter characterizing a Fabry-Perot filter is the *finesse*, the ratio of *FSR* to *HPBW*, which expresses the sharpness of the filter relative to the repeat period, and thus is closely related to the maximum number of channels that can be supported, N_{max}. The finesse is given by

$$F = \frac{FSR}{HPBW} = \frac{\pi\sqrt{R}}{1-R} \tag{4.13}$$

To tune the filter across one entire FSR it is only necessary to change the mirror spacing by $\lambda/2$. We may see from this that the tolerances on the spacing are very exacting. If we have N channels equally spaced within one FSR and wish to maintain the tuning to within, say, one-tenth of the interchannel spacing, then a tolerance on x to within $\lambda/20N$ must be maintained.

The height of the peak at resonance is

$$T(f)_{max} = \left[1 - \frac{A}{1-R}\right]^2 \tag{4.14}$$

and the *contrast*, the ratio of maximum to minimum of $T(f)$, is

$$C = \left(\frac{1+R}{1-R}\right)^2 = \frac{(1+R)^2 F^2}{\pi^2 R} \tag{4.15}$$

By examining the second bracketed term in Equation 4.10 we see how, as finesse increases, the HPBW gets narrower and narrower. It is clear that in order to get high finesse and thus a narrow peak and low sidelobe levels for the undesired channels, the mirror reflectivities R must be made high. Also, the mirrors must be made as plane and as parallel as possible so the reflections will add in phase. If the surfaces are not parallel, successive reflections will gradually *walk off* sideways and be lost.

One of the most attractive characteristics of etalons for lightwave network service is that they are usually completely polarization insensitive. As long as the mirrors are plane, and the intracavity material is not birefringent, an examination of Equation 3.51 confirms that there is nothing in the mirror reflection process that is polarization-dependent. Of course, if the cavity is filled with material whose velocity of propagation is polarization-dependent (and we shall discuss such a case later), then the entire device is polarization-dependent.

The tuning-range multiplication principle

It is most desirable to be able to tune over the entire 200-nm width of the the low-attenuation region at, say, 1.5 μ wavelength, representing a $\Delta\lambda/\lambda = \Delta f / f$ of 13 percent. Few physical phenomena are capable of this large a fractional change. For example, piezoelectric materials are limited to a $\Delta x/x$ of roughly 0.5 percent, and index change by current injection in semiconductors likewise shows about the same 0.5 percent limit on $\Delta n/n$.

There is a standard trick that is generally useful in expanding the tuning range beyond such limits. The basic idea is to use the *difference* between two large values of the parameter. Thus, even though each parameter may change by at most a fraction of one percent, the difference can be made to change by a much larger amount. This principle, which we shall arbitrarily call the *tuning-range multiplication* principle, will recur frequently in this chapter and in the following chapter when we discuss tunable lasers.

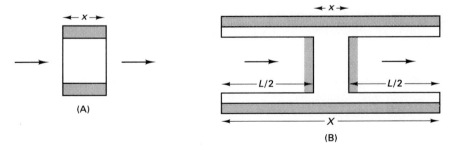

Figure   4-6. The use of mechanical advantage to increase tuning range. (A) Piezo elements placed between mirrors. (B) Piezo elements placed outside mirrors and spacers added.

The way this principle is used in piezoelectrically tuned etalons to overcome the 0.5 percent limitation is shown in Figure 4-6. At (A) we se that if a *short* piezo member is placed between the mirrors separated by x, then the tuning range $\Delta f / f = \Delta\lambda/\lambda = \Delta x/x$ will be limited to the 0.5 percent number. If, however a *long* such member of length X is placed *outside* as shown at (B), and constant-length spacers of length $L/2$ make up all of the spacing except x, then

$$\Delta f/f = \Delta \lambda / \lambda = \frac{X}{x} \frac{\Delta X}{X} \qquad (4.16)$$

The mechanical-advantage ratio X/x can easily be made in the thousands. Note that we have made the tuned variable x the *difference* between two large and almost identical quantities (X and the fixed distance L), so that when the first of these large quantities is varied (say by up to 0.5 percent), the difference exhibits a large fractional change (up to $X/x \times 0.5$ percent).

In Figure 4-6, the two quantities are distances; we shall encounter this idea again in connection with index-tuned etalons (this section), tunable acoustooptic and electrooptic phase-matching filters (Sections 4.9 and 4.10), and with tunable lasers (Section 5.12), and in all these cases the two quantities are two refractive indices.

Tunable etalon realizations

We now illustrate the tunable-etalon ideas introduced so far with several examples from the literature of product-level tunable filters. Tunable etalons for laboratory use at the near-infrared wavelengths of communication interest are available having finesses as high as 15,000 [4], but today such units require temperature-controlled ovens. Of greater interest are several devices that have been developed specifically for fiber optic communication applications. Although these have a more modest finesse, they are much smaller and can be provided with external means to make them highly stable, as will be discussed in Chapter 10.

The Queensgate Microfilter was developed some years ago [5] as an extension of stable bulk-optic etalon designs for astronomical scanning spectroscopy. Small versions of these have been evolved for communication-network use that are easily capable of finesses of 250 or more, and are between 2 and 3 cm on a side. Problems of drift, piezoelectric hysteresis, and vibration are dealt with by use of *capacitance micrometry*, to be described in Chapter 10. Measurements on early versions of Queensgate Microfilters [6] having an FSR of 80 nm (10 μ mirror spacing) have shown a measured finesse of 250, a temperature coefficient of tuning of 0.1 nm per °C, and an access time (time to for the filter to settle down after tuning between widely spaced channels) of about 1 millisecond. Insertion loss as low as 3.0 dB has been measured by the manufacturer.

Mallinson [7] at British Telecom Research Laboratory developed an etalon made entirely lithographically. This tunable filter design is particularly interesting because of its extreme mechanical simplicity and therefore its potential for low cost. The device, shown in Figure 4-7, takes advantage of two properties of silicon, its transparency at near-infrared wavelengths and the availability of well-developed low-cost techniques for chemically machining small precision structures in this material. Collimated light passes vertically through the device, which consists of two silicon chips a few millimeters in width whose facing surfaces have reflective coatings. One of these has a "moat" etched out so as to form a flexible diaphragm. The device is tuned by the application of a variable voltage (up to several tens of volts) between the two plates.

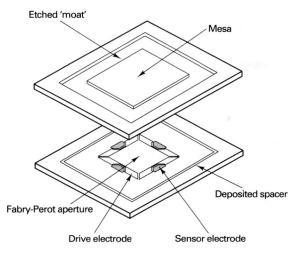

Figure 4-7. Exploded view of British Telecom Research Labs. silicon micromachined Fabry-Perot filter. (From [7], by permission IEE).

Tuning across more than one free spectral range has been demonstrated. As with the Queensgate device, there can be several capacitance electrodes placed around the periphery so that tilts may be corrected.

The spacing x and the applied voltage V are related by a nonlinear equation

$$V = Const \ x^2 \tag{4.17}$$

The first Mallinson filters exhibited finesses of about 100, and loss of 2.4 dB. Retuning time was not investigated. On the one hand we would expect the device to tune quite rapidly because of its small size; on the other hand, the restoring force is small.

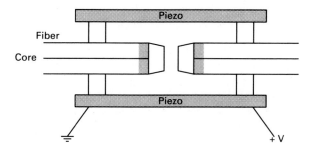

Figure 4-8. Physical configuration of the Fiber Fabry-Perot (FFP) tunable filter.

Probably the most successful system-usable etalon developed to date has been the *Fiber Fabry-Perot*, invented at AT&T Bell Laboratories [8], and further evolved by Micron Optics, Inc.. The basic premise of the design is that low cost and low attenuation are most easily achieved through reducing or eliminating the pigtailing problem by forming the mirrors directly on the butt end of the fiber core.

Tuning is by means of a voltage applied to an external piezoelectric member, as with the Queensgate and many other designs. This is shown in Figure 4-8. Finesses of up to 350 were measured by the Bell Laboratories group on selected laboratory prototypes. The commercial version, shown in Figure 4-9, exhibits a finesse of over 150, insertion loss of 1 to 2 dB, and good temperature stability gotten by a clever compensation scheme in which the negative temperature coefficient of the piezo material is played off against the positive coefficient of the metal parts [9]. These units are ideal for use in printed-circuit-board implementations, for example the IBM Rainbow system, described in Chapter 16. The time to retune across the entire free spectral range and then settle down is several milliseconds.

Several groups have developed automatic-frequency-control (AFC) loop systems to lock the filter to the incoming signal (e.g. [9]). This scheme is described in Chapter 10.

Figure   4-9. Commercial Fiber Fabry-Perot tunable filter. (Courtesy Micron Optics, Inc.).

Index-tuned etalons

There has been much interest in replacing the mechanical method of tuning a Fabry-Perot filter with a voltage-dependence of the cavity index, so as to gain faster tuning speed, while hopefully preserving the wide tuning range of the piezoelectrically tuned devices. The idea is to fill the cavity with an *electrooptic material*, one that has voltage-dependent indices of refraction. The electrooptic effect is discussed in more quantitative detail in Section 7.3. Again, there is a materials problem, since few materials exhibit a large electrooptic coefficient, and some of those that do change index very slowly.

Liquid crystal compounds offer some promising possibilities. They possess the property of solid materials (such as sheet Polaroid) that there is a parallel alignment of all the molecules along a certain axis. In the case of liquid crystal material, the alignment can be changed, since the molecules are suspended in a liquid solvent. Specifically, the molecules can be rotated by the application of an electric field.

Consider what happens when light is passed through a region in which all the molecules are aligned similarly, say perpendicularly to the propagation direction. The extreme anisotropy of the molecules means that light polarized along the molecule's axis sees an index that is different from that seen by light polarized normal to the axis. This results in these materials having extremely large birefringence, the relative index difference between ordinary and extraordinary waves being as much as 15-20 percent.

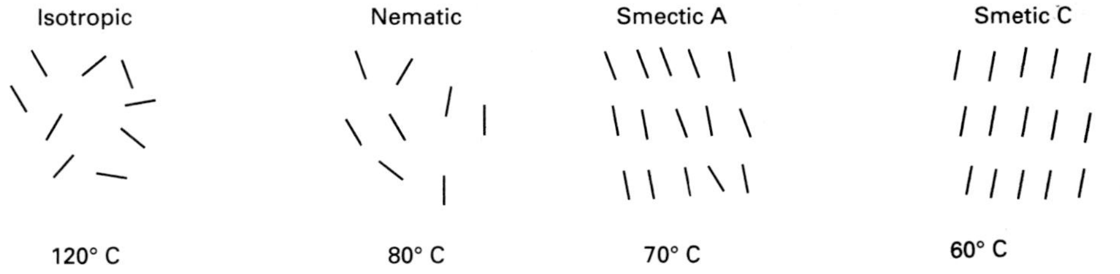

Isotropic	Nematic	Smectic A	Smetic C

| 120° C | 80° C | 70° C | 60° C |

Figure 4-10. Phases of a hypothetical liquid crystal material as the temperature is varied. Molecular orientation remains constant throughout the temperature range of the phase.

Figure 4-10 shows the various phases of the same idealized liquid crystal material, with the temperature decreasing from left to right [10]. (Not all materials go through all these phases). At high temperatures, the orientations of the molecules are completely disordered, i.e., isotropic. Then, as the temperature is reduced, the *nematic* phase takes over, and the molecules tend, on the average, to take up the same orientation. In a real device, this orientation can be preset by manufacturing roughness striations of the desired directionality into the adjacent glass surfaces. Going still lower in temperature, there is another phase-transition temperature below which the material enters the *smectic A* phase, and the molecules are oriented like books on a set of shelves. Finally, with still lower temperatures, there is the *smectic C* phase in which the molecules are oriented at some material-dependent tilt angle ψ from the orientation they would have had in the smectic A phase.

Thus the index seen by incident light depends on the relative orientation of the polarization plane to the orientation of the molecules, and the latter, in turn, depends on applied voltage. This is the basis of liquid-crystal flat-panel displays and of tunable filters using such materials. Rather than using the applied voltage to change either n_e or n_o directly, which amounts to a small fractional change, large effects can be gotten by reorienting the liquid crystal in such a way that the light sees an n_e or an n_o that is an indirect consequence of the fact that the molecules have a long shape.

Work on tunable filters for networking using liquid crystals has so far been confined to the nematic phase. For this phase, the retuning time is quite slow because of the absence of a strong restoring force. With nematic materials, the torque applied to the molecule is proportional to the square of the electric field strength. The molecules will change to a new orientation rapidly when a voltage is applied, but the only way to restore their original equilibrium orientation is to remove the voltage and let them drift back to it under the influence of thermal effects. Applying a reverse voltage would have no effect because of the squaring operation. An etalon developed at Bellcore [11], and using nematic liquid crystal material, exhibited a very wide tuning range of 175 nm, but with a tuning time of 20 msec.

Certain liquid crystal materials are not only smectic, but *ferroelectric* too, meaning that the molecules act like the electrical analogue of little magnets in that each has a net positive charge at one point in the molecule and a negative charge some distance away in another part. For nematics, this *polarization* (in the charge sense, not the light sense) had to be induced by the applied voltage; with ferroelectrics it is always there. With ferroelectric materials, the torque on the molecule is proportional to the voltage, not the square. This means, for example, that the original orientation of the molecules can be restored by applying a negative voltage. The response time of some materials to a reversal of applied voltage has been measured to be in the submicrosecond range, five orders of magnitude faster than nematics. They are therefore very promising materials for improving tuning speed.

While these tunable-filter directions are just projections at this time, it is quite possible that in the next several years, imaginative use of ferroelectric liquid crystal materials within the cavities of etalons will provide the rapid (nanosecond) tuning speed and the wide tuning range (200 nm) that are needed to build packet-switched networks.

Table 4-1.   Maximum number of nodes for worst-case (bit-synchronous) crosstalk in four Fabry-Perot filter arrangements. (From [1]).

Filter type	N_{max} formula	$F = 100$ case	$F = 300$ case
Single-cavity	$0.65F$	65	195
Two-pass	$1.4F$	140	420
Vernier two-cavity	$0.44F^2$	4400	39,600
Coarse-fine two-cavity	$0.25F^2$	2500	22,500

Crosstalk

The worst-case crosstalk (the bit-synchronous case) has been analyzed in [1] for the Fabry-Perot structures discussed in this chapter, with results such as those given in Table 4-1. The analyses are tedious and closed-form solutions are rarely possible. However, the method is exactly that given earlier in Section 4.3.

It is seen that there is a handy rule of thumb regarding the single-cavity etalon, namely, that to keep the detectability penalty below 1.0 dB, the maximum number of channels that can be crowded into one FSR is 65 percent of the etalon's finesse. Similar formulae are given in the table for other etalon arrangements.

4.5 Cascaded Multiple Fabry-Perot Filters

Multipass versus multicavity

As the need increases for more nodes per network, it is necessary to crowd more and more channels within one free spectral range. If an etalon has insufficient finesse, interchannel crosstalk will appear, due to insufficient adjacent-channel rejection, producing unacceptable error rates. However, it is possible to cascade several etalons, each having a modest finesse, in order to gain a higher overall *effective finesse*. This is often much simpler and cheaper than attempting very high values of finesse in a single cavity by carefully controlling mirror quality and parallelism. Certainly, finesses up into the thousands may be achieved in single etalons, but the cost makes the multiple-etalon idea an attractive alternative.

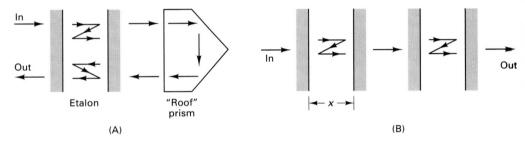

Figure 4-11. Filters made from two cascaded etalons (A) Two-pass. (B) Two-cavity.

Two approaches are possible, the *multipass* and the *multicavity* methods. The difference is illustrated in Figure 4-11 for two cavities. In the multipass scheme, an example of which is shown at the left, light passes twice through the same cavity. The composite power transfer function is then the square of the individual transfer functions that were given in Equation 4.8.

In the multicavity approach shown at the right, the FSRs of two cavities of roughly the same finesse F_o can be chosen to be in the ratio of two integers k and ℓ, and the result is an effective finesse that is roughly $F = \max (k, \ell) F_o$. Typical spectra are shown in Figure 4-12. At (A) is the frequency spectrum one would have had using a hypothetical single cavity of finesse F of roughly 10. At (B) is that of a cavity, also of finesse 10, but an FSR that is 1/3 the FSR in (A). The orders i of the peaks in (B) that line up with the first peak in (A) (order 1) are multiples of $k = 3$. Similarly (C) shows the response of a second cavity whose FSR is 1/4 that of (A), and for this cavity the aligned peaks are those that whose orders j are multiples of $\ell = 4$. If and only if k and ℓ are relatively prime, all other intervening peaks are misaligned.

It is helpful in keeping the two pairs of integers straight to note that k and ℓ are ratios of *FSR*, a quantity built into the filter, considered as fixed-tuned in the diagram.

The other integers *i* and *j*, on the other hand, depend on the frequency of the optical signal.

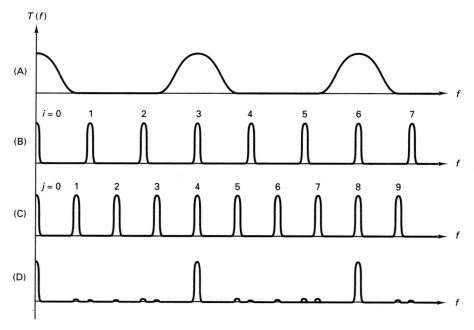

Figure 4-12. Spectra for a Vernier arrangement of two-cavity etalon. (A) Single cavity with the desired free spectral range. (B) First cavity. Same finesse as (A), but 1/3 the free spectral range (C) Second cavity. Same finesse as (A), but 1/4 the FSR. (D) Result of passing light sequentially through the two etalons. The finesse (ratio of FSR to peak width) has been effectively increased by 4.

The composite power transfer function of frequency is the product of the Airy functions of Equation 4.8. The result of passing light sequentially through the two etalons is shown at (D), and it is seen that we obtain the narrowness of peaks in (B) and (C), but with the *effective FSR* of (D) being $\ell = 4$ times that of (A). Since the *HPBW* of each of the two has not been changed, the finesse has been multiplied by 4. In addition, there is almost a factor of two additional increase in effective finesse of the cascade just due to the fact the the *HPBW* of the cascade is that of two cascaded filters of about the same *HPBW*.

Note that the actual band in use might be many *FSRs* off to the right in the diagram. For example, if we want to set our spread of wavelength channels $\Delta\lambda$ within one *FSR* = 200 nm at 1500 nm wavelength, we would be dealing with roughly the seventh multiple of the overall effective FSR shown at (D).

The advantage of the two-pass structure is that sidelobes away from the central peak are suppressed by the square of the Airy function, but the disadvantage is that the central peak is narrowed only by a factor of two. The advantage of the two-cavity approach is that the central peak is narrowed by the multiplier (3 or 4 in the example), and the disadvantage is that the secondary peaks are reduced only by the Airy function, not the Airy function squared. Picchi [12] has made an extensive comparison of the crosstalk due to the $N-1$ other channels at the output of multipass versus multicavity designs and has shown conclusively that the narrower peak of the multicavity design is much more important than the better suppression of distant sidelobes in the multipass design, a result that is corroborated by the data of Table 4-1. This is because most of the interchannel crosstalk comes from the two channels on either side of the desired channel, and very little from more remote channels.

It is necessary to provide some means of isolating the two cavities. At exact resonance, this is no problem, since the reflection spectrum from the second cavity has a null at exactly the frequency at which there is a peak of the transmission spectrum. At other wavelengths, however, this problem must be taken care of. If the two cavities are built into the same unit, the normal to the two mirrors of cavity 1 may be offset slightly (several degrees) from the normal to cavity 2, and the spurious reflections will walk off [3, 13], otherwise more elaborate measures [14] may be employed.

Vernier versus coarse-fine

The multicavity approach solves the problem of making a high-finesse structure out of several low-finesse structures. A number of variants are possible on the basic scheme of setting the two integers k and ℓ relatively prime. The vernier multicavity design of Figure 4-12 results when they are adjacent integers.

A second useful combination, which we shall meet shortly in connection with the *acoustically tuned double wedge* tunable filter, is to make one of the integers k and ℓ small and the other large. Then one cavity acts to give "coarse filtering" and the other "fine filtering," as illustrated in Figure 4-13 where $k = 1$ and $\ell = 4$. This means that every order i of the first cavity produces an aligned peak, as at (A), but that only every fourth one of the second cavity does so, as at (B).

The first etalon is thin and has a peak, shown at (A), only narrow enough to provide a modest level of rejection of unwanted maxima of the second, thicker etalon that are shown at (B). As the first etalon is tuned over its FSR, the second is tuned to keep up with it. This second one can be tuned continuously, or in a *sawtooth* pattern such that it tunes, say, only across a single FSR and then resets and tunes across its own FSR again, and so forth.

For the example in Figure 4-13, at the lowest-frequency output peak, the order for the first etalon is $i = 1$, and that of the second is $j = 4$, so that the effective finesse improvement of the combination over the single etalon of Figure 4-13(A) is $\ell = 4$, as with the vernier arrangement. Note, however, that the off-peak sidelobe level of the coarse-fine design can be slightly higher than for the vernier design, as evidenced by the numbers in Table 4-1.

Experimental use of multicavity etalons for gaining high effective finesse goes back almost thirty years. There have been a number of applications to lightwave net-

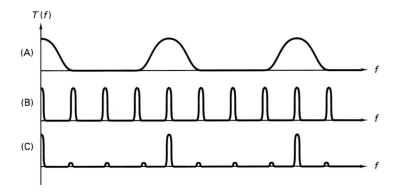

Figure   4-13. Spectra for coarse-fine form of two-cavity etalon, the cavities having widely different free spectral ranges but equal finesse (A) Rough-selectivity etalon (B) Fine-selectivity etalon (C) Cascade of the two.

works, several [15, 16] using two Fiber Fabry-Perots of the type shown in Figures 4-8 and 4-9.

Control

Multicavity designs, as used to date, have all posed problems of control complexity, since one applied control voltage no longer suffices; as the tuned frequency f changes, it is necessary to change the voltages applied to the cavities separately so that they will track each other.   There is a simple principle that can make it possible to get away with only one control voltage.   Consider, first, the two cavity lengths x (having order i) and y (having order j), which are

$$x = \frac{i\lambda}{2n} \qquad \text{and} \qquad \frac{j\lambda}{2n} \qquad (4.18)$$

Eliminating λ between these two equations (representing the condition that they are tuned to some common λ) gives

$$\frac{x}{y} = \frac{i}{j} \qquad (4.19)$$

All that is required, in order to make the two cavities track one another as λ changes, is to enforce this condition.   But this is just the condition that occurs anyhow when the two piezo-driven etalons are connected in electrical series, provided the piezo material used in the two has uniform resistivity and uniform stretching as a function of voltage gradient.   It is not even required that elongation as a function of voltage gradient be linear.   Since cavity 1 will be i/j as long as cavity 2 (3/4 in the example of Figure 4-12) it will have 3/4 as much voltage applied to it in a series connection, which is exactly the condition desired in order for the two cavities to track each other as the voltage is changed to tune to a new λ.

Crosstalk

As Table 4-1 shows, both the coarse-fine and vernier two-cavity etalons are capable of providing enough filtering selectivity for wavelength-division networks of several thousand nodes. Reference [16] reported achieving an effective finesse of over 2000 using two cascaded Fiber Fabry-Perots. This should be enough to support over 1000 WDM channels with adequately low crosstalk.

Advantages and disadvantages of etalons

Among the advantages of tunable etalons are their wide tuning range, the polarization independence often available, and the fact that the passbands can be made very narrow (either by improving single-cavity finesse or by using the multicavity scheme). Among the disadvantages are excessive access time of presently available designs, due to such effects as inertia or viscosity.

Also, the Airy function has no nulls, and so an etalon of a given bandwidth will have poorer crosstalk performance than certain other designs that do have nulls, for example, the Mach-Zehnder chain to be discussed next.

Introduction of electrooptic material in the cavity improves tuning time, but introduces polarization sensitivity. Several other methods of improving the tuning speed of Fabry-Perot filters exist. The acoustooptically deflected double-wedge etalon is one, and we shall describe it in Section 4.8 as part of the larger discussion of acoustooptic tuning possibilities.

4.6 The Mach-Zehnder Chain

Whereas the Fabry-Perot interferometer involves light interference by *many* repeated reflections, a single Mach-Zehnder interferometer (MZI) involves interference by only *two* versions of the same light traversing paths of slightly different length. (That is, the light path undergoes feedback in the etalon, but feedforward in the MZI).

Although this, by itself, will be seen to give very little wavelength-discriminating ability, we can get very good such capability by cascading a number of MZIs that are different from each other in a particular way. The process is somewhat analogous to the cascading of multiple etalons that we have just been discussing. We shall encounter this idea of cascading repeatedly as we discuss the wide variety of tunable filter options.

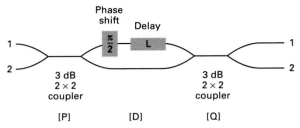

Figure   4-14. Structure of a single Mach-Zehnder interferometer (MZI).

Structure and fabrication

Figure 4-14 shows an idealized block diagram of such device. The light is split in a 3-dB coupler at the input and merged in another 3-dB coupler at the output. The two paths differ by delay τ. As indicated, the MZI can be thought of [17] as a cascade of three two-port black boxes having scattering matrices $[P]$, $[D]$, and $[Q]$. The scattering matrix was introduced in Section 3.11 in analyzing the 2×2 coupler.

We can write down the composite scattering matrix of the entire MZI as the product of the three individual scattering matrices.

$$
\begin{bmatrix} H_{11}(f) & H_{12}(f) \\ H_{21}(f) & H_{22}(f) \end{bmatrix} = [P][D][Q]
$$

$$
= \frac{1}{\sqrt{2}} \begin{bmatrix} 1 & -j \\ -j & 1 \end{bmatrix} \begin{bmatrix} \exp(-j2\pi f\tau) & 0 \\ 0 & 1 \end{bmatrix} \frac{1}{\sqrt{2}} \begin{bmatrix} 1 & -j \\ -j & 1 \end{bmatrix} \quad (4.20)
$$

$$
= \frac{1}{2j} \begin{bmatrix} j[\exp(-j2\pi f\tau) - 1] & \exp(-j2\pi f\tau) + 1 \\ \exp(-j2\pi f\tau) + 1 & j[(\exp(-j2\pi f\tau) - 1] \end{bmatrix}
$$

In many applications, the MZI is used as a 1-input device. If one of the inputs to the MZI (say input 2) is absent, the transfer function for the electric field vector is

$$
\begin{bmatrix} H_{11}(f) \\ H_{21}(f) \end{bmatrix} = \frac{1}{2j} \begin{bmatrix} j[\exp(-j2\pi f\tau) - 1] \\ \exp(-j2\pi f\tau) + 1 \end{bmatrix} \quad (4.21)
$$

and the power transfer function is

$$
\begin{bmatrix} |H_{11}(f)|^2 \\ |H_{21}(f)|^2 \end{bmatrix} = \begin{bmatrix} \sin^2(\pi f\tau) \\ \cos^2(\pi f\tau) \end{bmatrix} \quad (4.22)
$$

So both power transfer functions of frequency are raised sinusoids, 90° "out of phase" with one another.

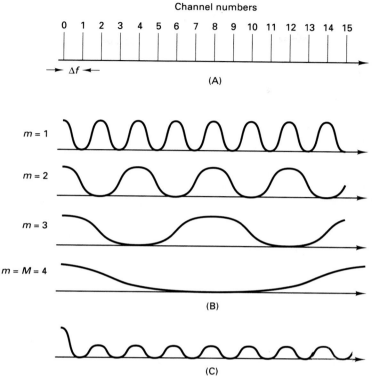

Figure **4-15.** Channel filtering action of a chain of M Mach-Zehnder interferometers (A) Input channels. It is desired to isolate Channel 0. (B) Transfer functions of the M successive stages. (C) Resultant overall transfer function.

Suppose we would like to build a filter to isolate one of $N = 2^M - 1$ channels equally spaced in frequency by δf, as illustrated in Figure 4-15(A) for the case $M = 4$. Suppose the frequency to be selected is the first one, number 0. A *Mach-Zehnder chain* of M stages will do just this [18], provided the successive stages have the correct path differential L_m, $m = 1, \ldots, M$ to make the individual transfer functions $T_m(f)$ have the pattern shown in Figure 4-15(B), so that the total overall transfer function comes out as shown in Figure 4-15(C). (With no loss of generality, we have added a $\pi/2$ phase shift in one of the arms in Figure 4-14, so as to get the $\cos^2$ rather than the $\sin^2$ form of transfer function). Such a filter is sometimes called a *periodic filter*.

We would like the first stage (having the longest L) to pass only channels 0, 2, 4, 6, 8, 10, ..., the second to pass only 0, 4, 8, the third to pass only 0, 8, 16, ..., etc. This requires that

$$L_m = \frac{v}{2^m \delta f} \qquad \text{or} \qquad \tau_m = \frac{1}{2^m \delta f} \qquad (4.23)$$

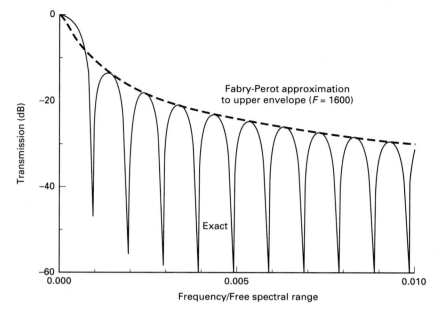

Figure 4-16. Ideal power transfer function of a 10-stage MZI chain, plotted on a logarithmic scale (solid curve). The transfer function for a Fabry-Perot filter (Airy function) for finesse $F = 1600$ is shown for comparison (dashed line). (From [2], courtesy SPIE).

If this is done, the overall transfer function is

$$T(f) = \prod_{m=1}^{M} \cos^2 \pi f \tau_m = \left(\frac{\sin(\pi f/\delta f)}{N \sin(\pi f/N\delta f)} \right)^2 \tag{4.24}$$

as shown on a logarithmic scale by the solid curve of Figure 4-16.

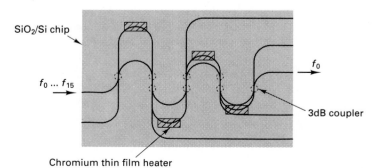

Figure 4-17. An MZI chain device that selects one of 16 channels ($M = 4$) fabricated at NTT Laboratories. (From [19], © 1990 IEEE).

It is clear that if the device is ideally constructed, infinitely deep nulls are produced at the undesired channel frequencies. This is to be compared with the single or multiple etalons (Sections 4.4 and 4.5), which have no nulls in their transfer functions.

Figure 4-17 shows an $M = 4$ MZI chain optical filter, as made by the NTT Transmission Systems and Optoelectronics Laboratories, who have pioneered the use of Mach-Zehnder chains for fiber optic networking. A 128-channel ($M = 7$) unit [20, 21] was fabricated on a single 5-cm by 6-cm substrate of silicon upon which a layer of silica had been deposited. The fibers were attached in V-grooves. Observed fiber-to-fiber attenuation was 6.7 dB, and crosstalk was $- 13$ dB. The path differential L_1 of the first MZI was about 1 cm.

To demultiplex all N channels, a tree structure consisting of $2^M - 1$ MZI units would be required. For a filter designed to output one channel only, a linear chain consisting of M of the MZIs is sufficient.

Control

To tune each MZI over its complete periodic range, it is only necessary to vary its differential path length by $\lambda/2$ at the most. The filters developed by NTT are completely tunable, and this is done by the means shown in the small shaded rectangles of Figure 4-17. These are patches of chromium plated on top of the planar waveguide to form resistive heaters having a time constant of about one millisecond, too slow for packet switching. There is great interest in finding an electrooptic technique that will produce a retardation of up to $\lambda/2$ in low-attenuation planar waveguide material, since this then provides sub-microsecond tuning.

To lock all the MZIs when receiving one channel, there is the complex approach [22] in which each of M AFC circuits (phase-lock loops) locks onto an acquired input signal.

For use in a practical network, digital addressing of the tunable element is always a desirable feature, and the MZI chain, with its binary tree structure, is particularly amenable to this. At first glance, you might think that simply supplying the various bits of the digital address word to the MZIs separately would do the trick, but a few moments' inspection shows that it is not quite so simple. This is certainly what happens with the first MZI, but the desired voltage to control the second MZI depends on the bit used for the first MZI, that of the third depends on the first two bits, and so forth. The voltage fed to the last Mth MZI depends on all $M - 1$ preceding bits in the address. A combinational logic circuit [17] that solves this problem is shown in Figure 4-18.

Crosstalk

Earlier, in connection with Fabry-Perot filters, the assumption was made that the signal channels could be represented in frequency as impulses that were exactly equally spaced by δf. Under those ideal circumstances, the crosstalk of an MZI chain will be

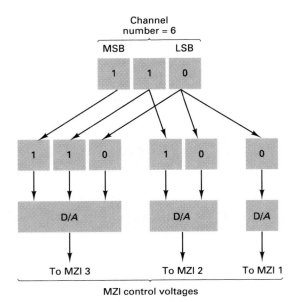

Figure 4-18. Combinational logic circuit used to convert the channel number expressed as a binary word into a set of voltages fed to the MZI stages. $M = 3$. The settings shown select frequency 6. MSB = most significant bit; LSB = least significant bit. (From [17]).

zero, since the transfer function has infinitely deep nulls. In practice, the amplitudes of the signals in the two arms of each 2×2 coupler will not be exactly equal, so each MZI stage will not have perfect signal cancellation at the null, which in turn means that the overall $T(f)$ will not have infinitely deep nulls either. Also, the channels will not usually be perfectly equally spaced, and they will each have a finite nonzero modulation bandwidth.

The calculation of crosstalk in an MZI is thus trivial for the ideal filter (it is zero), but fairly complex for a real filter, in which some of the transmitter frequencies may be misaligned with the resonant nulls and in which there will be spectral spread due to modulation. However, there is a most interesting upper bound on this real crosstalk [2], namely that the crosstalk for an M-stage MZI chain can be no worse than that of a single-stage Fabry-Perot etalon of finesse $F = (\pi/2)2^M$. For example, Figure 4-16 shows on a logarithmic scale the transfer function $T(f)$ for a ten-stage MZI chain (solid line), and the Airy function for an FPI having a finesse $F = (\pi/2)1024 = 1600$ (dashed line). It is seen that no matter how spread out or misaligned the different channels are, at worst they will see a transfer function $T(f)$ that can be only as high as the corresponding Airy function curve.

This bound comes about as follows:

$$\left(\frac{\sin \pi N/\delta f}{N \sin \pi/\delta f} \right)^2 \leq \frac{1}{N^2 \sin^2 (\pi/\delta f)} \qquad (4.25)$$

which is approximately

$$\frac{1}{1 + N^2 \sin^2 (\pi/\delta f)} \qquad (4.26)$$

which is the Airy function with the finesse expressed as $(\pi/2)N$.

Pros and cons of Mach-Zehnder chains

Perhaps the biggest advantage of Mach-Zehnder chains is the possibility of realizing the low fabrication costs that lithography offers. Also, by shaping the waveguide cross-section to be roughly square, these filters can be made almost completely polarization-insensitive. For channels spaced at exactly even intervals in frequency the crosstalk is very low, due to the nulls in the transfer function. Also, silicon has one of the lowest temperature coefficients of expansion of any metal (3×10^{-6} per $°C$).

Apart from the complexity of the tuning control, the principle disadvantage, as matters now stand, is the slow tuning speed, due to thermal inertia.

4.7 Interaction of Sound and Light - Phase Matching

We now turn to several tunable filters that exploit the interaction of sound waves and light waves in a solid. The sound wave produces an artificial diffraction grating and incident light interacts with this grating. (A grating is a region of many fine parallel ripples in the refractive index). For our purposes, the subject of sound-light interaction is of importance in itself because many tunable filter types are based on this inter-action. But beyond that, spending some time developing the subject is more broadly useful to us because it provides such a good understanding of all grating-related filter types, many of which do not involve acoustical effects.

As we said at the outset of this chapter, all practical tunable optical filters are based on some form of interference, constructively at the center of the passband(s) of the filter, and partially or completely destructively at other frequencies. In the acoustooptic filters we shall now describe, this is not so obvious; the fact that diffraction from a grating is really interference between hundreds or thousands of wavelets is not as effective a way to think of them as is the concept of *momentum con-servation*, also known as *phase matching*, or *mode coupling*.

When a sound wave travels through a solid transparent material [23, 24], the spatially periodic local compressions and rarefactions cause local increases and decreases in the refractive index by the action of the *photoelastic effect*. Therefore, by passing light through a material with a suitably high *photoelastic coefficient* it should be possible to form a periodic grating that will *diffract* the incident light beam to an angle that depends on the angle of incidence, the light wavelength λ and the sound wavelength Λ. Another way of saying this is that phase-matching devices can be built, ones in which that portion of the incoming light whose propagation vector **k** closely matches the propagation vector **K** of the sound in some particular way will reach the output by constructive interference, while incoming light at different enough **k**s will suffer partial or complete destructive interference, the same basic principle we saw earlier with the Fabry-Perot etalon. (Perhaps a more accurate term than "phase match" would have been "**k**-vector match").

We may construct a variety of devices in which only a certain narrow band of λs will be passed for a given Λ, and in which this band may be tuned in λ by varying Λ. The latter can be done simply by tuning the frequency of the electrical drive to the transducer that launches the acoustic wave. Since such RF drive frequencies can be derived accurately and reproducibly from quartz crystal oscillators to a harmonic specified by a digital input word, the idea of building highly stable and rapidly tunable filters using the acoustooptic interaction is an attractive one from the system point of view.

We shall use lower-case letters λ, v, f, ω, and k to denote wavelength, velocity, (temporal) frequency, angular frequency and propagation constant (i.e., wave number or spatial angular frequency), respectively, for the light, and upper-case letters (Λ, V, F, Ω, and K) for the corresponding acoustic variables. By convention, λ denotes the wavelength that light of the same ω would have had if travelling in free space, while Λ is sound wavelength in the medium.

To give a quantitative feel for things, consider light of $\lambda = 1.5\ \mu$, that is, frequency $f = 2 \times 10^{14}$ Hz, and sound of frequency $F = 100$ MHz, both travelling through a crystal of the commonly used material lead molybdate (PbMoO$_4$). This material has index $n = 2.3$ and sound velocity $V = 3.75$ mm/μsec, as compared with light velocity in the medium of $c/n = 1.3 \times 10^5$ mm/μsec. The acoustic wavelength $\Lambda = V/F$ is 37.5 μ, and the light wavelength in the medium is $\lambda/n = .65\ \mu$. So, even though the frequencies are different by six orders of magnitude in one direction, the velocities are different in the opposite direction by somewhat less than six orders of magnitude, and so the wavelengths in the medium usually come out within about one or two orders of magnitude of one another. This is one of the things that make the light-sound interaction practical.

One should note that, since the light velocity is orders of magnitude higher than the sound velocity, during the time the two interact, the grating created acoustically remains in essentially the same position as light passes through it.

To build a practical tunable acoustooptic device, it is desirable to pass into the diffracted beam the greatest possible fraction of the incident light with the least expenditure of RF drive power, to avoid excessive RF frequencies or RF tuning range, to minimize the time it takes to tune from one λ to another, and to resolve the maximum possible number of wavelength channels. In general, the acoustooptic properties necessary to achieve these objectives cannot be obtained from simple isotropic materials, in which all the variables to be discussed would be either scalars or vectors, but by propagating along carefully chosen axes of crystals, which, being anisotropic, require some of the variables to be represented as tensors. However, we shall be able to present the relevant ideas here by thinking of the various acoustooptic parameters as scalars and the parameters of the light and sound as scalars or vectors.

One practical consequence of the anisotropy of the acoustooptic parameters of crystals is that it requires some cleverness to use acoustooptic effects to build tunable optical filters that are completely polarization-independent. One way, which will be discussed in Section 4.10, is to build two parallel light paths into the middle of the device, one to handle one polarization and the other to handle the orthogonal polarization.

There are basically two types of acoustooptic devices, those for which the optical propagation vector and acoustic propagation vector are at a large angle from one another, and those for which they are almost collinear. Typically, the large-angle geometry is used for acoustooptic *deflectors* and the collinear one is used for *acoustooptic tunable filters*. Deflectors are of some interest, not because they necessarily make ideal tunable filters by themselves, but because they can be used as components in larger tunable-filter structures, because they can sometimes serve as modulators, and especially because they are useful for explaining basic principles.

The easiest way to see quantitatively what happens as light and sound interact in a solid and to see how wavelength tuning can be effected is to use the vector picture [24] in Figure 4-19. Instead of grinding through the geometry, it is much cleaner to explain the acoustooptic interaction by calling on the modern idea that light or sound has a dual personality, as a wave and as a quantized particle. For the light wave the particle is the familiar *photon*; for the sound wave the "particle" is the *phonon*.

Specifically, the light wave has an $\mathbf{E}$ field that can be represented as $\mathbf{E} = E \cos (\omega t - \mathbf{k}r)$, r being the magnitude of $\mathbf{r}$, the unit vector along the direction of propagation. This wave can be thought of as being composed of particles having

$$\text{Energy} = \frac{h}{2\pi} \, \omega \qquad \text{(a scalar)} \qquad (4.27)$$

and

$$\text{Momentum} = \frac{h}{2\pi} \, \mathbf{k} \qquad \text{(a vector)} \qquad (4.28)$$

and similarly for the acoustic phonon (angular frequency Ω radians per second and spatial frequency $\mathbf{K}$ radians per mm).

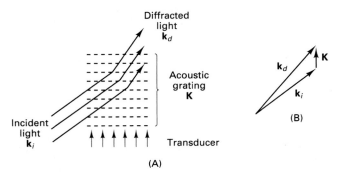

Figure   4-19. Interaction of light and sound. (A) Geometry of incident light ($\mathbf{k}_i$) and sound ($\mathbf{K}$) to produce diffracted light ($\mathbf{k}_d$). (B) Corresponding "conservation of momentum" triangle.

According to the particle view, the diffraction of light by the effective grating created by the acoustic wave is viewed as a lossless *collision* between an incident photon and a phonon in the medium. In such a collision, both energy and momentum must be conserved, which means that the diffracted light (subscript d) is related to the incident light (subscript i) and the acoustic field thus:

$$\omega_d = \omega_i \ \pm \ \Omega \qquad \text{(Conservation of energy)} \qquad (4.29)$$

and

$$\mathbf{k}_d = \mathbf{k}_i \ \pm \ \mathbf{K} \qquad \text{(Conservation of momentum)} \qquad (4.30)$$

The "collision" results in the annihilation of the incident photon and the phonon and the generation of a new photon that has different momentum and a different energy. If the light and sound are travelling in essentially the same directions ($\mathbf{k}_i \bullet \mathbf{K}$ positive), the + sign applies to both equations; the diffracted photons have a higher momentum (longer k-vector) and a slight up-doppler shift by the amount given up by the phonon. If they are propagating in roughly opposite directions, ($\mathbf{k}_i \bullet \mathbf{K}$ negative), then the − sign applies, and the new photon has a lower energy and a down-doppler by the amount given up by the phonon.

This model allows us to use Figure 4-19 to see what happens in a particularly simple way. At (A) is seen an incident beam with propagation vector $\mathbf{k}_i$ interacting with sound of propagation vector $\mathbf{K}$ producing the resultant $\mathbf{k}_d$ of the diffracted wave. At (B) is shown the equivalent *momentum triangle* representing the process by which all the wavelets partially reflected from the many peaks and troughs of the acoustic grating interfere constructively, i.e., diffraction, i.e., photon-phonon collision.

4.8 The Acoustically Tuned Double-Wedge Etalon

Figure 4-20 shows a tunable filter formed by combining two different technologies, the coarse-fine etalon of Section 4-5 and a two-axis acoustooptic deflector. We shall use this structure as a vehicle to illustrate acoustooptic effects.

The first wedge is thick and differs by $\lambda/2$ in thickness from end to end. Thus, referring to Figure 4-13(B), the fine resonance peak may be tuned by changing the vertical position at which light passes through this first wedge. Similarly, the second wedge, which is thinner and oriented at essentially $90°$ from the first wedge, provides tunability of the coarse peak of Figure 4-13(A) across its free spectral range according to the horizontal position. It is the function of the preceding acoustooptic deflector to position the beam of incoming light at the desired spot in the $X - Y$ plane in order to achieve a rapid tunability over many different resolvable wavelengths. We proceed now to a discussion of these deflectors. They can perform as parts of other structures and even as tunable filters themselves.

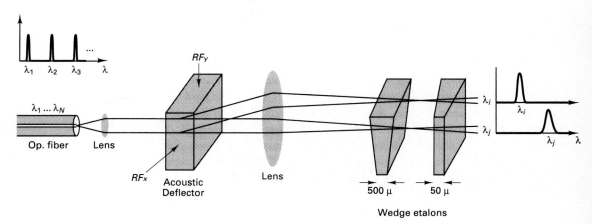

Figure 4-20. Acoustically tuned double-wedge etalon.

Both acoustooptic deflectors and the classical acoustooptic tunable filter (to be discussed in the next section) are devices that force exactly the condition just mentioned: that the device passes by constructive interference values of $\mathbf{k}_i$ that it is tuned to, the others interfering nonconstructively. The condition that is used in both devices is the *Bragg condition*.

The acoustooptic deflector

Consider the acoustooptic deflector shown in Figure 4-21(A). A collimated beam of acoustic energy of width L and height H propagates across a suitable crystal and is absorbed at the far end. By changing the acoustic frequency, the deflection angle can be changed.

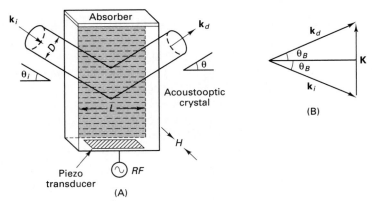

Figure 4-21. Acoustooptic deflector. The angles are exaggerated. (A) Geometry of the device, showing the interaction length L, the acoustic beam height H, the light beam diameter D, the transducer driven at frequency F, and the incident and diffracted k-vectors. (B) The Bragg condition.

In Figure 4-21(A), light enters from the left at an angle θ_i. Thinking of the grating as a series of partially reflecting mirrors, the angle of incidence and reflection must be equal (for situations in which the refractive indices of incident and diffracted light are equal). But what angle insures that all the elementary grating reflections will all add up in phase?

The question is easily answered from the vector model of diffraction using the diagram in Figure 4-21(B). Assuming that the incident and refracted waves see the same index of refraction, the elementary grating reflections will interfere constructively if and only if the angle of incidence (and of diffraction) equals the *Bragg* angle θ_B given by

$$\theta_d = \theta_i = \theta_B = \sin^{-1}\frac{K}{2k} = \sin^{-1}\frac{\lambda}{2n\Lambda} \qquad (4.31)$$

In other words, once given the acoustic propagation vector $\mathbf{K}$, the only way that the angle of incidence can equal the angle of diffraction and still form a closed momentum triangle is if this equation (*Bragg condition*) is satisfied. For practical devices θ will be very small. For example, in the earlier illustrative example $\theta = \sin^{-1}(0.5 \times .65/37.5) = 17$ milliradians $= 1°$. Replacing the sine by its argument,

$$\theta_B = \frac{\lambda}{2n\Lambda} = \frac{\lambda F}{2nV} \qquad (4.32)$$

where $F = \Omega/2\pi$.

There are several quantities of engineering interest to be determined. We would like to know the time it takes to deflect the beam (access time), how many spots can

Figure 4-22. Commercially available acoustooptic devices of interest for networking. (A) Single-axis and double-axis acoustooptic deflectors (Courtesy Isomet Corp.). (B) Bulk acoustooptic tunable filter (Courtesy Inrad Corp.).

be resolved, what fraction of light will be passed to the diffracted output beam, and the tuning range.

Access time

In practice light will be focussed into a narrow *beam waist* of diameter w positioned to embrace the interaction region of length L. After leaving the deflector, the beam suffers an irrecoverable divergence $\delta\theta$ by an amount given by the Fresnel diffraction limit

$$\delta\theta = \lambda / n\,w \tag{4.33}$$

(equivalent to considering the beam waist as an antenna of wn/λ wavelengths aperture). $\delta\theta$ is the angle between nulls. The time taken to set up a new deflection angle by changing the acoustic frequency F to a new value is clearly the time taken to form the new grating with a new spatial frequency K, i.e., the time it takes to fill up the width of the beam w with sound.

$$\tau = w/V \tag{4.34}$$

Number of resolvable spots

If the total allowed acoustic frequency swing is ΔF, then from the Bragg equation 4.32 and Figure 4-21(B), the total change of deflection angle is (counting the changes in both incident and diffracted angles)

$$\Delta\theta = \lambda\Delta F/nV \tag{4.35}$$

which, when divided through by the spot size $\lambda / nw = \lambda / \tau nV$ gives

$$N_s = \tau\Delta F \tag{4.36}$$

the number of resolvable spots.

It is seen that there is a conflict between keeping the beam waist w small, so as to improve the access time, and making it large, so as to minimize spot-size enlargement due to Fresnel diffraction.

The acoustooptic deflector as a tunable filter

It is tempting to think that an acoustooptic deflector followed by a slit or pinhole that passes only light of one wavelength deflected at one angle could serve as a practical tunable filter, but tunable filters built in this way do not turn out to be very interesting. The reason is that light of one wavelength isn't deflected at one angle, but at a smear of angles, the spot size. This smearing effect of diffraction limits the number of resolvable λs, as follows:

From the Bragg condition (4.32), the swing in wavelength of incident light to have produced the same $\delta\theta$ as that produced by diffraction (4.33) would be given by $\delta\theta = F\,\delta\lambda/2nV = \lambda/nw$ from which

$$HPBW = \delta\lambda = \frac{2V\lambda}{wF} \tag{4.37}$$

which for our illustrative example of $PbMoO_4$ material with 100-MHz drive and a 1-mm beam waist is 2.7 nm, too large to be interesting.

Deflector tuning range

The tuning range in optical frequency will be proportional to the range ΔF of permitted RF drive frequencies, as given in Equation 4.35. Several factors limit ΔF, the most important of which is *Bragg-angle mismatch*, in which the incoming θ_i, being unchanged, is not at the right tilt with the grating over the entire ΔF to satisfy the desired isosceles form of the Bragg triangle, Figure 4-21(B). This is discussed in detail in [23, 24], where it is also shown how this can be partially compensated by a variable phased array of transducers, in the style familiar to communication engineers as a phased-array antenna.

Other factors that limit ΔF include the achievable RF bandwidth of the matching network feeding the transducer. A good working figure for maximum practical fractional bandwidth $\Delta F/F$ is between 50 and 100 percent.

Figure 4-22 shows two commercially available acoustooptic deflectors, at (A) a single-axis one, such as described in Figure 4-21, and at (B) a device in which two acoustic signals propagate at two orthogonal angles, thus providing a two-dimensional deflection. Such a device is used in the acoustooptically tuned double wedge.

Performance of the double wedge etalon

The actual device of Figure 4-20 [25] used an RF frequency swing ΔF of 25 MHz on each axis, and the tuning time τ was 2 μsec, so that the number of resolvable spots $\tau\Delta F$ was 50 in each direction, and therefore there were 2500 resolvable spots. However, this is not the number of WDM channels that can be supported because they must be separated by several spot diameters in order to minimize crosstalk. A 60-nm optical tuning range Δf was achieved. The channel spacing was chosen so that the crosstalk penalty was 1.0 dB, and this led to a total number of channels around 500-600.

The problem with this tunable-filter solution is that the physical size (about one meter) is too large for successful use in small network nodes. This is an unavoidable consequence of the small deflection angles available from acoustooptic deflectors.

4.9 The Bulk Acoustooptic Tunable Filter

Structure

The bulk filter of Figure 4-22(C) uses an essentially collinear orientation of incident $\mathbf{k}_i$ vector and acoustic vector $\mathbf{K}$, as distinguished from the deflectors we have been discussing whose $\mathbf{k}_i$ and $\mathbf{K}$ vectors are orthogonal to within a few degrees. In the case shown, the light and sound propagate in the same direction, and

$$k_d - k_i = K \qquad (4.38)$$

but they could just as easily have propagated in opposite directions (plus sign in the equation). The device is schematized in Figure 4-23(A), where the Bragg momentum triangle of the earlier Figure 4-19(B) has been collapsed into three collinear vectors, as shown in Figure 4-23(B). Now the interaction length L is much longer. While this helps with wavelength selectivity and makes more efficient use of the applied acoustic power, the access time, given now by

$$\tau = L/V \tag{4.39}$$

is clearly much longer than it was for the acoustooptic deflector.

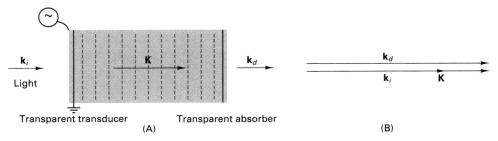

Figure   4-23. Bulk acoustooptic tunable filter (A) Geometry of the device. (B) Bragg triangle, which has now collapsed into a set of collinear vectors.

If we use an isotropic material for which the incident and diffracted light velocities are the same, then the momentum triangle of Fig. 4-19(B) cannot be closed, since this would require a lengthening of the **k**-vector (from the conservation-of-momentum equation 4.30) that cannot be fulfilled just by the much smaller up-doppler shift of the conservation-of-energy equation 4.29. However, suitable anisotropic materials can have acoustooptic properties of the diffracted light different from those of the incident light. In particular, if the refractive indices are different for two polarizations, this will do the trick. Materials whose birefringence depends on local compression have the property that, upon acoustooptic diffraction at exactly the Bragg condition, the indices of the ordinary and extraordinary wave will be different

$$n_o > n_e, \qquad\qquad n_o - n_e = \Delta n \tag{4.40}$$

where the subscripts o and e refer to the linearly polarized ordinary and extraordinary submodes, respectively.

From Equation 4.40 we have

$$k_o = \frac{2\pi}{\lambda}\, n_o = \frac{\omega}{c}\, n_o\,, \qquad\qquad k_e = \frac{2\pi}{\lambda}\, n_e = \frac{\omega}{c}\, n_e \tag{4.41}$$

and the triangle equation 4.30 is satisfied by

$$\frac{\omega}{c}\,\Delta n = \frac{\Omega}{V}\,, \qquad \text{i.e.,} \quad \lambda = \Lambda \Delta n \tag{4.42}$$

Note, once again, the use of the "tuning-range multiplication" principle, by using relative change of index differences rather than relative change of a single index.

There is an even simpler explanation [26] of this equation. Recall, from Section 3.13, that in a birefringent material, the two degenerate submodes, (say the E_x and E_y components of a z-propagating wave), have different propagation velocities. Therefore at periodic spaced intervals along the medium, called the *beat length*

$$L_{beat} = \frac{\lambda}{\Delta n} \tag{4.43}$$

they add, and halfwave in-between they cancel.

We saw in Section 3.13 that in an ordinary non-polarization-preserving fiber the beat length is usually one-half to several meters. But in a material whose photoeleastic properties are of the proper kind, (e.g., PbMoO sub 4), the beat length is so short that it can be matched by the acoustic wavelength ($\Lambda = L_{beat}$), so that the Bragg resonance occurs at $\lambda = \Lambda \, \Delta n$, i.e., Equation 4.42.

Very large tuning ranges can be gotten with materials having high birefringence. Note, once again, that this is because the tuning of the device is not being effected by varying an index relative to itself (which would only tune over $\Delta n/n = \Delta f/f < 0.005$), but by using the fractional difference Δn of two quantities that are almost equal.

Transfer function

The transfer function $H(f)$ of the bulk acoustooptic filter has been determined exactly by a lengthy wave-propagation analysis [24]; here we shall be satisfied with a more direct but approximate version.

We know that the power transfer function $T(f)$ will be centered at the optical frequency for which a phase match exists, but what is the mathematical form of $T(f)$ around that frequency? We can find out by the following argument. The effective delays seen by the ordinary and extraordinary submodes are

$$T_o = \frac{L}{v_o} = \frac{Ln_o}{c} \qquad \text{and} \qquad T_e = \frac{L}{v_e} = \frac{Ln_e}{c} \tag{4.44}$$

respectively, where v_o and v_e are phase velocities. The total difference in time delay experienced by the optical signal is therefore

$$T = T_o - T_e = \frac{L\Delta n}{c} \tag{4.45}$$

If we imagine an impulse applied to the input, the time difference between the earliest arriving signal at the output and the last arriving such signal will be T. Thus, the impulse response of the filter is a sinusoid with a square envelope of constant amplitude from 0 to T seconds, and zero amplitude thereafter.   The Fourier transform of such an impulse response of unit amplitude is

$$H(f) = T^2 \frac{\sin (\pi f T)}{(\pi f T)}$$
(4.46)

or, normalizing to unit response at the resonance peak and taking the squared magnitude,

$$T(f) = |H(f)|^2 = \frac{\sin^2 (\pi \Delta n L f / c)}{(\pi \Delta n L f / c)^2}$$
(4.47)

From this, it is readily seen that the larger the interaction length L and the birefringence Δn, the narrower the passband.   Roughly,

$$\frac{\Delta f}{f} = \frac{\Delta \lambda}{\lambda} \cong \frac{\Lambda}{L}$$
(4.48)

. Here Δf and $\Delta \lambda$ represent the 3-dB bandwidths in frequency and wavelength, respectively.

Crosstalk and number of channels

Bulk acoustooptic filters have a achieved very wide tuning range (e.g., from 1.3 to 1.56 μ) but with only modest wavelength selectivity (e.g., 4.5 nm) [27], even for usably long interaction lengths.   Moreover, they usually require several watts of RF power.   Access time is several microseconds.

A quantitative calculation of crosstalk can be made using Equation 4.47, but is unnecessary for present purposes because the maximum number of channels can be seen to be fairly modest.   We can see this by assuming that one spaces the channels roughly 1.65 times the half-power bandwidth apart.   The factor of 1.65 comes by reasoning that the shape of the curve near resonance will not be too different from that of an etalon, and we know from Section 4.4 and [1] that to keep the error-rate penalty less than one decibel, such an interchannel spacing is required.   If 200 nm of wavelength is available, then the number of channels is 200/2 × 4.5 = 22.

4.10 Filters Based on Separated Polarization Beam Splitters

Figure 4-24 schematizes the principle of operation of an entire class of polarization-independent tunable filters based on placing a resonant structure between two polarization beam splitters.   The resonant structure will do things to the state of polarization of light at the resonance frequency that it does not do to light at other frequencies. Thus a narrowband filter can be realized by discriminating at the output on the basis of polarization.

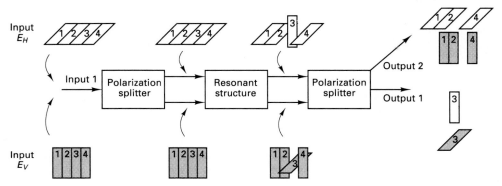

Figure   4-24. The general class of tunable filters consisting of two polarization beam splitters separated by a resonant structure. The diagram shows the filtering action schematically when the resonant section is tuned to the third of four wavelengths. The top row of polarization planes shows what happens to the incoming horizontally polarized components, and the lower row shows what happens to the incoming vertical component. (Courtesy David A. Smith, Bellcore).

Suppose arbitrarily polarized light enters input 1 at the left. The action of the left polarization splitter is to pass the horizontal component but to divert the vertical component to the other output branch. Two different things then happen, the first to the light that is at or near the phase-matched wavelength, and the second to light of other wavelengths. For phase-matched wavelengths, there is a 90-degree rotation between input and output planes of polarization. This rotation converts the horizontally polarized input to the filter section into vertically polarized light which is diverted by the second polarization splitter to output 1. Other wavelengths emerge from the grating region still horizontally polarized and pass through to output 2, never to be seen again (perhaps).

In a similar way, the vertical component at input 1 also has its phase-matched wavelengths rotated and passed to the output 1, while its off-resonance wavelengths are transferred by the second polarization splitter to output 2.

It is seen that, upon combining the outputs, no matter what the state of polarization of an incoming signal, this signal will reach the output similarly polarized. This is true for the resonance frequencies going to output 1 and for all the off-resonance light going to output 2.

This principle has been applied in devising acoustooptic filters and electrooptic filters, as we shall now discuss. It has also been implemented using liquid-crystal filters as the central resonant structure [28].

The surface-wave acoustooptic filter

Surface-acoustic-wave (SAW) technology [29] allows us to fabricate a wide variety of low-power consumption filtering and signal-processing devices lithographically. Most previous applications have involved electrical rather than optical inputs and outputs. Thus it has long been established that metallic transducers can be fabricated lithographically that launch an acoustic surface wave down the surface of a suitable material. The planar waveguide art allows waveguides and polarization beamsplitters to be similarly fabricated lithographically. Using this combination of ideas, devices such as the one shown in Figure 4-25 have been developed [30].

As mentioned in the last section, at the phase-matched condition the grating in the middle passes light strongly and rotates any linear polarization direction by 90 degrees so that it proceeds to output 1. Off-resonance wavelengths are not polarization-rotated and go to output 2.

The expression for the transfer function of the surface-wave acoustooptic tunable filter is, to first order, the same as Equation 4.47 for the bulk acoustooptic filter, by the same line of reasoning.

Such devices have been built [30] with the same extremely large tuning ranges (from 1.3 and 1.55 nm) as the bulk devices [27], but with about one-fourth the HPBW. As with the bulk devices, they have access times in the range of microseconds, the time for the acoustic wave to fill up the interaction length.

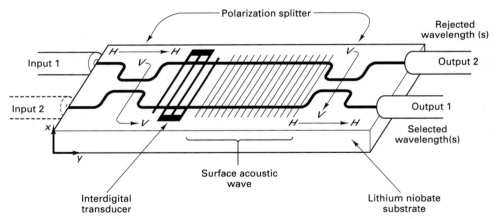

Figure   4-25. Surface wave acoustooptic tunable filter using polarization conversion. The labels show only what happens to signals at the resonance frequency. (From [30], © IEEE 1990).

Multi-frequency drive of acoustooptic devices

It should be pointed out that acoustooptic deflectors, bulk filters and surface wave filters all have the property that linear superposition of the RF drive pertains. This means that if several RF signals are supplied simultaneously, the deflector could be deflecting incident light of the same wavelength to several places at once and similarly the tunable filters could be sending light of several wavelengths to the filter output [30]. To date, the maximum number of wavelengths that can be separately resolve by using that number of RF drive signals is around seven for the surface wave acoustooptic filter.

This capability has some interesting system architecture consequences which will be mentioned in Chapter 11.

The electrooptic tunable filter

If electrooptic effects could be used to do the same thing as the acoustooptic tunable filters we have described, great reductions in access time could be obtained, since it would be limited by capacitance, rather than sound propagation velocity. Such devices have been built [31], and one is shown in Figure 4-26. Note the similarity in appearance to the surface-wave acoustooptic filter of Figure 4-25. In fact, the basic principles of the two are identical.

Both use polarization splitters at input and output, with a resonant polarization-rotating section in between, consisting essentially of a grating of wavelength Λ and interaction length L, so that the tuned wavelength and the transfer functions are as given in Equations 4.42 and 4.47, respectively. For the electrooptic filter, the behavior of the incoming horizontal and vertical components at resonance and off resonance, and the crossover or noncrossover before the resonant section or after, are all as just described for the acoustooptic filter. However, in the electrooptic device, the periodic grating of the refractive index is formed by finger electrodes lithographed permanently in place, whereas with the acoustooptic device, it is formed by an acoustic signal by means of the photoelastic effect.

Thus, while the acoustooptic filter is tuned by changing Λ by a change in the frequency of the acoustic drive, in the electrooptic device the spatial pitch Λ of the grating is fixed. According to Equation 4.32 (i.e., Equation 4.42) the tuning must therefore be done by changing the birefringence Δn that the grating sees.

This is called *birefringence tuning*, and in the earliest electrooptic grating [32] was done as follows: The lithographed grating extended uninterruptedly down the entire length of one side of the waveguide carrying the light, while on the other side there was a single electrode likewise extending down the entire length, and a voltage applied to this effected the tuning of Δn. A more effective scheme later proved to be one actually shown in Figure 4-26, in which the grating is placed directly over the waveguide for maximum coupling, and then is periodically interrupted to make room for fingers that apply the birefringence tuning voltage. The device described in [31] consisted of 46 grating sections of $\Lambda = 21\ \mu$ interspersed with 45 phase-shift sections. The 3-dB bandwidth was 1.2 nm at 1.52-μ wavelength.

The device shown in Figure 4-26 had about 3 dB of attenuation. A swing of the common voltage V_T applied to all 45 birefringence tuning sections from -100 to $+100$volts tuned the device over 11 nm in several tens of nanoseconds. In this way it

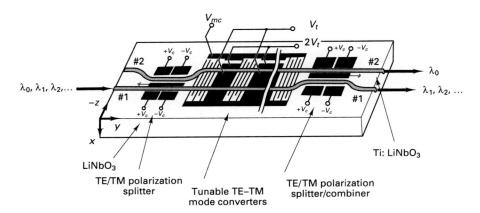

V_{mc} V_t

$2V_t$

$+V_c$ $-V_c$

#2 $+V_c$ $-V_c$ #2 λ_0

$\lambda_0, \lambda_1, \lambda_2, \ldots$ $\lambda_1, \lambda_2, \ldots$

$-z$ #1   #1

$+V_c$ $-V_c$ $+V_c$ $-V_c$

y

x Ti: LiNbO$_3$

LiNbO$_3$

TE/TM polarization splitter   Tunable TE–TM mode converters   TE/TM polarization splitter/combiner

Figure 4-26. Electrooptic tunable filter using polarization conversion. (From [31]).

was able to resolve about 10 channels. The interaction length could be lengthened to improve this, but at the expense of increased attenuation.

4.11 **Tunable Taps**

The class of filters discussed in the last section are really three-port devices. The optical signal at the selected wavelength exits from one output port and those at other wavelengths exit from the other.

Several filter types have this property. These include Fabry-Perot etalons. By simple conservation-of-energy principles the spectrum of light reflected back from the etalon must be roughly flat with a resonant notch in wavelength where the transmitted light has its resonant peak. In principle one can arrange to extract the energy reflected from the device and divert it to a second output port, for example by angling the axis of the device slightly with respect to the incoming propagation direction [33].

We can call all such three-port devices "tunable taps." They are of potential system architecture importance because, as is discussed in Section 3.11, a star topology using a star coupler suffers from a great waste of transmitted energy. Each transmission is broadcast to all receivers, and all but one of them throw it away. With a tunable tap, the energy is conserved and directed to a separate port where it can be used. In the future, if tunable filters of the three-port type can be made with low enough attenuation, they could be used in a bus geometry to avoid the energy wastage of the star geometry. Today, the attenuation of any available filter design is so much greater than that of an ordinary biconical tap (Section 3.11) that the idea is not yet practical.

4.12 Switched Gratings

In view of the difficulties we have encountered in finding physical effects that change an index rapidly and electronically, a different alternative suggests itself. Why not provide a dense array of N photodetectors and diffract all the received light onto this array in such a way that each detector sees a different narrow wavelength band? We can then tune by electrically switching the receiver input to accept only the output of that detector that sees the desired wavelength. This is the idea behind the *switched-grating* approach to wavelength tunability [34].

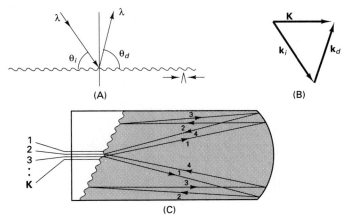

Figure 4-27. Ruled grating as a diffracting component. (A) Angles of incidence and diffraction. (B) Equivalent momentum triangle. (C) Stimax configuration.

Consider a classic diffraction grating, as shown in Figure 4-27(A). The *pitch* of the grating is the number of ruled lines per unit distance, usually between 100 and 5000 lines per millimeter. The spatial wavelength Λ is therefore between 10 and 0.2 μ, respectively. Incident light of (free-space) wavelength λ arrives at an angle θ_i from the plane of the grating and leaves at a wavelength-dependent angle θ_d. We can use the familiar momentum triangle of Figure 4-27(B), given by Equation 4.30, to find out what wavelength λ will be diffracted to what angle θ_d.

$$\mathbf{k}_d = \mathbf{k}_i + \mathbf{K} \qquad \text{i.e.,} \qquad \frac{\cos \theta_d}{\lambda} = \frac{\cos \theta_i}{\lambda} + \frac{1}{\Lambda} \qquad (4.49)$$

More generally, the condition will be satisfied if we replace λ with $m\lambda$, where m is the *order* of the diffracted light. $m = 1, 2, 3, \ldots$ corresponds to conditions in which the light sees the fundamental sinusoidal component of the grating, the spatial frequency second harmonic, third harmonic, and so forth. We shall usually be interested only in the first order, $m = 1$.

In making an optical filter, the angle of incidence θ_i is under the control of the designer, and can therefore be adjusted to any convenient value. A particularly interesting one is the *Littrow* configuration in which some particular wavelength is aimed back along the direction of incidence. The "Littrow condition" is thus

$$2 \cos \theta_i = \frac{\lambda}{\Lambda} \qquad (4.50)$$

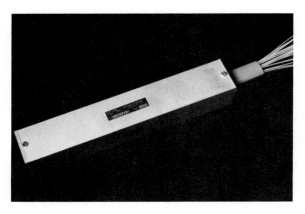

Figure 4-28. Commercial $N = 20$ Stimax diffraction grating (Courtesy of Instruments SA Division of Jobin-Yvon).

This Littrow condition has been exploited very cleverly by the Jobin-Yvon company in their *Stimax* configuration, shown in Figure 4-27(C) [35]. The cores of a number $K + 1$ of fibers are arranged in a row, side-by-side, in very precise V-grooves, the direction of the row being perpendicular to the grating lines. This row of fiber cores projects into the interior of the device through a tiny slot made in the grating, and one of these cores, that of the input fiber, radiates light into the device. The light makes two round trips through the length of the device, as shown in the figure. On the first round trip, it emerges as a cone (as given by the NA, discussed earlier in Section 3.8) aligned with the device axis, travels to the spherical mirror at the other end, where it is collimated into a parallel beam, and then travels back again to impinge on the grating. Now, according to Equation 4.49, it must leave the grating at an angle that is tipped by a wavelength-dependent amount from the axis. It travels back to the mirror as a collimated beam, is reconverged by the mirror and strikes one of the output fiber ends.

Such devices have been built with K up to 32, and exhibit an adjacent-channel crosstalk level at least 50 dB below the in-channel signal. Attenuations are typically 5 dB or less for such devices. One thing that helps minimize light loss is to use *blazed* gratings. This term refers to the process of fabricating the grating not in the form of a sinusoid or a triangular wave, but as a stairstep, where the long side of the step is normal to the light. This is visible in Figure 4-27(C). The expression "blazed" comes from the fact that as you look at the grating at various angles, the light gets suddenly very bright for angles θ_d near the normal to the long side of the steps.

The usefulness of diffraction gratings for optical networks is not limited to tunable filters alone. We shall see in Chapter 11 how they can also be used for such things as *wavelength routing*, a scheme by which signals are directed along different paths in a network according to their wavelength.

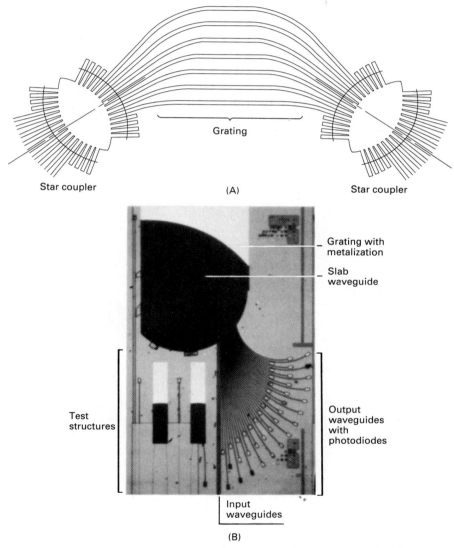

Figure 4-29. Planar-waveguide grating implementations. (A) 16×16 grating-based wavelength-routing component made from two planar star couplers. (From [36]). (B) $K = 77$ grating with integrated photodetector array. (From [37], courtesy C. Cremer, Siemens AG).

Figure 4-28 shows a commercial Stimax unit with $K = 20$. Clearly, a tunable filter can be formed by attaching a photodetector to each output fiber and then switching the correct photodetector output to the receiver circuitry. An even more satisfactory solution is to integrate the photodetector array directly into the grating device without going through long fiber pigtails.

Recently, planar waveguide technology has advanced to the point where not only have gratings been realized having K greater than that achievable with bulk devices, but some have been made with the array of photodetectors integrated into the structure. Figure 4-29(A) shows a planar grating developed at Bellcore [38] and (B) shows a Bell Labs device using two of the planar couplers of Figure 3-26 to form a grating with multiple outputs, each shifted in wavelength from its neighbor [36]. Such a component can be used as a wavelength-router [39]. Figure 4-29(C) shows a combined grating-photodiode array from Siemens [37].

Such devices are evidence that planar waveguide technology is beginning to displace bulk devices and take its place in photonics just as LSI did thirty years ago with bulk electronics.

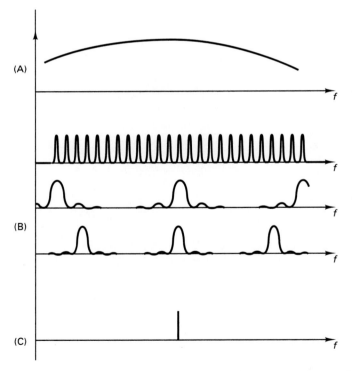

Figure   4-30. Frequency transfer functions of an active filter. (A) Gain curve of the medium. (B) Various resonating sections. (C) Composite frequency transfer function.

4.13 Active Filters

In the next chapter we shall look at tunable lasers. These all involve some kind of index-tunable structure, some of them exactly those described in this chapter. If the bias current that pumps the laser is insufficient to support lasing, the device can be used as an *active tunable filter* (see, e.g., [40]) with very narrow passband and even modest gain. Because of the *side-mode rejection* action of lasers, the selectivity is greatly enhanced; mistuning of the input light that would produce only a moderate diminution of output if the filter were of the passive type we have been describing can cause a very large fall-off of output power in an active filter.

We shall defer the discussion of the principles that make active filters possible to the next chapter, but meanwhile it is instructive to point out that many of the same principles of cascading separate filter sections are used. We saw this in connection with the Mach-Zehnder chain and the multiple Fabry-Perot interferometer. Here, in Figure 4-30, is a preview of some ideas from Chapter 5, showing a number of concatenated effects, in the same style as Figures 4-12, 4-13, and 4-15. The principles are the same, except that the top curve is that of an active process; it is the *gain curve* of the lasing material. All the other transfer functions (B) in the cascade are of the type we have seen in this chapter, namely transfer functions of linear systems. The net effect is that the side-mode rejection of the lasing causes the composite transfer function (C) of the overall device to be much narrower than the composite product of curves (A) and all of (B) would have led us to expect. Research on active tunable filters is in its earliest stages.

Table 4-2. Achievable tunable filter characteristics. (PBS = polarization beam splitter).

Type of filter	Number of nodes in 200 nm	Tuning time	Atten. at peak (dB)	Remarks
Micron FFP	100	1 ms	1	Temp. compensated
Queensgate FPI	100	1 ms	3	Capacitance-stabilized
MZI chain	128	1 ms	5	Lithographic
A/O double wedge	600	2 μs	9	Excessive size
Bulk acoustooptic	22	3 μs	5	
Acoustooptic-PBS	90	6μs	3	
Electrooptic-PBS	80	0.1 μs	4	
Switched gratings	70	0.1 μs	5	

4.14 What We Have Learned

Usable tunable filters have been invented and built in great variety. They are all based on interference effects. In some designs, a great many wavelets interfere constructively only at resonance, in some a great many successive reflections are in phase only at resonance, and in some many light paths merge in phase only at resonance.

We have discussed the detailed principles of those options that appear most promising for practical system use. It should be noted, however, that this is a very rapidly moving field with new innovations occurring every few months. It is interesting to compare surveys of the tunable optical filter art made five years apart, [41] and [42], respectively, and observe that most of the designs that were earlier considered promising fell by the wayside in five years. While not yet full stable, the tunable filter art has finally produced in the last two years a number of usable commercial products.

Table 4-2 is a tabular summary of the various parameters of the tunable filter types that have actually been built and tested.

4.15 Problems

4.1 (*Typical laser diode resonant cavity*) The active region of a certain semiconductor laser diode is 200 microns long and is composed of an indium-gallium-arsenic-phosphide (InGaAsP) alloy with a refractive index of 3.5. The "end facets" constituting the cavity mirrors are parallel plane surfaces exposed to air. (A) What is the power reflectivity of each facet? (B) What is the finesse of the cavity? (C) What is the spacing in frequency between the Fabry-Perot resonance lines of this laser? (D) What is the corresponding wavelength difference, for an operating wavelength of 1.5 μ? (E) For 1.3 μ?

4.2 Rewrite Equation 4.10 in a more convenient form that involves only frequency, finesse, and free spectral range.

4.3 (A) What is the transfer function $T_R(f) = |E_R / E_i|^2$ corresponding to Equation 4.10, but for the light reflected back toward the source of a Fabry-Perot etalon? (See Figure 4-4(B)). Assume that the absorption passing through the mirror supporting structure $A = 0$. (B) Give a word description, based on Figure 4-5.

4.4 A wavelength-division multiple-access (WDMA) system is to have 1000 channels, equally spaced in frequency over the 200-nm window centered at 1.5 μ wavelength. (A) Roughly, what is the width of this window in GHz? (B) How would you choose the FSR of a tunable etalon for the receivers of this WDMA system? (C) What would be the minimum finesse you would require for these components? (D) What mirror reflectivity would they have to have? (E) What would be the cavity length, assuming air as the intracavity medium? (F) What is the minimum length of the device if piezo material of $(\Delta x/x)_{max} = 0.5$ percent is used? (G) How far would the mirrors have to move with respect to one another to tune over the 200 nm range if the Fabry-Perot order $i = 1$? If the order were $i = 10$?

4.5 Given two etalons of finesse 100, by roughly what factor is the maximum number of WDM channels improved using the two-cavity cascaded connection with $i = 9$ and $j = 10$ relative to the two-pass use of just one of the etalons?

4.6 (*Ring resonators*) The fiber ring resonator, shown at the left of the figure below, can be made tunable over a small range by wrapping the fiber around a piezoelectrically enlarged cylinder. (A) Using the model of a 2×2 coupler introduced in Section 3.11, compute the transfer function $T(f)$ of the ring resonator. (B) How does this relate to the Fabry-Perot etalon whose transfer function $T(f)$ is given in Equation 4.9 or 4.10 ? *Hint*: Consider Problem 4-3.

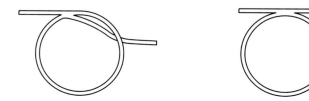

(C) The alternative at the right looks attractive because it can be made from a single length of fiber without splices by fashioning the 2×2 coupler as shown in Figure 3-22. Compute its transfer function.

4.7 Show that the required percent tolerance on the mirror spacing for a vernier two-cavity filter of a given finesse is just as severe as for a single-cavity device having the same finesse.

4.8 By comparison of Figures 4-4 and 4-14, what can you say qualitatively about the relative durations of the impulse responses of a Fabry-Perot interferometer and a Mach-Zehnder interferometer of equal HPBW?

4.9 The spectral resolving power of a ruled grating is determined by the *pitch*, the number of lines per mm. A common value of pitch for gratings made either by a mechanical ruling engine or by interference (holographic) means is 1200 lines per mm. What would the RF drive frequency of an acoustooptic device made of $PbMoO_4$ have to be in order to achieve this grating pitch?

4.10 Consider a tunable filter based on an acoustooptic deflector using $PbMoO_4$ with an optical beam waist 1.0 mm wide. (A) How fast can it tune? (B) How many spots could be resolved by swinging the RF frequency by 50 MHz?

4.11 Describe the action of the surface-wave acoustooptic filter when light is applied to input 2 instead of input 1.

4.12 In the same surface-wave acoustooptic filter, what happens when left circular polarized light is applied to input 1 and right circular polarized light of exactly the same frequency and phase is applied to input 2 ?

4.13 In the Stimax grating multiplexor of Figure 4.27(C), the first fibers are equally spaced. Fiber 1 receives the input signal, a mixture of signals at different wavelengths, Fiber 2 is Output 1 at λ_1, Fiber 3 is the Output 2 at $\lambda_2 = \lambda_1 + \delta\lambda$, and so forth. The last fiber, the $N + 1$st, is

Output N for the *N*th channel. Suppose it was the third fiber in the sequence that received the input, all the others being output fibers. What wavelengths would appear at the outputs?

4.16 References

1. P. A. Humblet and W. M. Hamdy, "Crosstalk analysis and filter optimization of single- and double-cavity Fabry-Perot filters," *IEEE Jour. Sel. Areas in Comm.*, vol. 8, no. 6, August, 1990.

2. W. Hamdy, "Crosstalk in direct-detection optical fiber FDMA networks," *PhD Thesis, M.I.T. Elec. Engrg. Dept.*, June, 1991.

3. J. M. Vaughn, *The Fabry-Perot Interferometer*, Am. Inst. of Physics, 1989.

4. Supercavity–High-Finesse Tunable Fabry-Perot Filter, Newport Corp., Fountain Valley, CA 92728, 1990.

5. H. R. Hicks, N. K. Reay, and P. D. Atherton, "The application of capacitance micrometry to the control of Fabry-Perot etalons," *J. Phys. E, Sci. Instrum.*, vol. 17, pp. 49-55, 1984.

6. N. R. Dono, P. E. Green, K. Liu, R. Ramaswami, and F. F. Tong, "Wavelength division multiple access networks for computer communication," *IEEE Jour. Sel. Areas in Comm.*, vol. 8, no. 6, 1990.

7. S. R. Mallinson and J. H. Jerman, "Miniature micromachined Fabry-Perot interferometers in silicon," *Electron. Lett.*, vol. 23, no. 20, pp. 1041-1043, September, 1987.

8. J. Stone and L. W. Stulz, "Pigtailed high-finesse tunable fiber Fabry-Perot interferometers with large, medium and small free spectral range," *Electron. Lett.*, vol. 23, pp. 781-783, 1987.

9. C. M. Miller and F. J. Janniello, "Passively temperature-compensated fiber Fabry-Perot filter and its application in wavelength division multiple access computer network," *Elect. Ltrs.*, vol. 26, no. 25, pp. 2122-2123, December, 1990.

10. D. A. Jared and K. M. Johnson, "Ferroelectric liquid crystal spatial light modulators," *SPIE Critical Reviews Series*, vol. 1150, pp. 46-60, 1990.

11. J. S. Patel, M. A. Saifi, D. W. Berreman, C. Lin, N. Andreadakis, and S. D. Lee, "Electrically tunable optical filter for infrared wavelength using liquid crystals in a Fabry-Perot etalon," *Appl. Phys. Ltrs.*, vol. 57, no. 17, pp. 1718-1720, October, 1990.

12. G. Picchi, "Double-cavity vs. multipass Fabry-Perot filters for channel selection in optical WDMA networks," *Res. note, IBM Research, Hawthorne*, September, 1988.

13. J. E. Mack, D. P. McNutt, F. L. Roesler, and R. Chabbal, "The PEPSIOS purely interferometric high-resolution scanning spectrometer," *Applied Optics*, vol. 2, no. 2, pp. 873-885, February, 1963.

14. A. A. Saleh and J. Stone, "Two-stage Fabry-Perot filters as demultiplexers in optical FDMA LANs," *IEEE/OSA Jour. Lightwave Tech.*, vol. 7, pp. 323-330, February, 1989.

15. I. P. Kaminow, P. P. Iannone, J. Stone, and L. W. Stulz, "A tunable vernier fiber Fabry Perot filter for FDM demultiplexing and detection," *IEEE Phot. Tech. Ltrs.*, vol. 1, no. 1, pp. 24-25, January, 1989.

16. C. M. Miller and J. W. Miller, "Wavelength-locked two-stage fibre Fabry-Perot filter for dense wavelength division demultiplexing in an erbium-doped fibre amplifier spectrum," *Electronic Ltrs.*, vol. 28, no. 3, pp. 216-217, January, 1992.

17. K. Sivarajan, "Digital control of the tunable Mach-Zehnder optical filter chain," *IBM Research Report RC-16157*, September, 1990.

18. K. Nosu, H. Toba, and K. Iwashita, "Optical FDM transmission technique," *IEEE/OSA Jour. Lightwave Tech.*, vol. 5, no. 9, pp. 1301-1308, September, 1987.

19. H. Toba, K. Oda, K. Nosu, and N. Takato, "Factors affecting the design of optical FDM information distributing systems," *IEEE Jour. Selected Areas in Commun.*, vol. 8, no. 6, pp. 965-972, 1990.

20. H. Toba, K. Oda, K. Nakanishi, N. Shibata, K. Nosu, N. Takato, and K. Sato, "100-channel optical FDM transmission/distribution at 622 Mb/s over 50 km using a waveguide frequency selection switch," *Elect. Ltrs.*, vol. 26, no. 6, pp. 376-377, March, 1990.

21. N. Takato, A. Sugita, K. Onose, K. Okazaki, M. Okuno, M. Kawachi, and K. Oda, "128-channel polarization-insensitive frequency-selection-switch using high-silica waveguides on Si," *IEEE Phot. Tech. Ltrs.*, vol. 2, no. 6, pp. 441-443, 1990.

22. K. Oda, N. Takato, T. Kominato, and H. Toba, "Channel selection and stabilization technique for waveguide-type 16-channel frequency selection switch for optical FDM distribution systems," *IEEE Phot. Tech. Ltrs.*, vol. 1, no. 6, pp. 1-2, June, 1989.

23. J. Sapriel, *Acousto-Optics*, Wiley, 1979.

24. Y. Yariv and P. Yeh, *Optical Waves in Crystals*, Wiley, 1984.

25. F. F. Tong, D. F. Bowen, and P. A. Humblet, "A high-speed tunable optical filter for wavelength division multiaccess networks," *Submitted to IEEE/OSA Jour. Lightwave Tech.*, November, 1991.

26. D. A. Smith, K. W. Cheung, J. E. Baran, and J. J. Johnson, "Acoustically-tunable integrated-optical filters for WDM networks," *Conf. Record, LEOS Topical Meeting on Optical Multiaccess Networks*, Monterey, CA, July, 1990.

27. K. W. Cheung, M. M. Choy, and H. Kobrinsky, "Electronic wavelength tuning using acoustooptic tunable filter with broad continuous tuning range and narrow channel spacing," *IEEE Phot. Tech. Ltrs.*, vol. 1, no. 2, February, 1989.

28. K. M. Johnson and G. D. Sharp, "Chiral smectic liquid crystal optical modulator," *Patent pending*, 1991.

29. C. K. Campbell, "Applications of surface acoustic and shallow bulk acoustic wave devices," *Proc. IEEE*, vol. 77, no. 10, pp. 1453-1484, October, 1989.

30. D. A. Smith, J. E. Baran, J. J. Johnson, and K. W. Cheung, "Integrated-optic acoustically-tunable filters for WDM networks," *IEEE Jour. on Selected Areas in Commun.*, vol. 8, no. 6, pp. 1151-1159, 1990.

31. W. Warzanskyj, F. Heismann, and R. C. Alferness, "Polarization-independent electro-optically tunable narrow-band wavelength filter," *Appl. Phys. Ltrs.*, vol. 53, no. 1, pp. 13-15, July, 1988.

32. R. C. Alferness and L. L. Buhl, "Tunable electro-optic waveguide TE−TM converter/wavelength filter," *Appl. Phys. Ltrs.*, vol. 40, no. 10, pp. 861-2, May, 1982.

33. K. Liu, "Tunable tap proposal," *Private communication*, December, 1990.

34. P. A. Kirby, "Multichannel wavelength-switched transmitters and receivers−New component concepts for broad-band networks and distributed switching systems," *IEEE/OSA Jour. of Lightwave Tech.*, vol. 8, no. 2, pp. 202-211, 1990.

35. J. P. Laude and J. M. Lerner, "Wavelength division multiplexing/demultiplexing (WDM) using diffraction gratings," *SPIE−Application, Theory and Fabrication of Periodic Structures*, vol. 503, pp. 22-28, 1984.

36. C. Dragone, C. A. Edwards, and R. C. Kistler, "Integrated optics N x N multiplexor in silicon," *IEEE Photonics Tech. Ltrs.*, vol. 3, no. 10, pp. 896-899, 1991.

37. C. Cremer, N. Emeis, M. Schier, G. Heise, G. Ebbinghaus, and L. Stoll, "Grating spectrograph integrated with photodiode array in InGaAsP/InGaAs/InP," *IEEE Photonics Tech. Ltrs.*, vol. 4, no. 1, pp. 108-110, 1992.

38. J. B. Soole, A. Scherer, H. P. LeBlanc, R. Bhat, and M. A. Koza, "Spectrometer on chip: a monolithic WDM component," *Conf. Record, Optical Fiber Commun. Conference*, p. 123, February, 1992.

39. C. Dragone, "An N x N optical multiplexer using a planar arrangement of two star couplers," *IEEE Photonics Tech. Ltrs.*, vol. 3, no. 9, pp. 812-815, 1991.

40. L. G. Kazovsky, M. Stern, S. G. Menocal, and C. E. Zah, "DBR active optical filters: Transfer function and noise characteristics," *IEEE/OSA Jour. Lightwave Tech.*, vol. 8, no. 10, 1990.

41. H. Ishio, "Review and status of wavelength-division multiplexing," *IEEE/OSA Jour. Lightwave Tech.*, vol. 2, no. 4, pp. 448-463, 1984.

42. H. Kobrinski and K. W. Cheung, "Wavelength-tunable optical filters: Applications and Technologies," *IEEE Commun. Mag.*, vol. 27, no. 10, pp. 53-63, October, 1989.

CHAPTER 5

Laser Diodes

5.1 Overview

Sources of light

Genesis I:3 tells us that "there was light," but it does not say what kind. The source of photons might have been either a thermal *incandescent* source or a nonthermal *fluorescent* one, the latter type being subdivided in turn into the *phosphorescent* kind (in which the radiation persists long after the stimulus is removed), and the *fluorescent* kind. Fluorescence is a term applied until recently to all sources in which the disappearance of response after removal of the stimulus is essentially instantaneous. Since Einstein's work in 1916, we know that there are two ways that this kind of light can be produced, *spontaneous emission*: the effect exploited in a fluorescent light bulb and in light-emitting diodes (LEDs), and *stimulated emission*, as in lasers. We have no idea whether the light mentioned in Genesis was stimulated emission, since until 1916, only God knew about it.

The words "stimulated emission" are what the letters "SE" in *laser* stand for, the complete acronym being "light amplifier using stimulated emission of radiation." In stimulated emission, an incoming photon of just the right wavelength triggers off the emission of a second photon, which, being of the same frequency, phase, propagation direction and polarization as the incident one, results in an increase in light intensity.

As we shall see in Chapter 6, it turns out not to be particularly easy to make a pure light amplifier using stimulated emission, but nevertheless it is this form of amplification that makes a laser work by balancing out the losses. In a laser, the material chosen to exhibit stimulated emission is placed between two mirrors, thus forming a Fabry-Perot cavity. The situation is exactly analogous to that of an electronic oscillator in that amplification by stimulated emission provides the gain element to overcome losses, and the reflections inside the laser cavity provide the feedback mechanism that starts and sustains the oscillations.

In this chapter, we shall gradually build up a working model of what goes on inside a laser that has not been specifically designed for tunability and at the end present the state of the tunable laser diode art. We start at the simplest level of what the device looks like externally, both electrically and physically. The electrical characteristics are embodied in the curve of light output as a function of current input. The various physical phenomena that explain this curve are introduced quantitatively, both for ordinary "bulk" lasers and for those based on quantum confinement. Going beyond the simple input-output curve, we examine the steady-state and transient external characteristics before finally addressing tunability.

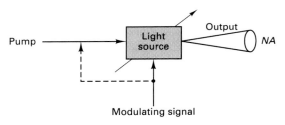

Figure 5-1. Idealized light source.

Figure 5-1 indicates the kind of device we seek for use in a lightwave network. Input energy, called the *pump* energy, is supplied in the form of input light or input electric current, and light energy comes out at the output. Data is impressed on the light signal by modulating its amplitude, frequency, phase or polarization. If binary data is to be impressed by on-off modulation, then this can be done by simply varying the pump between two values that insure that the light is on or off (dashed line).

There are several things we would like our light source to do. We would like the conversion of pump photons or electrons into output photons to be as efficient as possible, and modulation at high data rates to be as easy as possible. The light that comes out should be as *monochromatic* (single wavelength) as possible, so as to minimize mode partition noise and the effects of fiber dispersion, to allow wavelength multiaccess channels to be closely spaced, and to allow the use of coherent receivers. We would like the light output to be a linear function of the input signal, so as to minimize harmonic distortion. As the arrow indicates, it is desirable in some applications that this wavelength be tunable, preferably over the entire low-attenuation band of the fiber (say, 200 nm) and within nanoseconds, if possible, so as to support packet-switching protocols, such as those to be described in Chapter 13. The directivity pattern of the output light should relate appropriately to the angular acceptance angle of the fiber into which the light is to be fed; in other words, the NA of the laser should be no greater than that of the fiber. And finally, all these properties should be minimally sensitive to temperature, vibration and other environmental effects, the device should have a very long lifetime, and should sell for peanuts.

The semiconductor laser diode is a most astonishing device. When the original bulky and elaborate gas laser was invented in 1960, no one could possibly have predicted that lasers would eventually take the form of tiny units small enough to be integrated by the hundreds or thousands on a single chip, easily built into a convenient electronic-circuit environment of a few volts, attachable to a revolutionary new communication substitute for copper, tunable like a radio transmitter or receiver, and, thanks to the compact audio disc, more plentiful than phonograph needles.

For an accessible tutorial introduction to classical nontunable laser diodes, there are several good books [1, 2]. For more advanced and comprehensive discussions, the books [3, 4] and the review article [5] are recommended.

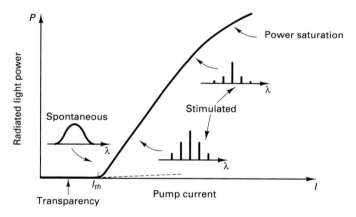

Figure 5-2. The *P-I* curve, light output power from one side of the device versus pump current. The changes in output spectrum with increasing current are indicated.

The P-I curve

Some orientation about the role that spontaneous and stimulated emission play in LEDs and laser diodes can be gotten from looking at a typical laser *P-I curve*, such as the one in Figure 5-2. *P* is the light power exiting one end of the device, and *I* is the pump current. As the current increases from zero, there is a small amount of spontaneous emission whose power level increases slowly and linearly with *I*. In this region, the device is acting strictly as an LED, emitting incoherent radiation whose power spectrum is a broad smear, the *fluorescence spectrum* of the material. At some point, there is enough pump current that the amount of emitted light equals the amount of absorbed light and *transparency* is achieved. Transparency (sometimes called *bleaching*) means that the amount of light emitted and the amount absorbed exactly balance out. Eventually, the pump current reaches the lasing *threshold* current I_{th}, stimulated emission takes over, and the light output P rises rapidly with further increases of I. The character of the radiated spectrum changes from a broad smear into a series of lines, the Fabry-Perot cavity resonances, whose envelope around threshold is roughly the same as the fluorescence spectrum. This spectrum, the *gain curve*, expresses the wavelength dependence of the lasing material inside the device as the the pump current increases beyond the threshold value I_{th}.

As the pump current continues to increase above threshold, the envelope of the line spectrum becomes narrower than the original gain curve by the phenomena associated with *side-mode suppression*. The main mode shifts toward shorter wavelengths as the pump current increases.

It is the presence of more than one emission line that is responsible for the mode partition noise mentioned in Chapter 3; the radiated power can jump around from one frequency to another, thus exciting the fiber in a a strongly time-varying way. Since single-mode fibers are an imperative for all but the most modest values of bitrate-distance product, lasers with a high degree of side-mode suppression are very desirable. For wavelength-division systems with close channel spacing, the ratio of main-mode to side-mode must be especially large. The exact amount can be deduced from the crosstalk analyses given in the last chapter.

Note that the behavior of the device, as expressed by the linear *P-I* curve, is essentially quadratic. That is, above threshold, current in translates into power out, not square root of power. In this sense a laser diode is very different from more conventional forms of communication transmitters where the output power and the power drawn from the power supply are roughly proportional.

Eventually, further increases of I produce diminishing increases of P. Because of this, and because protracted use at excessive instantaneous power levels can limit the device lifetime, laser diodes and LEDs must be considered by system designers as peak- power-limited devices rather than average-power-limited devices.

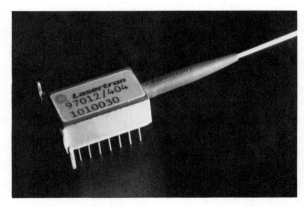

Figure 5-3. A contemporary laser diode package with attached (pigtailed) fiber. (Courtesy Lasertron, Inc.).

Figure 5-3 shows a typical commercial laser diode. Inside the can are not only the laser diode itself, a small chunk of semiconductor material only a few hundreds of microns on a side, but some method of coupling light into the fiber pigtail, plus several electrical pinout connections. These include not only those supplying the pump current, but also pins carrying the output of a small *monitor photodetector* that can be used for failure detection and to assist in adjusting the bias current properly with relation to I_{th}. Since operating characteristics and lifetime are strong functions of temperature, a thermistor temperature sensor is often included, along with a *thermoelectric cooler* element that uses the Peltier effect to turn current into heat removal. These two are connected by a simple external servo system that quite easily keeps the temperature of the laser chip itself to within 0.1°C. over a wide range of ambient conditions outside the can.

For on-off keying, the current is adjusted, or *biased*, to be near I_{th} for a "0" bit and significantly above I_{th} for a "1" bit. Since biasing I too low leads to a random delay in the onset of lasing during a "1" bit (due to the *turn-on delay* to be discussed in Section 5.6), it is customary not to bias the device so that the light is completely turned off during a "0" bit. The ratio of light outputs P for a "0" bit and a "1" bit is called the *extinction ratio r*.

Device geometries

Figure 5-4 shows the geometries of several forms of semiconductor laser diodes. The meanings of the letters p and n will be discussed in Section 5.2. In the *homostructure* device of (A), there is a single material, with the *active region* being one interface between parts of this material that are doped with two different impurities. Light is emitted edge-on from a narrow line at that interface. In the *double heterostructure* devices of (B) and (C), there are three kinds of material, and the active region is now a slab (B) or strip (C). The medium surrounding the active region of such devices is often called the *cladding*, because it usually has a lower index than the surrounding region, as with a fiber.

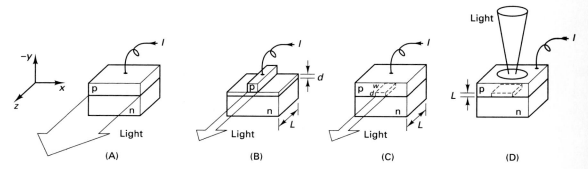

Figure   5-4. Idealized view of several laser diode structures. (A) Homostructure. (B) Double heterostructure with planar active region and electrode in the form of a ridge. (C) Buried double heterostructure. (D) Surface emitter.

These are just schematic representations; real devices bearing the names in the figure have quite complex structures. The literature on these variegated structures is fairly intimidating, with sentences such as "Minimum spectral linewidth and the linewidth power products for the SCH-MQW-DFB-DC-PBH-LD were 2.0 MHz and 5 MHz, respectively" (an actual quote).

For best coupling of light from the source into the fiber, the source should have equal height and width, but it turns out to be very difficult to achieve this. It is particularly important not to generate more than one transverse mode of light radiation across the emitting surface, because that would misdirect some of the light output that should go into the fiber instead. To ensure that there is only one transverse optical mode radiating from the laser, the width w in the x-direction is made small. Another reason for making w small is to keep it less than $2a$, the diameter of the fiber, so as to simplify the laser-fiber coupling elements. Actually these are minor reasons for making w small compared to the need for spatial confinement of the injected current, because the gain inside the active region is higher the higher the density of atoms that are engaged in stimulated emission. The required reduction of w can be effected by causing only a strip of the material in the otherwise uniform middle layer to emit. This can be done either by using a stripe electrode as at (B), or, even more effectively, by *gain guiding* or *index guiding* using a *double heterostructure*, as shown at (C). Gain guiding involves forcing the injection current into a narrow active region, whereas index guiding involves tailoring the lateral variations of refractive indices of the various regions, as is done in fibers (Chapter 3), so that the light tends to be confined laterally.

Values of $w = 10\,\mu$, $d = 0.2\,\mu$, and $L = 100 - 400\,\mu$ are typical for the x, y, and z dimensions of the active region, respectively. Since d is so much smaller than w, not only is the directivity pattern narrower in x than in y, but also the light from most laser diodes is strongly horizontally polarized.

The front and back x-y surfaces are called the *facets*, and simply by cleaving the block of semiconductor material, a facet results that is almost ideally plane (due to the highly uniform crystal structure), and has a reflectivity of around 30 percent, which is enough to provide the feedback necessary for lasing.

Under the conditions of low facet reflectivity, continued increase of the pump current I does not lead to lasing, but to continued action as an LED, that is to continued emission of more and more spontaneous radiation only, so that we have the linear P-I curve given by the dashed line in Figure 5-2. Eventually, it, too, saturates. LEDs in which antireflection coatings and other artifices are used, together with tight confinement and high current densities, in order to generate high light output power P are called *superluminescent diodes (SLDs)*.

If the field varies across the emitting surface with random phase (as it would with the spontaneous emission of an LED, SLD, or fluorescent light bulb), the light is emitted with a highly spread-out directivity pattern

$$P \propto \cos \theta \qquad (5.1)$$

called a *Lambertian* pattern, where θ is angular deviation from normal, and this leads to relatively poor coupling to multimode fibers and even worse coupling to single-mode fiber. In a laser diode, on the other hand, the light is in phase across the emitting facet, which gives a much more directive beam than a Lambertian pattern does.

Poor directivity is just one reason that LEDs have been superseded by lasers as communication light sources. An even more important reason is readily seen from Figure 5-2, namely, the high power available from laser diodes. Commercially available LEDs emit less than about 100 microwatts, whereas laser diodes of ten to one hundred times that output power are readily available. Equally important is the chromatic dispersion that long lengths of fiber introduce if the source bandwidth is too large (Section 3.6). Still another disadvantage of LEDs is that the speed with which they can be turned on and off is controlled by slower-acting internal processes than those of laser diodes. Because of this basic limitation, LEDs cannot be modulated much more rapidly than about one-tenth the equivalent figure for laser diodes. For example, although research LEDs have achieved speeds of 1 Gb/s and research laser diodes over 10 Gb/s, commercial LEDs are unable to support speeds higher than 100 Mb/s, whereas laser diodes of more than 1 Gb/s speed are widely available.

Threshold condition

The condition for onset of laser action can be gotten from Figure 5-5 in terms of the (intensity) reflectivity R of the facets (assumed identical), the gain per unit length g supplied by the stimulated emission effect, the internal attenuation per unit length α, the cavity length L, the index of the cavity material n, and the free-space wavelength λ.

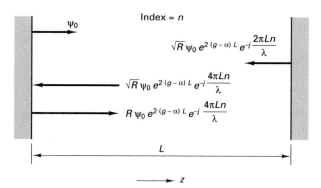

Figure 5-5. The threshold condition exists when there is just enough internal gain per unit length g to offset the internal losses α per unit length and the escape of useful light from the two mirrors, each having reflectance R.

Recall that the field strength of a plane wave propagating in the z-direction can be represented as

$$\psi(z) = \psi \exp(-kz) = \psi \exp[-(\alpha + j\beta)z] \tag{5.2}$$

where ψ is either the electric or the magnetic field strength (Equation 3.31). The quantity β is $2\pi n/\lambda$ (Equation 3.28) for an ideal lossles, nondispersive medium. (Actually, the medium inside the laser is dispersive, but for the purposes of this chapter we may assume it is not). In the case of a laser, where gain is present, we replace the attenuation per unit length α by $(\alpha - \Gamma g)$, where g is the gain per unit length of the material and Γ is the *confinement factor*, the fraction of the optical power that lies within the active region. Γ is closely related to the quantity given in Equation 3.79, the ratio of power in a fiber core to the total power in core and cladding. The index n of semiconductor laser materials ranges from 3.5 to 4.5, and actually has a slight (less than one percent) negative dependency on current I.

If the field strength of the light leaving the left mirror is ψ_o, then when it hits the right mirror the incident field strength is

$$\psi_o \exp\left[(\Gamma g - \alpha) L - j \frac{2\pi L n}{\lambda}\right] \tag{5.3}$$

That reflected back toward the first mirror is $\sqrt{R}$ times this, and that rereflected from the first mirror is

$$R \psi_o \exp\left[2(\Gamma g - \alpha) L - j \frac{4\pi L n}{\lambda}\right] \tag{5.4}$$

Lasing will occur when two conditions are met, (i) the magnitude of this quantity equals or exceeds that of ψ_o,

$$R \exp\left[2(\Gamma g - \alpha) L\right] \geq 1 \qquad \text{i.e.,} \qquad \Gamma g \geq \alpha - \frac{\ln R}{2L} \tag{5.5}$$

and (ii) the phases match

$$\lambda = \frac{2nL}{i} \tag{5.6}$$

the familiar condition for Fabry-Perot resonances of order i, the number of standing half-waves in the cavity.

The cavity resonances are the lines in the spectra shown in Figure 5-2. As the pump current increases, the index n inside the cavity decreases, and this decreases the wavelength separation between the cavity resonances. The fact that the fractional change of index over the range of pump currents used is less than one percent severely limits the use of current injection alone as a mechanism for tuning the frequency of a laser diode.

We shall see that gain g is a function of wavelength. So we can say that the real and imaginary parts of $k = (\alpha - \Gamma g + j\beta)$, the propagation constant in Equation 5.2, relate to Figure 5-2 in the following way: The real part defines the *P-I* curve and the envelope of the spectrum as a function of the pump current I (Equation 5.5). The imaginary part defines the spacing and location in λ of the modes underneath that envelope (Equation 5.6).

Population inversion and lasing

The easiest way to explain how the necessary gain is produced by stimulated emission is to go back in history to the first lasers of the early 1960s, the *gas lasers*. In gases, an outer electron of an atom does not see the electrons of any nearby other atoms (as it would in a solid material), and therefore when the electron is driven into a higher energy orbit (by the absorption of a photon), or goes to a lower-energy orbit (by the emission of a photon), it does so in isolation, and each orbit is characterized by one value of energy.

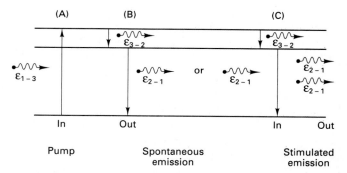

Figure 5-6. A three level system. (A) Pumping to a higher level. (B) Spontaneous decay to level 2 followed by spontaneous decay to level 1. (C) Same but with stimulated emission from 2 to 1.

Figure 5-6 shows how both spontaneous and stimulated emission can occur in a *three-level* system. The absorption or emission of energy by the transition of an electron between any two levels having energy difference $\mathcal{E}$ will be accompanied by the absorption or emission of a single photon of frequency f, given by the famous Planck equation

$$\mathcal{E} = hf \tag{5.7}$$

where h is Planck's constant (6.63×10^{-34} joules $\times$ second). The energy levels that the electron can occupy include the *ground state*, or minimum energy level for that electron (level 1), and two upper levels, 2 and 3. If the gas is at some temperature T, there is a simple exponential probability density function, the *Boltzmann distribution* (Figure 5-7(A))

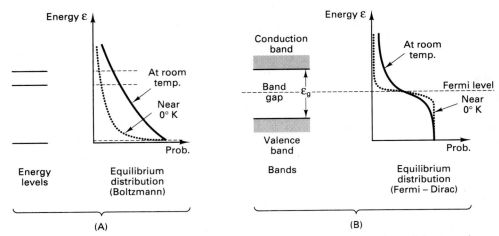

Figure   5-7. Energy levels and probability of their occupancy for a gas (A) and for a semiconductor (B). For a gas, the probability versus energy level is the Boltzmann distribution, and for a semiconductor, the Fermi-Dirac distribution, each sketched for two temperatures.

$$P(\mathcal{E}) = \exp[-(\mathcal{E} - \mathcal{E}_{ground})]/KY \qquad (5.8)$$

that gives the probability that the particular electron will be at one of the three permitted levels, as shown by the dashed lines in Figure 5-7(A). This is the *equilibrium* state that describes an ensemble view of a volume of gas at temperature We shall use the notation Y instead of T for absolute temperature, reserving the latter to indicate the length of one bit period. When we get to detectability calculations in Chapter 8, you will see why, since temperature and bit duration often appear together in the same equation.

Suppose now that pump photons of exactly the correct energy $\mathcal{E}_{1-3}$ are incident on the gas. That is, suppose they have frequency $\mathcal{E}_{1-3}/h$. They will drive many electrons into the third state, whereupon, looking at the volume of gas as a whole, it will be observed that many more atoms have the electron at level 3 than are supposed to be there according to the Boltzmann distribution, while at level 1 there are many fewer. If the number of atoms having the electron in the higher-level state is actually larger than the number with the lower state, the condition of *population inversion* has taken place.

When a population inversion exists, two things can happen to electrons that have been raised to a higher level than the normal equilibrium state. To regain the equilibrium distribution, they can either revert to lower states spontaneously after some mean *lifetime* τ, or they can be stimulated to do so by some incident photons of exactly the energy corresponding to the drop to a lower state. Spontaneous emission produces *incoherent light*, light that has no identifiable sinusoidal nature, since it consists of a stream of photons having independent phase and polarization. This sort of light is characteristic of a thermal source, such as a light bulb, or a spontaneous source such as an LED. The stimulated emission of a laser, on the other hand produces *coherent light*, either at one frequency (monochromatic coherent light), or at several frequencies at once. Such light has one or more identifiable values of frequency, phase and polarization; these quantities may be randomly time varying, but the variation is slow compared to the center frequency.

In the three-level gas laser, light of a short wavelength (high energy) is used as the pump from ground state to a third level that has been chosen for its short lifetime in decaying spontaneously to level two. There the electron sits until either it, too, decays to the ground state spontaneously or does so via a stimulated-emission interaction when an incident photon of the right frequency happens along, whereupon there are now two photons of the same frequency, phase, propagation direction and polarization, the incident photon and the stimulated one, and light amplification has taken place. This scheme is made to work by choosing the particular gas and the particular pump wavelength so that τ for the 3-to-2 transition is much shorter than the τ for the 2-to-1 transition, causing the electron hangs around for a long time in level 2.

If now, the gas is constrained between two mirrors, as the intensity of the pump light at λ_{1-3} is increased, first the point of transparency is passed, and then, when most of the emitted light at λ_{2-1} is stimulated rather than spontaneous emission, the device is said to be *lasing*. Since, with the coherent light of stimulated emission all emitted photons are working together in phase and polarization state, whereas with the incoherent light of stimulated emission they are adding with random phases, the optical power grows rapidly with increased pump above I_{th}. To say it another way, the power of a coherent addition goes as the square of the sum of all the individual contributions, whereas with incoherent addition the power goes as the sum of the squares.

The curve of light output versus pump input for a gas laser looks like that of a solid-state laser (Figure 5-2), even though the pumping is being done by photons rather than electrons.

5.2 Conventional (Bulk) Laser Diodes and Photodetectors

Intrinsic and extrinsic semiconductors

Figure 5-7(B) shows what happens when, instead of a gas, we are dealing with a solid. Figure 5-8 shows in a most graphic way how, as the spacing between atoms gets smaller, what started out as the atomic lines in a gas become bands in a solid.

For the solid, the simple picture of energy levels that are representable as essentially infinitely narrow lines with a Boltzmann distribution of probability of level occupancy is replaced by one in which there are two bands of energies, separated by a

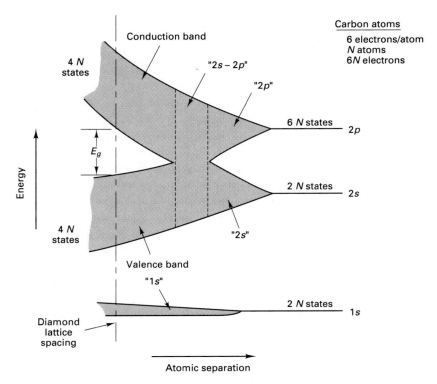

Figure   5-8. As the atoms get closer together, the wave functions of the various electrons begin to feel the effect of each other and the atomic lines of the gas begin to smear into the bands of the solid. (From [6]).

bandgap.   Also, we must replace the Boltzmann probability of occupancy versus energy (Equation 5.8) by the *Fermi-Dirac* probability density distribution

$$P(\mathcal{E}) = \frac{1}{1 + \exp \ (\mathcal{E} - \mathcal{E}_f) \, / \, KY} \tag{5.9}$$

K being the Boltzmann constant (1.38×10^{-23} joules per $^{\circ}K$), and Y the absolute temperature.

The convention is to refer to this probability distribution simply as the "Fermi function." The lower band is the *valence* band, so called because it embraces the energy levels the electron normally has when it participates in chemical reactions. The upper band is the *conduction* band, because normally, when the electron is in this region of energy, it has broken free of the parent atom and can participate in current flow through the solid material. In between these two bands is the forbidden region of the band gap. The energy difference between the bottom of the conduction band and the top of the valence band is the *bandgap energy* $\mathcal{E}_g$ whose value differs for different materials.

As Equation 5.9 and Figure 5-7(B) indicate, at absolute zero all the electrons are in the valence band, $P(\mathcal{E})$ being a step function at $\mathcal{E}_f$, but when $Y > 0$, some of the electrons will have enough energy to make it across the bandgap into the conduction band, each leaving behind it a vacancy called a *hole*. Holes are not real particles, but fictitious positively charged particles, each corresponding to a location where an electron has been taken away from an electrically neutral atom.

To save complicated bookkeeping and clutter in depicting what is going on, it is convenient to abbreviate the entire Fermi function by simply referring to $\mathcal{E}_f$, the point where the probability density function has dropped to half its maximum value. This value is called the *Fermi level*, and is not just a calculational convenience, but is the energy at which the distribution becomes a step function when $Y = 0$.

What makes transistors, semiconductor diodes, as well as photonic sources and detectors possible is the presence in a semiconductor of a small enough bandgap that at room temperature electrons may cross it into the conduction band with significant probability. The conduction band has the property that any unbound electrons in this energy region are effectively free to move around physically. Conversely, the valence band has the property that unbound holes in this energy region are free to move around physically. In the gas laser, the operative particle was an electron. In semiconductor lasers and LEDs, it is a *hole-electron pair*, often referred to as a *carrier*.

Notice in Figure 5-4 that the different pieces of material making up the sandwich of the light source device have been labelled "p" or "n." There is "i"-type material that we shall meet in connection with photodetectors. "i" stands for *intrinsic* semiconductor material, that in which no particular impurities or *dopants* have been included. The "n" and "p" denote the two kinds of *extrinsic* semiconductor material, those doped with *donor* impurities that donate electrons or with *acceptor* impurities that take them away, respectively. For n-material, the electrons are said to be the *majority carriers* and the holes the *minority carriers*; for p-materials the converse is true.

We now have two easily confused uses of the same word "carrier." In speaking of the semiconductor material there can be majority and minority carriers in the form of electrons or holes. In speaking of lasers and photodetectors, the word usually refers to hole-electron pairs. When we get to modulation and detection there will be another use of the word "carrier" to mean a sinusoid at some optical frequency that has information modulated onto it by changing its phase, frequency, amplitude or polarization. Which one of these meanings is intended should be clear from the context.

Figure 5-7(B) shows the band structure of an intrinsic semiconductor material, and Figure 5-9 shows the bandgap, expressed in terms of the free-space wavelength $\lambda = c/f$ of light of energy $\mathcal{E} = hf$ that will be emitted or absorbed by the laser, LED, or photodetector devices made from the III-V materials indium and gallium from column III of the periodic table and arsenic and phosphorus from column V. The variables x and y range between zero and unity and denote the range of *mole fractions* of the different elements. At the corners of the diagram are the four choices of *binary* alloys, and in the interior the various possible *quaternary* alloys.

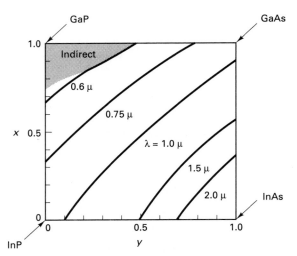

Figure 5-9. Free space wavelength corresponding to the bandgap energy for quaternary III-V semiconductor alloys whose composition is $In_{1-x}Ga_xAs_yP_{1-y}$. (From [4]).

For example, in the lower left-hand corner ($x = y = 0$), we have pure indium phosphide, and in the upper right pure gallium arsenide. It is seen that the operating wavelength of the device can in principle be adjusted to lie anywhere in the range 0.6 to 2.0 μ by controlling the composition of the active region.

The notation often employed to describe the material composition of a semiconductor laser is to state the composition of the laser material of bandgap $\mathcal{E}_g$, followed, after a "/," by that of the substrate material. For example, most lasers at 1.3 and 1.5 μ are InGaAsP/InP devices.

An important fabrication question has been swept under the rug, and will stay there for the elementary system-oriented discussions to which this book is limited, and that is the question of *lattice matching*. In fabricating a device in which the different layers are fashioned out of different materials, if the *lattice constant* of one material, the physical spacing between atoms in the crystal, does not match that of the adjacent material, physical strain is produced which results in the formation of defects at or near the boundary. These act somewhat like unwanted impurities in that they absorb carriers that would otherwise go to producing useful radiation at the desired wavelength. The lattice matching requirement strongly limits the combinations of materials that can be used for the substrate and the various other physical regions of the device. Recently, people have learned to gain certain advantages by introducing a controlled amount of strain by means of a controlled lattice mismatch.

Figure 5-10(A), (B), and (C) show energy versus position along the device for the three types of material. Note carefully that this figure and others like it that we will be discussing do not show device shape; if they show the location of holes and electrons, for example, they show these locations horizontally only. The vertical axis is energy. In Figure 5-10 is seen what happens to the Fermi level for intrinsic, n, and p semiconductors, respectively. They abbreviate Figure 5-7(B) by showing just the

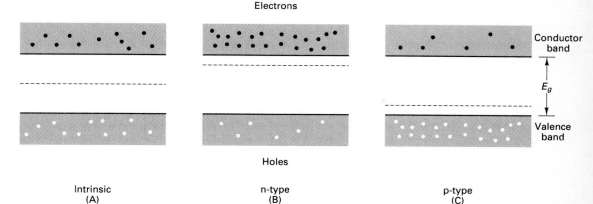

Figure 5-10. Simplified representation of the valence and conduction bands, Fermi levels (dashed lines), electrons (black dots) and holes (white dots) for (A) Intrinsic, (B) n-type, and (C) p-type semiconductor materials.

Fermi level rather than the entire probability distribution. For intrinsic semiconductor material of Figures 5-7(B) and 5-10(A), the Fermi level is in the middle of the gap. Free electrons may move into the conduction band, and since the material as a whole is electrically neutral, for those electrons that do so, a corresponding number of holes will appear in the valence band, thus forming an electron-hole pair, or in photonics lingo, a carrier.

If now we introduce donor impurities, the electrons donated will be free to move only within the conduction band, where there will be an excess of them (compared to the number of holes in the conduction band), and the Fermi level will move up, as shown in Figure 5-10(B). Similarly, if acceptor impurities are added, the Fermi level will move down, as with (C).

These diagrams will now be used to explain how spontaneous and stimulated emission occur in the semiconductor devices of Figure 5-4, first for a PN homostructure, then for a heterostructure.

The p-n homojunction as emitter and detector

Consider the homojunction between p and n materials, as shown in Figure 5-11(A), and assume for the moment that no current is applied. The diagram reads from left to right in the y-direction (Figure 5-4(A)). When the two materials are first brought into contact, the Fermi levels do not line up, but very quickly, free electrons from the n side try to flow leftward to the p side to even out the density of electrons laterally (and the converse happens with holes travelling rightward from the p side to the n side). However, as these electrons and holes pass across the junction, many of them combine instantly. This leaves the p region populated with a few less holes, and similarly for electrons in the n region, but in the middle there is a *depletion region* where there are almost no free electrons or holes. This is shown in Figure 5-11(B). Any further

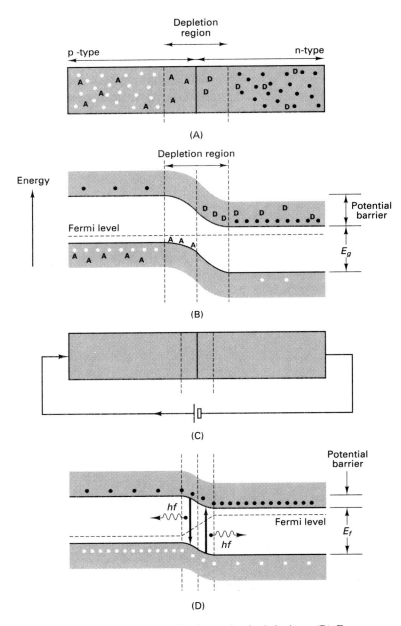

Figure 5-11. Homostructure light emitter (A) Open-circuited device. (B) Energy vs. position diagram for open-circuited homojunction. (C) Forward biasing. (D) Energy vs. position for forward biasing.

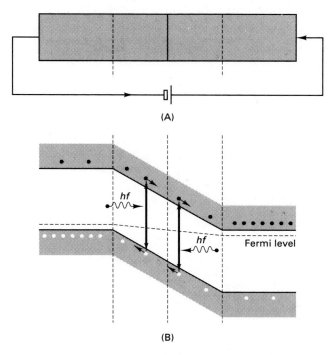

Figure 5-12. Homostructure photodetector (A) Reverse biasing. (B) Energy vs. position for reverse biasing.

leftward electron movement from the n region is opposed by the coulomb repulsion between the donor atoms on the right side of the junction and the free electrons on the n side. Similarly, the acceptor atoms immediately to the left side of the junction repel the holes on the p side from moving to the right. In this way, an energy barrier is produced that the electrons are unable to climb over and the holes (for which increasing energy is downward) are unable to get under. The effect of all this is that the Fermi level runs continuously across the device, as shown by the dashed line. The height of the energy barrier, when expressed as a voltage, is called the *contact potential*.

This standoff condition at the depletion layer is modified when a *forward bias* current is applied to the device, as in Figure 5-11(C). The injection of electrons, shown by the arrow, reduces the built-in potential difference, which reduces the coulomb repulsion force that was holding the electrons on one side and the holes on the other. More of them are now able to cross the narrowed depletion. When they get to the other side, they greatly increase the concentration of minority carriers, which are now available to recombine with the majority carriers already there due to the doping. Many of the electron-hole pairs will combine radiatively and produce spontaneous emission of frequency $f = \mathcal{E}_g / h$, and some will combine unproductively by some form of nonradiative recombination. If the current is insufficient, many of those photons that were generated will be absorbed in a reverse process that generates electron-hole pairs. As the current is increased, the potential barrier is reduced further, these two unproductive (loss) processes lose out to the process of photon generation, and we have population inversion, analogous to that of the gas laser; there are more electrons producing light by combining with holes than there are supposed to be for that material at that temperature.

So what happens is exactly as was shown earlier in Figure 5-2 and illustrated with the gas-laser example. As I is increased, there is a greater population inversion. If mirror facets exist at the ends of the active region, the transparency point is passed, and, as the threshold current I_{th} is exceeded, stimulated emission rapidly predominates over spontaneous emission, and lasing starts taking place. If there are no mirrors, the spontaneous emission continues to grow in intensity with injected current.

It is worth making a digression at this point to see how sending the current through the device in the opposite direction turns it from an emitter into a photodetector. This will lay some groundwork for Chapter 8. If the device is reverse-biased, as in Figure 5-12, the depletion region becomes even wider than in the unbiased case of Figure 5-11(A) and (B). The potential barrier becomes even taller, that is, the voltage difference across the depletion region becomes even stronger, and every time an electron-hole pair is created by an incident photon of $f = \mathcal{E}_g$, the electron will move rapidly to the right and the hole to the left. In this way a stream of photons is converted into a current. Again, the behavior of the device is quadratic; milliwatts of light are converted into milliamperes of current.

Summarizing the behavior of the p-n junction as an emitter and a detector, the device is forward and reverse biased, respectively, and carriers are converted to light or from light, respectively.

P-I curve equations

At a given point in the active region, the gain per unit length g is actually a function of carrier density N

$$g(N) = a(N - N_o) \tag{5.10}$$

where N_o is the carrier density at transparency (somewhat below threshold), the point where the active region makes a transition from being opaque to light of the bandgap wavelength to a condition of luminosity due to spontaneous emission. The quantity $a = dg/dN$ is called the *differential gain constant*, and has the dimensions of square centimeters. The gain per unit length g is also a function of λ, as will be discussed in Section 5.5.

Further increases of pump current I above I_{th} do not increase N, since all of the pump electrons are either being converted to photons or are being lost to such processes as ohmic leakage across the device. In other words, above threshold, the carrier density is "clamped" or "pinned" at a value N_{th}, and thus the gain is clamped too. This is consistent with the linear nature of the *P-I* curve above threshold.

Several useful relations can be gotten, starting with Equation 5.10, by a line of reasoning that is clearly laid out in Section 2.6 of Reference [4].   There is one equation for the threshold current,

$$I_{th} = \frac{qdwLN_{th}}{\tau_e} \qquad \text{(amperes)}$$ (5.11)

and one for the the slope of the *P-I* curve above threshold

$$\frac{dP}{dI} = \frac{hf}{2q} \frac{\eta_d}{\eta_i} \qquad \text{(watts per ampere)}$$ (5.12)

Here d, w, L are the height, width, and length, respectively of the active region, q is the electronic charge (1.6×10^{-19} coulombs), τ_e is the *carrier recombination lifetime*, analogous to the lifetimes τ_{i-j} characterizing the transitions between energy levels in the gas-laser example, and η_d and η_i are to be defined shortly.   When we get to the rate equations in Section 5.6 an easier way to derive Equations 5.11 and 5.12 will be presented.

From Equations 5.5 and 5.10, the threshold carrier density N_{th} is given by

$$N_{th} = N_o + \frac{\alpha_m + \alpha}{\Gamma a}$$ (5.13)

where N_o is the transparency carrier density, and Γ is the confinement factor of the transverse optical waveguide mode set up across the active region.   It represents the fraction of the optical mode that lies within the active region, and thus ranges from 0 to 1.   The higher the confinement, the fewer injected electrons are needed to achieve threshold.   The quantity α is the actual attenuation (absorption) per unit length inside the material (used earlier in Equation 5.5), and α_m is a fictitious equivalent loss per unit length due to useful escape of light through one of the two cavity mirrors, given by

$$\alpha_m = \frac{1}{2L} \ln \frac{1}{R}$$ (5.14)

by analogy with Equation 5.5.

It is clear from Figure 5-2 that low threshold current is one of the keys to high light output with low pump current.   Equations 5.11 and 5.13 say that I_{th} can be reduced by having a short, narrow, and low active region, tight confinement of the field to within that region, and highly reflective facets at the ends.

The quantity that expresses the slope of the *P-I* curve is η_d, the *external quantum efficiency* of the entire device, the fraction of additionally injected electrons that get converted to useful output photons.   The quantity η_i is the *internal quantum efficiency*, the fraction of injected electrons that produce stimulated photons, not all of which will reach the output of the device. Above threshold, we can approximate η_i by unity. These two ηs can be related to the internal loss per unit length α and the mirror reflectivity R by

$$\eta_d = \eta_i \frac{\text{photon escape rate}}{\text{photon generation rate}}$$

$$= \eta_i \frac{\alpha_m}{\alpha_m + \alpha} = \frac{\eta_i \ln 1/R}{\ln 1/R + 2\alpha L} \quad \text{photon per electron} \tag{5.15}$$

The threshold current is very temperature-dependent, more so for long-wavelength InGaAsP lasers than for short-wavelength GaAs lasers [4]. This is one of the principal reasons that long-wavelength lasers usually include an internal thermoelectric cooler. Figure 5-13 depicts the situation for a long wavelength laser that is particularly insensitive to temperature. Many theoretical explanations have been advanced for the detailed temperature behavior of semiconductor lasers, but the total picture is somewhat unclear. As for I_{th}, a good fit to the observable facts can be gotten by writing

$$I_{th}(Y) = I_o \exp(Y/Y_o) \tag{5.16}$$

where I_0 is a hypothetical threshold current at absolute zero and Y_o is some fictitious temperature constituting a metric for temperature-dependence. It is desired to maximize Y_o, which has proved to be easier for short-wavelength lasers (Y_o of around $120°K$) than for those at 1.3 or 1.5 μ ($Y_o = 50 - 70°K$). Long-wavelength multiple-quantum-well lasers (next section) have been made that have Y_o values that are higher. The multiple-quantum-well device whose performance is shown in Figure 5-13 has a Y of $96°K$.

The double-heterojunction laser

The greater the optical mode confinement factor Γ, and the smaller the dimensions w, d, and L of the region through which the current is concentrated, the more pump electrons will be converted into light, according to Equations 5.11 through 5.15. However, the earlier laser diodes, which were PN homostructure devices, had an insufficient N_o to lase continuously at room temperature, and even then the threshold current was very high, more than an ampere.

A heterojunction is a junction between two materials of different bandgaps. We saw in conjuction with the homojunction that various regions could be doped differently so as to force an overpopulation of one kind of carrier or another, but the bandgap remained constant from one region to another because the mole fractions (Figure 5-9) were unchanged. This is clearly seen in Figure 5-11, where the energy

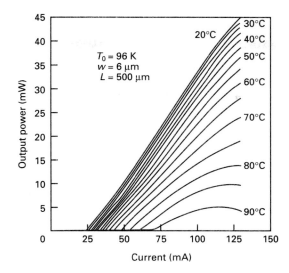

Figure   5-13. *P − I* curves as a function of absolute temperature Y for a long-wavelength quantum-well laser diode. (Courtesy Peter Unger, IBM Research Laboratory, Zurich).

levels may change, but the valence and conduction bands are a constant distance apart in energy.

In a single heterojunction, there are two materials whose mole fractions *x* and *y* in Figure 5-9 are actually different, so that the gap is different in the two materials. A *double heterostructure* is just two such structures back-to-back, as shown in Figure 5-14.

The double heterostructure device, proposed by Woodall and first implemented by Hayashi and Panish in 1970, solved the part of the density problem due to carriers not being adequately confined in the *y*-direction in Figure 5-4 (the direction normal to the plane of the active region). Also, if the index of the active region can be made higher than that of the surrounding medium, improved optical confinement also takes place. The evolution of various index- and gain-guiding schemes solved the problem of confinement in the *x*-direction. Today's double-heterostructure laser diodes have values of I_{th} as low as 5-10 milliamperes, and quantum-well laser diodes (to be described in the next section) have attained values of less than 1.0 mA [7].

Very low threshold currents have important practical consequences. Not only is power saved and device lifetime increased, but the control circuitry is much simpler. Normally, for on-off modulation, there is a *bias-current* supply that must be stabilized to lie very close to I_{th}, which depends on such variables as temperature. If I_{th} is a small fraction of the maximum drive current, an essentially biasless mode of operation can be used.

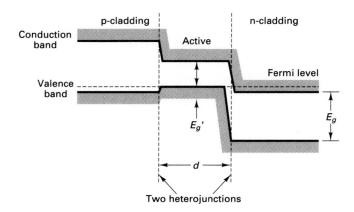

Figure 5-14. Modification of the energy barrier of Figure 5-11(B) by replacing the homostructure with a double heterostructure of depth d.

The way in which the heterostructure improves carrier confinement in the y-direction by making d small is shown in Figure 5-14. Problem 5-1 gives some typical parameters of such laser diodes.

Again, the diagram reads left to right in the y-direction. The extrinsic materials of the p-region and the n-region are shown having the same bandgap, whereas the region in the middle has a smaller bandgap than either. Long-wavelength (1.3 and 1.5 μ) lasers can be made by using intrinsic InGaAsP material for the active region and p-doped and n-doped InGaAsP for the two cladding regions, respectively. InP would be used for the substrate. When no bias current is applied, once again the Fermi levels line up across the device, as shown in Figure 5-14. But this time, when the electrons move leftward in the conduction band in the n-region and the holes move rightward in the valence band in the p-region, once they get into the active region, they are resisted from going further, because there is another potential barrier to climb. The result is that the electron and hole concentrations build up to much higher values in the active region than they would have in the depletion region of the simple homojunction of Figure 5-11(A). Then, as forward or reverse bias currents are applied, analogously to Figure 5-11(C) and (D), light of frequency $f = \mathcal{E}_{g\ active}\ /\ h$ is emitted.

In addition to confining the carriers tightly within the active region, the heterostructure has another neat property. By a lucky coincidence, the refractive index of the active-region material is slightly greater than that of the extrinsic materials of the cladding, just as in the fibers discussed in Chapter 3, so that not only is the injected current concentrated, but the light, too, is concentrated. These two forms of confinement are gain guiding and index guiding, respectively.

5.3 State Density and State Confinement

There is another totally different kind of confinement of the carriers other than geometrically, and this has its own advantages. This is a confinement of the energy *states* that the electrons or holes are allowed to occupy. This is what *quantum-confinement lasers* do. The many benefits include lower I_{th}, lower linewidth, higher differential gain $a = dg/dN$ (as in Equation 5.10), more control over the gain curve, and lower temperature sensitivity [8].

We shall now discuss quantum states, both to give a more accurate picture of some of the phenomena already discussed and to prepare the way for a discussion of lasers using quantum confinement.

Lasers that use quantum confinement include those having a single quantum well (QW), multiple quantum wells (MQW), quantum wires (QWi), or quantum dots (QD), sometimes called quantum boxes (QB). Closely coupled multiple quantum wells are called *superlattices*.

We have glossed over this question of what states the electrons are allowed to occupy in a gas laser or the carriers are allowed to occupy in a solid-state laser, and it is time now to correct this deficiency. Note in Figure 5-7(A) that, for a gas, there are whole regions in energy $\mathcal{E}$ that the electron may not occupy; it must lie on one of the infinitely narrow lines representing a quantized energy level. In the case of the solid-state semiconductor, the electron may not occupy the forbidden energy bandgap. What has been swept under the rug is that there are even more restrictions on where the electron can and cannot be. There are just so many *states* at any energy that the electron can legitimately occupy. The *Pauli exclusion principle* says that if there is already an electron occupying a certain state, no second electron anywhere in the vicinity can occupy it. In all the diagrams like Figures 5-10, 5-11, 5-12 and 5-14, if there are no empty states at one of the energy levels shown, the electron will not be permitted to occupy it. It is all very well for the Boltzmann or Fermi-Dirac probability distributions to say that there is a certain probability that the electron will find itself at a certain energy, but it will actually only do so with that probability if there is a state available for it to occupy at that energy level. The mean number of states $v(\mathcal{E})\,d\mathcal{E}$ actually occupied in an incremental energy range $d\mathcal{E}$ is

$$v(\mathcal{E})\,d\mathcal{E} = \rho(\mathcal{E})\,P(\mathcal{E})\,d\mathcal{E} \tag{5.17}$$

where $P(\mathcal{E})$ is given by Equations 5.8 or 5.9, and $\rho(\mathcal{E})$ is called the the *density of (permitted) states*.

The quantum-well class of devices all force a narrower choice of energy levels at which permitted states exist than are given by the valence and conduction bands of Figures 5-7(B), 5-10, and 5-11. This is because $\rho(\mathcal{E})$ is more concentrated for quantum-well devices than for the conventional bulk devices we have discussed up to now.

The situation is schematized in Figure 5-15, where the upper row of diagrams shows the shape of the device and the bottom row a rough idea of $\rho(\mathcal{E})$, the number of states available as a function of energy on each side of the bandgap. (A) is the conventional bulk heterostructure device, (B) is the multiple quantum-well device where the active layer consists of a succession of quantum wells, forming a multilayer sandwich of alternating layers of two different semiconductor materials. These layers are only about 0.005 to 0.010 μ thick (50-100 Angstroms), which is only 7 to 15 atomic diameters. (C) and (D) carry the two-dimensional confinement idea of (B) further by confining to one dimension with quantum wires, or to essentially zero dimensions with quantum dots (quantum boxes), respectively.

Density of states

Perhaps the reader will recall, from our discussion of interacting acoustic and light waves in the last chapter, that each has a dual personality, as a wave and as a particle. This is reflected in the fact that in order to describe an electron, photon, or phonon completely, we must talk not only about its energy (as we have up to now in this chapter), but also about its momentum. So far, all the diagrams showing bandgaps, Fermi levels, and so forth have been in terms of energy $\mathcal{E}$, a scalar. Momentum $\mathbf{p}$, which we have been leaving out of the picture, is a vector, directed along the propagation direction of the equivalent wave.

For completeness, the simple picture of levels in Figure 5-7(A) and of the bandgap in Figure 5-7(B) must be extended to include the momentum of the particle. The result is the *energy-momentum* representation or $\mathcal{E}-k$ diagram which we shall now explain, using Figure 5-16. Particularly clear tutorial discussions of the ideas involved are given in Chapter 11 of [3] and Chapter 3 of [9], with more detailed discussion in [10], Chapter 15.

The fact that k is shown at right angles to $\mathcal{E}$ does not mean that there are two things propagating in perpendicular directions. It means that we are using a two-dimensional coordinate system to portray two things about the same particle, the energy and another scalar, the magnitude k of the propagation vector $\mathbf{k}$, which in turn is proportional to the momentum vector $\mathbf{p}$ by

$$\mathbf{k} = 2\pi\mathbf{p} / h \qquad (5.18)$$

which we have seen before as Equation 4.28. In discussing the $\mathcal{E}-k$ diagram, it is customary to speak of k as the (scalar) momentum, although actually it is something proportional to momentum.

Again, the case of an electron in a gas proves a useful starting point in understanding where the permissible states lie in $\mathcal{E}-k$ space. Thinking of the electron as a wave, it has energy

$$\mathcal{E} = hf \qquad (5.19)$$

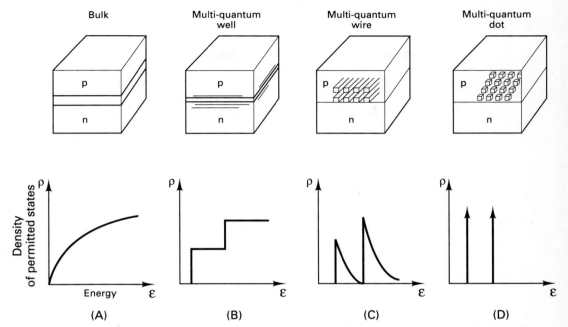

Figure   5-15. Comparing the physical device shape and the energy dependence of permitted state density for (A) conventional bulk laser diode, (B) Multiple quantum well device, (C) Quantum wire device, and (D) Quantum dot device.

and as a particle it has momentum

$$\mathbf{p} = m\,\mathbf{v} \tag{5.20}$$

and energy

$$\mathcal{E} = \frac{p^2}{2m} = \frac{h^2\,k^2}{8\pi^2 m} \tag{5.21}$$

from the familiar $\mathcal{E} = m\,v^2/2$, where m is the effective mass (which, in general can be different from the rest mass of the particle). From Equations 5.20 and 5.21, it is clear that in $\mathcal{E}-k$ space, the available states must lie along the parabola of Figure 5-16(A). The two heavy dots show that at a specified energy $\mathcal{E}$, there are only two permissible states.

Why two? The exclusion principle says "only one." The answer is that the electron can have two equal and opposite spins, giving plus and minus values of (angular) momentum. What about all the other atoms in the gas? Can no other atoms have their corresponding electron occupy either of these states? The answer is "yes," because in a gas the atoms are so far apart that they do not interact, and so the exclu-

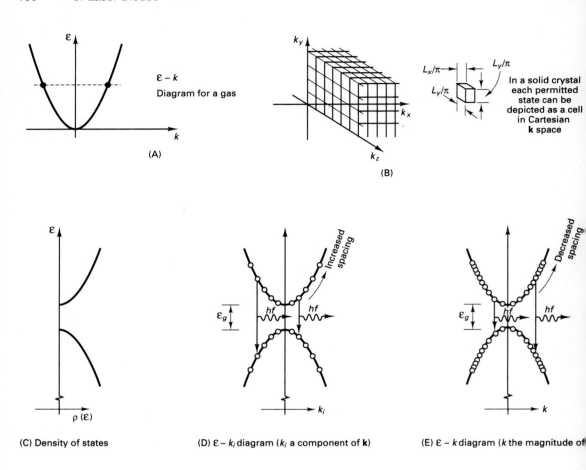

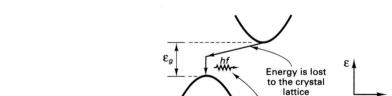

(F) ℰ – k diagram for an indirect semiconductor

Figure 5-16. The energy-momentum (ℰ–k) diagram explained. (A) ℰ − k diagram for an electron in a gas. (B) Countable cells in **k** space for an electron in a bulk solid. (C) Density of permitted states $\rho(ℰ)$. (D) ℰ–k_i diagram. (E) ℰ–k diagram. (F) The ℰ–k diagram of an indirect bandgap material.

sion principle is not violated. (Specifically, the tails of their Schrödinger wave functions have negligible overlap).

It is an interesting consequence of the wave-particle duality of light that the two values of particle spin can be identified with the two degenerate polarization states discussed in Sections 3.9 and 3.13.

In a solid crystal, such as the semiconductor material used in laser diodes, the tails of the wave function of a given electron lying in either the valence or the conduction band see many nearby atoms in the crystal lattice, and so a different treatment is necessary. The following discussion ([3], Section 11.2) leads to Figure 5-16(E), the more complete version of Figure 5-10(A).

When an electron's parent atom is tightly confined in a crystal of dimensions L_x, L_y, and L_z centimeters, quantum theory says that each of the three components, k_x, k_y, and k_z, of $\mathbf{k} = \mathbf{u}_x k_x + \mathbf{u}_y k_y + \mathbf{u}_z k_z$, the electron's momentum, is quantized to values

$$k_x = \ell\pi/L_x \qquad k_y = p\pi/L_y \qquad k_z = q\pi/L_z \qquad (5.22)$$

Each of these quantizations (by the integers ℓ, p, and q) is analogous to the various orders of Fabry-Perot cavity modes of the preceding chapter, and we can think of this set of relationships as expressing a set of standing cavity waves, where we are not talking about a classical propagating electromagnetic wave, but the wave functions of the particle, the solutions to the Schrödinger equation for the electron.

We can now proceed to develop all the $\mathcal{E}-k$ diagram ideas of Figure 5-16, and also gain considerable insight about quantum-well, -wire and -dot devices, simply by counting states. Figure 5-16(B) shows one octant of a cartesian space of $\mathbf{k}$, the one corresponding to positive values of the k_i, i = 1, 2, or 3. The momentum $\mathbf{k}$ is a vector from the origin to the center of one of the little cells. There is one state within each of the countable cells whose dimensions are as shown in the figure. These little cells are piled up like a child's toy blocks. Their density in $\mathbf{k}$ space is $L_x L_y L_z/\pi^3$, the reciprocal of the volume of each in $\mathbf{k}$ space. Consider a thin incremental shell dk thick on a sphere of radius k. The part of its volume contained in the octant is

$$\frac{1}{8}\frac{d}{dk}\frac{4\pi k^3}{3} = \pi k^2/2 \qquad (5.23)$$

Multiplying this by the density in $\mathbf{k}$ space gives the number of states $N_k\,dk$ swept through by letting k increase from k to $k + dk$.

$$N_k\,dk = \frac{1}{2\pi^2}L_x L_y L_z k^2\,dk \qquad (5.24)$$

Dividing through by the unit spatial volume of the piece of material ($L_x L_y L_z$ cm^3), and multiplying by two to account for the two possible values of spin for each state, we have the density per cubic centimeter of states as a function of momentum

$$\rho_k(k)dk = \frac{k^2}{\pi^2} \, dk \qquad (5.25)$$

Inverting Equation 5.21 and finding $dk/d\mathcal{E}$ gives

$$dk = \frac{\pi\sqrt{2m}}{h\sqrt{\mathcal{E}}} \, d\mathcal{E} \qquad (5.26)$$

so, substituting this for dk and Equation 5.21 for k in Equation 5.25 gives the density of permitted states per unit spatial volume of the material as a function of energy $\mathcal{E}$, all ready to be used in Equation 5.17

$$\rho(\mathcal{E})d\,\mathcal{E} = \frac{8\pi m\sqrt{2m}}{h^3} \sqrt{\mathcal{E}} \, d\mathcal{E} \qquad (5.27)$$

This is plotted in Figure 5-16(C), where we have now identified $\mathcal{E}$ with $(\mathcal{E} - \mathcal{E}_c)$ for the conduction band and $(\mathcal{E}_v - \mathcal{E})$ for the valence band, respectively.

To see where the individual states are in energy-momentum space, first look at the plot of $\mathcal{E}$ as a function of one component, say the x-component. The plot of $\mathcal{E}$ versus k_x must be a parabola, by Equation 5.21, but according to the depiction of Figure 5-16(B), the states (shown as little dots) must be spaced equally in k_x Therefore the dots must exhibit the steady increase of spacing shown.

It is more customary to plot $\mathcal{E}$ as a function of the magnitude k, and here Equation 5.26 indicates that the spacing of the dots must get tighter with increasing $\mathcal{E}$. These two situations of increasing and decreasing spacing are easily confused if a careful distinction is not made as to which is the independent variable in the diagram, a slice along one coordinate of the crystal (e.g., k_i) or the total magnitude k.

The reason that materials such as GaAs, InP, and indeed suitable compositions of InGaAsP are so easily able to generate and detect photons is that they are *direct-bandgap* materials; the noses of the two parabolas line up vertically, that is, they are at exactly the same value of momentum. Thus, light will be emitted when a hole and electron combine to form a photon via purely radiative transitions like the ones shown by the purely vertical arrows.

For many semiconductors like silicon, the noses of the two parabolas do not have the same k-value (Figure 5-16(F)), so not only is a radiative energy change (vertically) involved in a transition, but a momentum change (sideways) happens, too, so that part of the energy goes into the crystal lattice vibration as heat. This is why silicon does not work as a laser material. Materials having this sideways property are called *indirect bandgap* materials. (Note from the upper left-hand corner of Figure 5-9 that this can also occur with III-V alloys as the pure GaP condition is approached).

Figure 5-17 shows the effect of using Equation 5.17 to get the actual occupancy of states at any particular temperature. We multiply $\rho(\mathcal{E})$, the density of permitted states, by the Fermi function $P(\mathcal{E})$ of Figure 5-7(B) to get $\nu(\mathcal{E})$, the average number of states per unit volume occupied in an incremental energy range $d\mathcal{E}$ at energy $\mathcal{E}$, shown in the rightmost diagrams of Figure 5-17.

5.4 Quantum-Well, -Wire and -Dot Lasers

We have given a bit of quantification and physical plausibility to the parabolic depiction of the permitted (but not necessarily occupied) electron and hole energy levels schematized for the conventional or "bulk" device by the smooth parabola of Figure 5-15(A). How do the analogous curves of the quantum-well devices [7] of (B), (C), and (D) come about? The answer is by *quantum-mechanical confinement*. We have discussed the single-valued discrete permitted states of a gas, and a continuum of permitted states in the valence and conduction bands. Quantum-mechanical confinement provides another mechanism for confining the permitted states of electrons and holes to discrete values that are more concentrated than the occupancy of an entire octant of **k** space discussed in connection with Figure 5-16(B) for the bulk laser diode.

Figure 5-18 shows at the top left the active region of a bulk laser diode in the form of a block of material with the dimensions L_x, L_y, and L_z, as before. On the right is the pile of children's blocks that was discussed before, the counting cells that were used for deducing the density of states in either an increment of a component k_i or an increment of the magnitude k. The representation of each state is a little rectangular solid, because there is confinement in all three directions.

Suppose, by using modern MBE (molecular beam epitaxy) or MOCVD (metallo-organic chemical vapor deposition) we are able to control the deposition of the material so as to make the active layer thickness d in Figure 5-4(B) and (C) only a few tens of Angstroms thick, as shown for the quantum well example of Figure 5-18. Such distances are comparable to or less than the reach of the electron's wave function, its *de Broglie wavelength*. (Recall that we have the option of thinking of the electron as either a particle with a certain momentum **p** or a wave with a certain wavelength, and we can see that the latter is h/mv by combining Equations 5.18 and 5.20).

The result in **k** space is shown at the right of the second row of Figure 5-18. The cells are very long in the y-direction. Instead of a pile of children's blocks filling up the octant of positive k_x-, k_y-, and k_z-directions, we how have a big mass of mailing tubes, side by side and end on end. Carrying this idea one step further, we can have the quantum wire along z, with the confinement now being less tight along the k_z-axis to give us a volume full of flat pizza boxes standing on their edges, and finally, the quantum dot, where we are back to children's blocks again, except that each child is now at least 50 feet tall, with toy blocks to match.

Taking the quantum-well case, and doing the state-count bookkeeping again (as in Section 11.6 of [3]), leads to the $\mathcal{E}$–k diagram of Figure 5-19(A) and the state

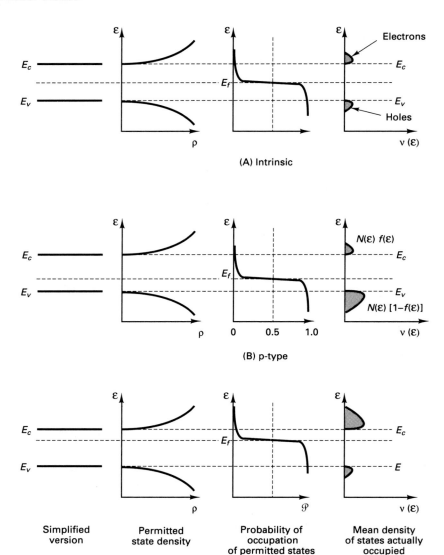

Figure   5-17. The energy dependence of permitted state spatial density ρ is multiplied by P the probability density that an electron or hole is a candidate for a state of that energy to form v the overall mean state density. (From p. 75 of [9]).

density function $\rho(\mathcal{E})$ of (B). Each increasing value of the quantum number q corresponds to another parabola of Figure 5-19(A). Recall from Equation 5.22 that $k_z = q\pi/L_z$. It can be demonstrated fairly simply (e.g., Section 11.6 of [3]) that the corners of the double staircase of ρ in Figure 5-19(B) have the previous parabolas in Figure 5-16(C) as their envelope.

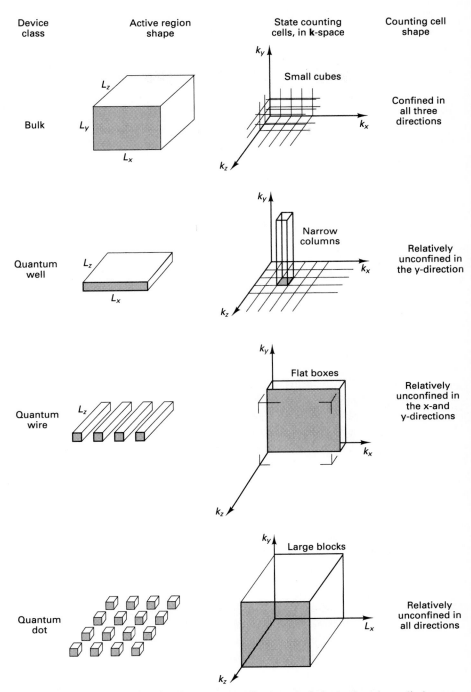

Figure 5-18. Active region shape and quantum confinement in **k** for various laser diode types.

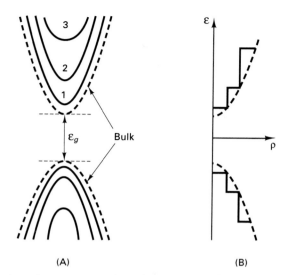

Figure   5-19. Behavior of a quantum well device.   (A) The $\mathcal{E}-k$ diagram, and (B) density of permitted states $\rho(\mathcal{E})$.

If the active region contains Q quantum wells that are more or less isolated from one another, then the density of states is just the same stairstep, but Q times as great in the $\mathcal{E}$-coordinate.   If they are not isolated, the superlattice situation, then the stairsteps get smeared, thus providing some control over the $\mathcal{E}-k$ function.

In order to achieve practical confinement of states into the narrower exponentially decaying bands of quantum wires (Figure 5-15(C)) or the essentially impulse-like behavior of the quantum dot of Figure 5-15(D), improved methods of controlling growth in lateral directions must be developed [11].

The fact that, in a quantum-well laser diode, the top of the valence band and the bottom of the conduction band are now squared-off steps instead of rising parabolas has a number of practical consequences.   Generally the gain curve of a bulk laser is a smooth bell-shaped curve, reflecting the smooth onset of the noses of the two parabolas, as smoothed by the Fermi function (rightmost column of Figure 5-17).   But for a quantum-well device, the gain curve has a much sharper onset, reading downward in wavelength (upward in $\mathcal{E}_g$).

The sharpness of the band edge means, among other things, that there is less temperature sensitivity of the width of the actual density of states occupied, as was shown in the rightmost column of Figure 5-17.   For higher temperatures, the Fermi function has higher values at these steep band edges, but the temperature-dependent tails on the Fermi function have a much weaker effect than with a bulk laser diode. Thus, as the temperature changes, the center of the gain curve is less affected (as will be discussed in the next section) and the same goes for the threshold current (Equations 5.11 and 5.13).

This sharpness of the band edge will also prove useful in the external modulators to be discussed in Chapter 7.

As mentioned before, the quantum-well approach offers not only lower threshold current, but also lower temperature dependence of threshold current, higher differential gain $a = dg/dN$, and other advantages. A good tutorial survey of these advantages is given by [8].

5.5 Steady-State Spectral Characteristics

In this section, we shall look more closely at the overall gain curve as a function of pump current and also the character of the individual Fabry-Perot laser radiation lines, before going on to transient effects and then to the special *distributed Bragg reflector (DBR)* and *distributed feedback (DFB)* structures. These are aimed at getting the narrow spectral line that we want for our systems.

The gain curve of bulk laser diodes

According to Equation 5.17 and Figure 5-17, we may compute the overall mean number $v(\mathcal{E})$ of occupied states in an incremental energy range at $\mathcal{E}$, and from this it should be possible to deduce the gain as a function of optical frequency using Equation 5-10. However, by the time all the relevant factors are included, no closed-form solution has proved possible. The result of a numerical calculation [4] for a particular InGaAsP material for 1.3 μ lasers is given in Figure 5-20. It shows several interesting things. First, as we already know, there is no net gain at any wavelength until the current I has gotten high enough for the gain to exceed the loss at some wavelength. Second, the gain curve for high carrier density has a 3-dB bandwidth of about 30-40 nm, which tells us something about the bandwidth available for wavelength-division systems using fixed-tuned or tunable lasers made out of this particular material. Third, the gain peak is seen to shift toward shorter wavelengths with increasing bias current, i.e., carrier concentration. This is because, as predicted by the Fermi function, it is the states near the noses of the $\mathcal{E}-k$ curves of Figure 5-16(E) that get occupied with the highest probability by the injected electrons and the corresponding holes, and the electron-hole pairs are thus characterized by a smaller energy (longer wavelength). As more of these states become occupied, the states farther away from the noses of the curves become stronger candidates.

This downward wavelength shift of the gain peak with pump current is in the same direction as the smaller negative shift of the cavity modes due to index decrease. It seems counterintuitive that the index would decrease with pump current   We are used to thinking that an increase in "density" of anything will slow down the propagating wave and thus increase the free-space wavelength at which the cavity resonates. While this would be true if the actual density of the semiconducting material increased, that is not what is increasing; it is the carrier density. That this causes a decrease in index can be proved by a complicated set of arguments (see Section 2.2 of [4]) that starts from the Maxwell equations, as modified to take into account dielectric

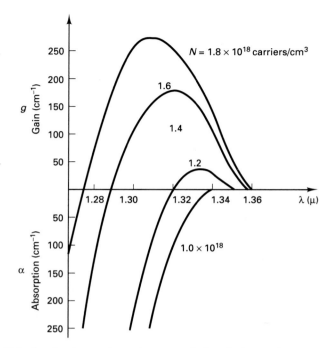

Figure   5-20. Calculated gain or absorption per unit length (α or g) versus λ of the emitted radiation for various values of the carrier density N. (From [4]).

polarization, the flow of charges proportional to the static electric field. Following the consequences of this generalization leads to the correct result.

The gain curve also has a considerable temperature dependence. From the character of the Fermi function, it is clear that at higher temperatures, the carriers are distributed over a wider range of energy differences, so the gain curve is broader and the peak gain is not as high. The fact that gain decreases with temperature can be seen from the decrease in slope in Figure 5-13. The peak gain is usually the quantity to be maximized, which is still another reason to use a thermoelectric cooler to keep the operating temperature low.

Since the index is temperature dependent too, the cavity mode frequencies have a temperature dependency, specifically about 0.1 nanometer per degree centigrade.

As explained earlier, the temperature sensitivity of peak gain and gain-curve width are considerably smaller with quantum-well devices than with bulk devices.

Broadening the gain curve

It is worth returning briefly to the question of gain-curve width and speculating about what can be done to broaden it out beyond the typical 30-40 nm values represented in Figure 5-20. To be able to do so is important for making widely tunable lasers.

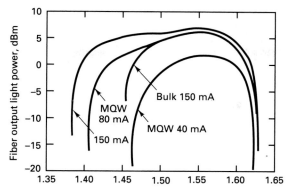

Figure   5-21. Gain curve broadening by heavy pumping of a multiple quantum well laser diode. (From [12], by permission IEE).

(Recall that at 1.3 and 1.5 μ, the useful spectral range over which the fiber attenuation is less than twice the minimum number of dB per kilometer is around 200 nm).

Quantum-well devices offer the most attractive possibility for doing this broadening. The approach is to pump the device with such high current that the tails of the Fermi function pick up not only each first step in the staircase of Figure 5-19(B) but also significant carrier density in the second step. Figure 5-21 shows how this artifice has been used to broaden out the gain curve of an InGaAsP/InP multiple-quantum-well laser to 240 nm, more than 16 percent of its center wavelength [12].

Side-mode suppression

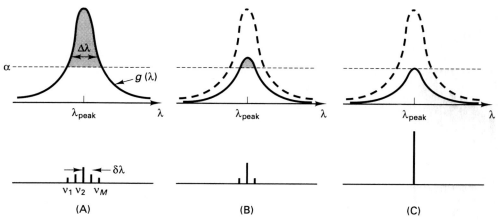

Figure   5-22. Time history of gain clamping (pinning), with the portion of the gain curve $g(\lambda)$ that exceeds the loss a by the amount shown shaded at the top. At the bottom are the corresponding spectra of radiation in the cavity modes. (From [10], John Wiley & Sons, by permission).

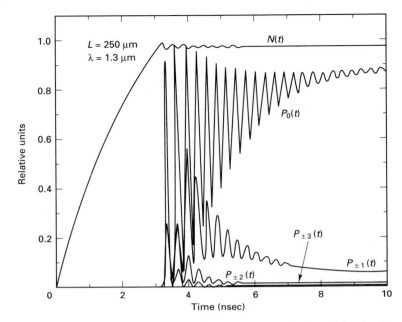

Figure 5-23. Calculated waveforms underlying Figure 5-22. N is the carrier density, P_o is the photon density in the main mode, $P_{\pm 1}$ the adjacent ones, $P_{\pm 2}$ the next ones, and so forth. (From [13], ©, 1983 IEEE).

It was stated at the beginning of this chapter that as the injection current is increased, the side modes are suppressed more strongly (Figure 5-2). A fairly realistic picture of the behavior of the side-mode suppression mechanism above threshold can be gotten by approximating the appropriate gain curve of Figure 5-20 by a gaussian function and the absorption α as a constant, independent of λ. (For details, see Section 14-2-B of [10]). Figure 5-22 shows what happens immediately after a laser diode, whose pump current I has been well below threshold I_{th}, suddenly experiences a step increase to a value well above I_{th}. The top row of diagrams show the gaussian gain curve of width $\Delta\lambda$, and the bottom row the cavity modes of separation $\delta\lambda$. At any given instant, light will be emitted at all those Fabry-Perot cavity mode wavelengths that have a gain exceeding the loss, i.e., those that satisfy Equation 5.5. As light begins to be radiated, it depletes the carriers and the gain curve falls. The radiation spectrum shown at the bottom of Figure 5-22(A) persists only for a brief instant after the laser is turned on. The growth of stimulated emission is strongest for that mode that is closest to λ_{peak}, and quickly the overall gain curve is lowered, as shown at (B), because so many of the carriers have been turned into light. The end result is shown at (C); all the gain is going into λ_{peak}.

Another view of this history is shown in Figure 5-23. The injected current I was assumed to go from zero to about 1.5 I_{th} at $t = 0$. For several nanoseconds the carrier density rises at a steady rate (to be discussed in the next section). Suddenly, light starts being emitted at all cavity resonances for which Equation 5.5 is satisfied. The light power in the main mode is P_0, that in the two nearest side modes is $P_{\pm 1}$, in the next two side modes $P_{\pm 2}$, and so forth. Eventually, only P_o is left. The oscillatory behavior is the *relaxation oscillation* in which energy is exchanged back and forth for a while between carriers and photon emission until eventually the oscillations die out.

This phenomenon of one wavelength taking over all the carriers is an idealization. Actually, there still remains significant side-mode output at high pump current (as shown in Figure 5-2) due to *spatial hole burning*. Two kinds of hole burning are defined, spectral and spatial. The word "hole," as used here, has nothing to do with the holes constituting part of each carrier. *Spectral hole burning* happens in a gas laser for cases in which the overall gain curve might consist of the overlap of the narrow spectra of many classes of different atoms, each class having its own center wavelength (an *inhomogeneously broadened* overall spectrum). For example, there can be many classes of atoms, each class having a narrow range of doppler velocities. Then, if all the atoms in that class are depleted, a hole or notch appears in the gain spectrum at that wavelength. In a semiconductor laser, this division of atoms into classes is usually absent (*homogeneous broadening*), so there is no spectral hole burning. However, there can be the spatial analog of this. The Fabry-Perot mode closest to the gain peak certainly depletes the carriers in those places along the cavity length L (i.e., burns holes) corresponding to peaks of field strength, but that leaves plenty of carriers elsewhere, and this allows the existence of significant side-mode power. As the pump current is increased, less and less of the active region length has not been burned out, but even at high pump levels, as shown in Figure 5-2, there is still enough side-mode power to prevent adequately tight channel spacing in dense wavelength-division networks, and to disrupt high-speed links by introducing dispersion-induced intersymbol interference and mode partition noise. Whatever incidental spontaneous emission is going on helps to stimulate coherent radiation at these side modes.

One can define the *mode suppression ratio* [4], a quantitative measure of side-mode suppression, as

$$\text{MSR} = \frac{\text{Power in the main mode} \ = \ P_o}{\text{Power in the strongest side mode} \ = \ P_{\pm}1}$$
$$= 1 + \frac{\Pi_o}{\tau_p R_{sp}} \left(\frac{\delta\lambda}{\Delta\lambda} \right)^2 \tag{5.28}$$

where τ_p is the photon lifetime, typically picoseconds, and not to be confused with the carrier (electron-hole pair) recombination lifetime τ_e (several nanoseconds), which we have met before with Equation 5.11. Π_0 is the photon density in the cavity. The quantity R_{sp} is the rate, per unit volume, at which spontaneously emitted photons are added to the active-region cavity. In general R_{sp} is a function of wavelength whose shape is identical to the gain curve, and can therefore be considered constant over a few cavity modes. It is these spontaneously emitted photons that play the dominant role in exciting the side modes by stimulating radiation in physical regions of the cavity that have been depleted of carrier density by spatial hole burning.

Numbers like 20 to 30 are typical of MSR for ordinary *Fabry-Perot lasers*, that is, the kinds we have been talking about in which no special tricks (e.g., DBR or DFB techniques) have been used to get monochromatic performance. Laser diodes that use only the cavity resonances as frequency filters are therefore quite insufficient for long haul-links and for networks using dense wavelength division.

Linewidth, phase noise and intensity noise

Up to now, we have considered the linewidth of an individual Fabry-Perot cavity mode to be negligible, but in real devices this is not so. The consequences of nonzero linewidth for signal detectability and the architecture of the network are substantial. In wavelength-division systems with low modulation rates and tight channel crowding, the individual linewidth can control how many nodes can exist on the network. For any form of signal detection, particularly those involving phase, seemingly modest values of linewidth can produce severe performance penalties, as will be discussed in Section 8.13.

The major cause of nonzero linewidth is randomness in the laser output phase, caused principally by spontaneous emission [14-16], and this makes these devices different from most other sources with finite linewidths with which the communication engineer might be familiar–for example, gaussian noise passed through a single-pole *RLC* filter.

A conventional filtered gaussian process will have random amplitude as well as phase, and this can be modelled as a sinusoid at the band center frequency, plus narrowband gaussian noise, as shown in Figure 5-24(A). The sinusoid has constant amplitude and phase. The noise can be represented as the sum of an in-phase component and a quadrature component of equal power but uncorrelated, and therefore, in the phasor representation there is just as much variance in the amplitude as there is in the phase, as shown by the dashed circle in the figure. As time goes on, the phase fluctuations result in the process "forgetting" its original value of phase, and eventually it will have occupied all possible values of phase from 0 to 2π uniformly. Such a random process is called a *Wiener process*, a form of random walk in which the probability distribution of phase is at any instant gaussian, but with a variance that grows linearly with time, as shown in Figure 5-24(B).

The early analyses of the linewidth to be expected from the physical processes involved considered the random noise to be equipartitioned between amplitude and phase, as in Figure 5-24(A). The resulting spectrum had the *Lorentzian* shape (Figure 5-24(C)) of a classical *RLC* resonant electric circuit, given in this case by

$$P(f) = P_o \left[1 + \left(\frac{f - f_o}{\Delta f / 2} \right)^2 \right]^{-1}$$

(5.29)

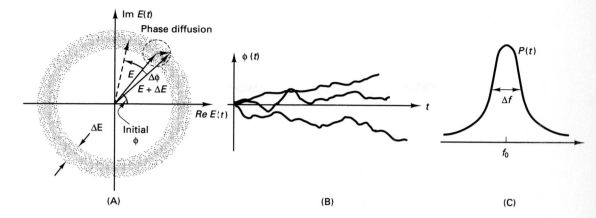

Figure 5-24. The classical narrow band noise, the sum of a sinusoid at the band center and a low-pass gaussian noise. (A) Phasor representation, showing diffusion of phase away from initial value. (From [16]). (B) Three sample time histories of phase $\phi(t)$. (C) Lorentzian spectrum.

where the width Δf is

$$\Delta f = \frac{v_g^2 \, h f \, n_{sp} \, g \, \alpha_m}{8 \pi P_o} \qquad \text{Hz} \qquad (5.30)$$

v_g being the group velocity in the cavity, α_m being the equivalent mirror loss (Equation 5.14), P_o being the power, and n_{sp} being the *spontaneous-emission factor*, a very important quantity that expresses the ratio of downward radiative energy transitions that are spontaneous to the total of spontaneous plus stimulated. It is defined as

$$n_{sp} = N_{upper} / (N_{upper} - N_{lower}) \qquad n_{sp} \geq 1 \qquad (5.31)$$

expressing how complete a population inversion has been achieved between two energy levels. As the inversion becomes more complete, n_{sp} approaches unity.

When an attempt was made to reconcile the theoretical result of Equation 5.30 with observation, it was found that the observed spectral shape was Lorentzian all right, but had a bandwidth that was higher by a factor of 30 to 40 than that of Equation 5.30. The more complete analysis [14] showed that the randomness in phase is caused not only by its own direct contribution to linewidth, but also by an additional indirect effect due to the change in index that a change of carrier density produces. The additional noise was therefore all in the phase, and the linewidth Δf in Equations 5.29 and 5.30 had to be multiplied by $(1 + \beta^2)$, where β is the *linewidth enhancement factor*

$$\beta = \frac{d(\text{Re } index)/dN}{d(\text{Im } index)/dN} = \frac{dn/dN}{dg/dN} \tag{5.32}$$

where N is the carrier density, n is the (real part of the) dielectric constant, and dg/dN is the differential gain a of Equation 5.10 (related to the imaginary part of the index).

This last equation expresses the fact that the random fluctuations of intensity (amplitude squared) don't just remain random fluctuations of intensity only; they produce fluctuations of refractive index which then produce (after a cycle or two of damped relaxation oscillation) an amount of phase fluctuation that adds incoherently to that already present from the more classical model. This coupling between the process of amplitude change and phase change is embodied in the quantity β. Quantum-well devices have a linewidth that is about a third that of equivalent bulk devices, because of a higher dg/dN.

Spontaneous-emission components cause amplitude fluctuations which, when converted to phase noise, act to broaden the linewidth. For this reason, it is necessary to minimize any light reflected back into the laser from splices, connectors, etc., since these will act to increase the level of spontaneous emission.

As we shall see in Chapter 8, for some forms of detector the preoccupation is with linewidth and phase noise, while for others it is with the amplitude fluctuations. The quantity *relative intensity noise* (*RIN*) is a form of noise-to-signal power ratio, defined as the variance of the light-output amplitude divided by the average power. As a spectral power density in watts per hertz, *RIN* is usually expressed in dB/Hz.

For systems whose receivers are particularly sensitive to intensity noise (for example the coherent receivers of Section 8.10 or the subcarrier modulation systems of Chapter 9), it may be necessary to provide isolators (Section 3.13). These can attenuate reflections from splices and connectors by as much as 50 dB. From experiments reviewed in [17], *RIN* may be expected to be as low as -150 dB/Hz if reflections are kept below $-50-60$ dB, but as bad as -110dB/Hz for reflected power around -20 dB.

Laser linewidth can be greatly reduced by the use of an external cavity, as will be described in Section 5.9.

5.6 Transient Effects and the Rate Equations

In recent years, as long-haul time-division links between telephone central offices have been going to higher and higher speeds, and as interest has increased in dense wavelength-division links and networks, more and more attention has been given to the modulation characteristics of laser diodes, particularly any undesired effect that will spread the transmitted spectrum much further than a fraction of the modulation bandwidth. What that fraction is will be considered in Section 8.13.

The discussions of Figures 5-22 and 5-23 have already introduced many of the transient phenomena that occur with high-speed on-off keying (OOK). In particular, we have seen that when a laser is on-off modulated, the side-mode content is much worse than the several percent steady-state side-mode level of the CW case, (1/*MSR*) of Equation 5.28. The additional transient effects that must be discussed include

- **Turn-on delay**. If the diode is biased too low, for example, below threshold, then when I suddenly goes well above I_{th}, it takes some time before there is a great enough density of injected carriers $N(t)$ to start the stimulated emission process. Since the onset of lasing is triggered by spontaneous emission, the turn-on time is a random variable.

- **Relaxation oscillations**. Once stimulated emission commences, it steals carriers from the overall $N(t)$, the light level drops, carriers build up sufficiently for strong light emission to recommence (Figure 5-22(B)), and so forth in a ping-pong effect between photon density and carrier density. This oscillation eventually dies out, steady-state carrier density given by Equation 5.10 and Figure 5-22(C) persists, and radiation tends to be in that cavity mode whose frequency is nearest the gain peak.

- **Instantaneous (transient) chirp.** Instantaneous chirp is a direct manifestation of relaxation oscillation. We have mentioned before how the frequency of the Fabry-Perot cavity modes depend on index, which in turn depends on carrier density. As the latter swings back and forth during relaxation oscillation, this produces corresponding frequency modulation of the laser output until the frequency eventually settles down to a constant value.

- **Adiabatic chirp.** This final constant value depends on pump current. With growth of the carrier density, there is a bulk frequency shift due to index change. So, for example, changing the bias current between two current values, both above threshold, produces a frequency shift. This is often used as a simple way of doing frequency modulation or a modest amount of tuning.

- **Mode hopping.** When there is more than one cavity mode with frequencies lying very near the top of the gain curve, random fluctuations of various kinds can cause one mode to capture the gain for a while and then another. A high *MSR* does away with this effect.

- **Thermal effects.** At lower modulation frequencies, below about 10 MHz, heating and cooling can keep up with the tiny thermal time constant of around 0.1 μsec typical of the laser-diode structure.

Let us dispose of the last of these effects first. Figure 5-25 schematizes the *FM response* of a typical laser diode [18]. The frequency deviation Δf is defined as the observed width in GHz/mA of the frequency swing as the laser frequency is deliberately frequency-modulated by sinusoidal modulation (at frequency F) of the injected current I. At the lowest frequencies, thermal effects dominate. As the injected current increases and decreases, the laser is heated and cooled, whereupon, as Equation 5.9 for

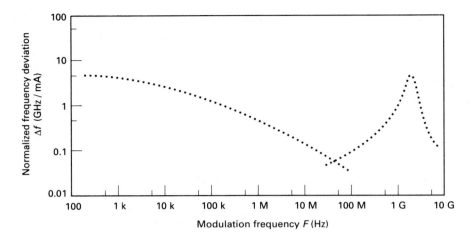

Figure 5-25. FM response of a typical laser diode. (From [18], © 1982, IEEE).

the Fermi distribution says, larger and then smaller values of $\mathcal{E}$ are favored, respectively.

As F gets too high for heating and cooling to keep up, there is a gradual roll-off of the amplitude of the frequency swing. The frequency swing is now due only to decrease of index with increase of carrier density. At extremely high frequencies, F resonates with the relaxation frequency, and the curve rises again. The relaxation oscillations [3] have a frequency

$$f_o \;=\; \frac{m}{4\pi\tau_e} \sqrt{\frac{4\tau_e(m-1)}{m^2\tau_p} - 1} \tag{5.33}$$

and a decay time

$$\tau_d \;=\; \frac{2\tau_e}{m} \tag{5.34}$$

where m is the ratio of I/I_{th} , τ_e is the carrier lifetime, and τ_p is the photon lifetime. The relaxation frequency f_o lies typically in the range 1 to 4 GHz. Above $F = f_o$, the radiated frequency is constant; nothing in the laser can keep up with F. As a practical matter, the upper modulation-frequency limit of a laser diode is as likely to be set by the inductance of the electrical connection and the parasitic capacitances as by f_o, although careful attention to such details, as well as to f_o , has allowed commercial laser diodes to operate at modulation bandwidths exceeding 10 GHz.

All the effects shown in Figures 5-23 and 5-25, the relaxation frequency and the time constant, as well as the explanation of phase noise embodied in Equation 5.32, and even the *P-I* curve, can be derived from a single pair of coupled equations, the

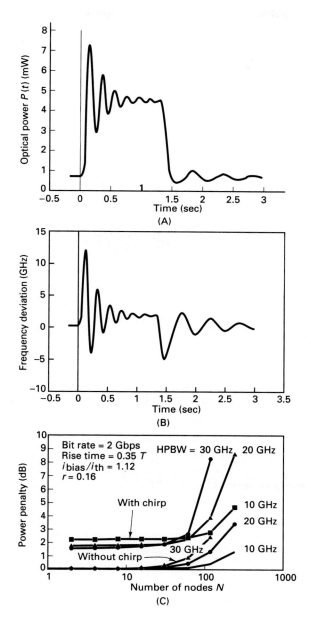

Figure   5-26. The effect of chirp on a typical ASK link consisting of a laser diode, a 50 km. link, doped fiber amplifier, Fabry-Perot filter and photodetector.   (A) Optical power.   (B) Corresponding optical frequency deviation, (C) Optical power penalty vs. number of stations for various filter bandwidths, with and without chirp. (From [20], © 1991, IEEE).

rate equations, one for the time derivative of the carrier density N, and the other for that of the photon density Π [4, 19] .

The carrier-density rate equation is

$$\dot{N}(t) = \frac{I(t)}{qLwd} - \frac{N(t)}{\tau_e} - \frac{\Gamma ac}{n} [N(t) - N_o] \Pi(t) \tag{5.35}$$

where Π is the photon density in the cavity, I is the pump-current input, n is the effective refractive index within the cavity, q is the electronic charge, a is the gain constant, representing the rate at which the gain per unit length grows with the carrier density N, and L, w, and d are the cavity dimensions. N_o is carrier density at transparency, and τ_e is the carrier recombination lifetime. Γ is the confinement factor of the transverse optical modes.

The first term on the right side simply gives the number of carriers being generated per unit volume by current injection. In the second term, carriers of density N are being taken away by spontaneous emission (and some nonradiative processes) at a rate $(1/\tau_e)$ per carrier. The third term expresses the depletion of carrier density by the radiation of useful stimulated light.

The photon power-density rate equation is

$$\dot{\Pi}(t) = \frac{\Gamma ac}{n} [N(t) - N_o]\Pi(t) - \frac{\Pi(t)}{\tau_p} + \frac{R_{sp}}{Lwd} \tag{5.36}$$

where R_{sp} is the same as with Equation 5.28. The first term on the right is obviously the rate at which carriers are being turned into useful photons (last term from the carrier rate equation, but with the sign reversed). The second term expresses the rate at which photons are being lost to the external world, either usefully as emitted light or by any of several loss mechanisms. The third term expresses the spontaneous-emission contribution. These contribute negligibly to the useful light at useful output levels, so are represented here as a constant, independent of Π.

From these equations, or the more detailed forms that take into account other slight dependencies, essentially everything about the static and dynamic behavior of the device can be deduced, including the *P-I* curve Equations 5.11 through 5.15. For example, the value of the threshold current (Equation 5.11) follows immediately from Equation 5.35 by setting both the time derivative and the photon density Π to zero.

The quantity of practical interest is the output power P, which is related to the (spatially uniform) photon density Π by

$$P(t) = (hfc/n) (1 - R) wd \, \Pi(t) \tag{5.37}$$

where R is the reflectivity of the output facet. This equation can be explained as follows. ΠLwd is the number of photons in the cavity. Each photon carries hf joules of energy, and $((1 - R) \times c/nL)$ is the rate at which the photons escape through the output facet, because the photons are travelling at velocity c/n, and it takes Ln/c seconds to lose a fraction $(1 - R)$ of them.

A complete characterization of laser behavior from the rate equations is complicated by the fact that the stimulated photon emission is shared across the several Fabry-Perot cavity modes. We shall break off the discussion of these questions here after simply showing how they explain the turn-on time seen in Figure 5-23 and also chirp. Detailed explorations of the rate equations and their consequences are given in [3, 4, 19].

From the carrier rate equation 5.35 it is seen that the instant that I goes from 0 to a large value, N starts a linear ramp increase (first term) that begins to bend over increasingly as soon as there are more and more carriers that can emit spontaneously (second term). Finally, stimulated photons are strongly emitted (third term) and the carrier density clamps. As for the delay for this to happen, Equation 5.35 says that, starting from zero light output, nothing but a slow increase of spontaneous emission will occur (third term), and even some of that is radiated away (second term), until the first term takes over. The condition for this to happen is the threshold condition. In other words, not until carrier density N reaches N_{th} (Equation 5.13) is there any significant stimulated light output. It is seen that spontaneous photons play an important role in getting the process going.

An expression for the amount of chirp as a function of time can be derived from the two rate equations to be [4]

$$\Delta f(t) = \frac{\beta}{4\pi} \left[\frac{\dot{\Pi}(t)}{\Pi(t)} + K\,\Pi(t) \right] \tag{5.38}$$

where β is the linewidth enhancement factor, and K is a constant. The first term is the dynamic chirp (and therefore present only during the relaxation oscillations) and the second term is the adiabatic chirp.

Another transient effect must be at least mentioned, namely the heating and cooling of the active region when ASK is used and there is a long string of "1"s or "0"s in the data stream, respectively. The magnitude of the effect can be deduced from Figure 5-25. Fortunately, almost any digital communication equipment today has run-length-constrained line codes (usually built in to the front-end electronic chips), which artificially add overhead bits at the transmitter and delete them at the receiver, these bits forcing a significant break in what would otherwise be long strings of "1"s or "0"s.

There have been several analytical and simulation studies of laser transient effects, either focussing only on the laser itself, treating the combination of laser and fiber, or, in one case [20], the complete cascaded connection of laser, fiber, doped-fiber amplifier, Fabry-Perot optical receiver filter and photodetector. We conclude this discussion of transient effects by presenting several results from this last-named study. The device parameters assumed were typical of a commercial DFB laser and photodetector-amplifier. The optical filter was a Fabry-Perot filter whose bandwidth was an adjustable parameter.

Figure 5-26(A) and (B) show the laser output optical power P and the optical frequency as a function of time when the input was a 3.0 Gb/s signal consisting of four "1" bits preceded and followed by a long string of "0"s. The dynamic chirp and the adiabatic frequency offset are clearly visible in (B). Figure 5-26(C) summarizes the complete system results in terms of the number of channels that can be crowded within one free spectral range of the Fabry-Perot filter when the crowding is limited by chirp. The FSR was assumed equal to a doped-fiber amplifier bandwidth of 30 nm. The ordinate is the detectability penalty in dB, that is, the number of dB the signal-to-noise ratio would have to degrade to produce an equal effect on the bit error rate. Results are given both with and without chirp being present, and it is seen that about 100 channels at 2 Gb/s can be accommodated before some measure to reduce the chirp has to be introduced.

5.7 Single-Frequency Laser Diodes

Up to this point, we have explored the things that go on inside a laser when the passive resonant structure consists only of the mirrors of a simple Fabry-Perot cavity, such as that which began our discussion of tunable filters in the last chapter.

The term *single-frequency laser diodes* is applied to the modification of the basic two-facet Fabry-Perot laser structure we have been discussing to somehow filter out all the side modes by an amount greater than the factor of several tens afforded by the side-mode suppression effects discussed earlier.

The remainder of this chapter will focus on the different types of resonant structures and deal no longer with the physics of emission inside these structures. To a first approximation, most of the characteristics we have been discussing (*P-I* curve, threshold condition, temperature and pump dependencies of mode frequencies, and even linewidth) carry over to the cases where further resonances are added to suppress side modes, to decrease linewidth, and to widen the tunability beyond what can be done by altering the pump current or the temperature.

There are two basic approaches to fashioning a single-frequency laser: to couple the active region to a resonant structure that is either outside it or effectively inside it. In the next section the external-cavity approach will be discussed, and attention will be devoted meanwhile to the two principal forms of single-frequency semiconductor laser diode that imbed the frequency-selective mechanism inside the device, the DBR (distributed Bragg region) devices and the DFB (distributed feedback) devices. The DFB laser is the form of single-frequency laser diode that is in the widest use today. The device shown in Figure 5-3 is a DFB laser.

Consider the dielectric slab of Figure 5-27(A). Along either the top or the bottom of a length L_g of the dielectric region is a corrugation that forms a grating of pitch Λ length per cycle, whose strength depends on the contrast of indices and on the lithographed physical depth. Recall, from Chapter 3, that there are two permitted solutions of the wave equation, two waves propagating in opposite directions. A sideways grating such as the one shown has the property that it will couple one into the other (i.e., act as a mirror) if the pitch of the grating and the wavelength of the light match up properly. At other light wavelengths, successive wavelets that are scattered from the corrugations do not add up constructively.

Alternatively, if the light makes just one pass through the grating, it can be considered a filter, which is just the situation we met earlier in connection with the tunable acoustooptic and electrooptic filters of Chapter 4. There we found that the outlines of an elaborate coupled mode analysis could be understood in terms of the momentum triangle (Bragg triangle). Using that same artifice again, one can represent ψ^+, the wave travelling in the $+z$ direction, by the k-vector (in radians per unit distance) $k = (2\pi p/h) = 2\pi n/\lambda$, and the counterpropagating one ψ^- as the oppositely directed vector. The Bragg condition requires that momentum has to be conserved by forcing the momentum triangle to be closed. This resonance condition

$$\Lambda = \lambda/2n \tag{5.39}$$

is depicted in Figure 5-27(B). The grating pitch Λ must be twice as tight as that which matches the light wavelength in the medium λ/n.

The coupling of ψ^+ into ψ^- is shown in Figure 5-27(C) for the case where there is no gain in the region of the grating, and in Figure 5-27(D) where there is. These two situations correspond to the DBR and DFB lasers of Figures 5-27(E) and (F), respectively. In the DBR case, the gain is outside the grating, so ψ^+ gradually attenuates, while coupling into ψ^-, which is then strongest near the entrance at $z = 0$. In other words, the grating forms a frequency-selective mirror at an effective distance $z = L_{dbr} < L_g$. In the DFB laser diode, where the grating is actually along the entire length L of the active region, a wave of the right frequency in the $+z$-direction gets amplified and so does one in the $-z$-direction.

In both these cases, the transfer function in frequency can be gotten by Fourier transforming the impulse response in time. Therefore, the power transfer function should be (within a constant multiplier)

$$T(f) = \frac{\sin^2 \pi f \tau}{(\pi f \tau)^2} \tag{5.40}$$

in which $\tau = L'/v_g$, where L' is either $L_g = L$ for the DFB case or $L_{dbr} < L_g$ for the DBR case. However, matters are in reality more complicated than this; the index contrast and lithographed depth of the grating must be taken into account too. Equation 5.40 is accurate only when the grating is infinitely shallow and of small index contrast. For the more realistic case, the successive scattering centers represented by the steps in the grating gradually weaken the signal exponentially so that the ideal $\sin^2 x/x$ curve

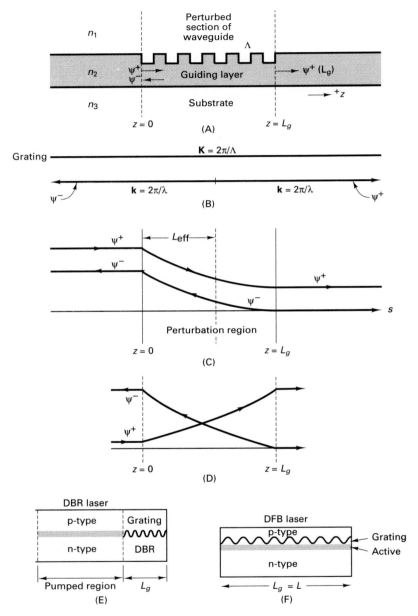

Figure   5-27. Grating structures for single-frequency lasers.   (A) Grating in a slab waveguide. (B) Bragg momentum triangle.   (C) Field strengths of forward and counterpropagating waves; no gain.   (D) Same with gain (E) Structure of a DBR laser diode. (F) Structure of a DFB laser diode.

gets broadened somewhat by being multiplied by the Fourier transform of a decaying exponential, namely a Lorentzian curve [15].

There is one complication concerning the DFB laser, in which the grating is used as a resonant filter, not a mirror. In this situation, the light makes makes many passes back and forth through the grating region. If a grating is used either as a single mirror (as in the DBR laser) or as a simple filter (where the light makes one pass through the device, as in the acoustooptic and electrooptic filters of Chapter 4), it does not need to be modified from the form shown in Figures 5-27(A) and 5-28(A). However, if it is to be used as a resonant filter (either with or without gain) so that there are many passes back and forth through the grating, in order to have maximum transmission occur exactly at the Bragg resonance, the grating must be modified by adding a $\lambda/4$ phase shift, as shown at (B).

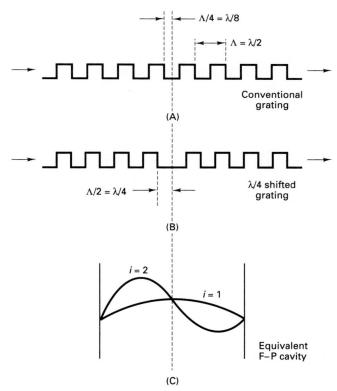

Figure   5-28. Grating resonator with a $\lambda/4$ shift compared with a conventional grating. (A) Conventional periodic grating. (B) $\lambda/4$ shifted grating. (C) Equivalent Fabry-Perot mirrors.

This can be seen as follows: To use it as a resonator is equivalent to considering it as two mirrors. (They do not form a true classical Fabry-Perot cavity, since they are narrowband mirrors.) This in turn means that an observer at the center (dashed line) looking to the right must see all the little wavelets from the positive-going grating steps add coherently. Looking to the left he must see the same thing, and both the left and the right composite ones must arrive at him in phase with one another. In the classical Fabry-Perot cavity of Figure 5-28(C), the observer in the middle is i times 90° from either mirror (i being the order), but in (A) he is seen to be 45° from either mirror. The fix of (B) puts him back at 90°.

Thus, the resonator of Figure 5-28(A) will not resonate exactly at the peak value of Equation 5.39 that is given by Equation 5.38, but at one of two frequencies slightly on either side of the grating peak. Therefore, a DFB laser made with an uninterrupted grating may exhibit sporadic frequency hopping between these two modes, and this can happen whether only the grating forms the two mirrors (low facet reflectivity), or the grating acts as a filter of the facet Fabry-Perot modes (high facet reflectivity).

Today's commercial DFB lasers may use the quarter-wave-shifted grating, but often there are enough asymmetries to cause one of the two side modes to be slightly stronger (although not as strong as would be the case if the trouble had been taken to include a quarter-wave-shifted grating); those lasers that do not have enough asymmetry are discarded early in the manufacturing process.

Using either quarter-wave-shifted gratings or component selection, plus the right choice of the strength of the grating, DFB lasers with sidemode suppression ratios *MSR* of 30-35 dB and linewidths of 5 MHz are routinely available. Again, reflections into the laser from couplers, splices, etc. have to be carefully controlled or the same undesired spontaneous emission that was of concern for linewidth will reduce the *MSR* as stated in Equation 5.28.

Complete coupled mode analyses of DFB and DBR structures are contained in [4], [19], and [3].

5.8 Tunability Review

We shall focus on tunability for the remainder of this chapter. To review a few numbers:

- At 1.3 or 1.5 μ, the spectral width of the fiber between points where the attenuation in dB per km doubles is 200 nm (25,000 GHz), which is a 13 percent fractional bandwidth $\Delta f/f = \Delta\lambda/\lambda$ at 1.5 μ.

- The spectral width of the gain curve of an InGaAsP/InP bulk laser is about 35 nm, some 2.3 percent.

- A gain-curve width of 240 nm at 1.5 μ is available using a strongly pumped quantum-well laser, about 16 percent.

- Index change by current injection leads to only a maximum 1 percent $\Delta n/n$. $d\lambda/dn$ is negative. This effect is identical to the adiabatic chirp mentioned earlier in Section 5.6. $d\lambda/dI$ varies between devices, but is typically -1 nm/mA [21].

- Temperature tuning can be used, but the sinusoidal tuning rate (or frequency modulation by injection current variation) is limited by thermal inertia to no more than 10 MHz (Figure 5-24). The fractional tuning range is $d\lambda/dY = -1.0 \text{ Å}/°C = -0.1 \text{ nm}/°C = -13 \text{ GHz}/°C$ at $\lambda = 1.5 \mu$. It is probably not advisable to temperature-tune a laser diode over much more than $\pm 10°C$ because of lifetime considerations at the upper temperatures and loss of output (Figure 5-13) at the lower temperatures, thus giving a total practical temperature-tuning range of some 2 nm at the very most.

- As we shall see in the next chapter, doped-fiber amplifiers are available at 1.5 μ using erbium, and they have a bandwidth of about 35 nm. Laser-diode amplifiers have about the same bandwidth of the gain spectrum, but are available at both 1.3 and 1.5 μ. Gain-curve broadening of quantum well structures by heavy biasing is a potential option for laser-diode amplifiers but not for doped-fiber amplifiers.

It is clear that what system architects and implementers need ideally is a single-frequency laser diode that can span up to 200 nm and is rapidly tunable. As mentioned in Section 1.6, until fast-tunable filters become available, tunable lasers are the only known practical method of getting submicrosecond tuning times, which is the key to gigabit fast packet-switching, to be discussed in Chapter 13.

In the remainder of this chapter we shall discuss what has been achieved so far in providing tuning over as much of the bandwidth of the available gain curve as possible. An easily understood review of this area, as it existed in 1988, is available in [22].

5.9 External-Cavity Tunable Lasers

The first thing we would think of in order to make a laser tunable over a wider range than can be provided by temperature tuning or index tuning is to add an external tunable filter in place of one of the mirror facets. This can be done by providing an *antireflection (AR) coating* on one facet of a Fabry-Perot laser diode and then putting any of the tunable filters that have been discussed in Chapter 4 between the active laser region and some external mirror reflection. Alternatively, the filter itself can be used in a mode where it is the reflection rather than the transmission that is narrowly frequency-selective.

Both these classes of structures are referred to generically as *external-cavity* lasers, a slight misnomer because the external structure can be something other than a simple Fabry-Perot mirror. In fact, if the laser diode itself is a Fabry-Perot laser (without any internal frequency-filtering structure), the external structure must be more selective than a simple mirror, since the spectral lines would otherwise be spaced too closely to give adequate side-mode suppression and to prevent mode hopping.

There is a second important reason for adding an external filter in place of one of the mirror facets, namely to reduce the linewidth. A fixed external reflector or

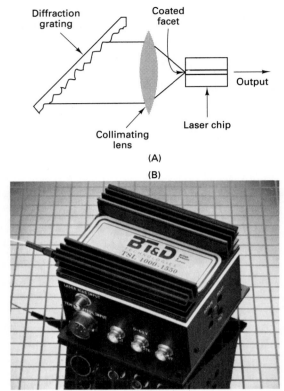

Figure   5-29. External cavity laser diodes using rotatable gratings (A) Basic structure, (B) A commercially available external-cavity tunable laser (Courtesy BTD Technologies).

tuned filter is often used for that purpose alone. The replacement of one facet with an external reflector reduces the width of any line by a very large factor because the effect goes as the *square* of L_t / L, where L_t is the overall effective cavity length and L is the length of the active region [23]. Two factors contribute to this decrease of linewidth (Equation 5.30). First, the fraction of the overall effective laser length L_t over which spontaneous emission acts is L / L_t. Second, the fictitious quantity α_m that prorates the mirror loss on a per-unit-length basis (Equation 5.14) mathematically has an L_t in the denominator. Linewidths of a few kilohertz have quite frequently been obtained using external gratings. This is one part in 10^{11}. Without the external grating the linewidth would have been three orders of magnitude higher.

Returning to the tunability problem, we must define several forms of tunability.

- *Overall tuning range*, the difference between the highest and lowest limit of tuning,

- *Continuous tuning range*, the width of the largest range all of which can be reached, and

- *Range of continuous control,* the range over which all control currents and voltages vary monotonically without jumps.

Accepted (but misleading) practice is to say that if the device can be tuned somehow to any wavelength between two limits, then it is "continuously tunable" over that range. This does not necessarily mean that in order to tune over this range the several currents or voltages applied to the device vary monotonically with no jumps. In fact, it is often necessary to have discontinuous control in order to achieve continuous tuning.

Many different forms of external filters have been employed. The one most widely used is the tiltable grating, used as shown in Figure 5-29(A), its limitation being the slow tuning speed of seconds imposed by the large mass that must be moved. Using external gratings, tuning over the entire 240-nm range of a strongly pumped multiple-quantum-well laser has been demonstrated by several groups, including the group that devised the laser with the gain curve shown in Figure 5-21 These devices have continuous tuning, but discontinuous control, because of mode hopping. [12]. A packaged commercial external-cavity laser diode using a grating is shown in Figure 5-29(B) [24]. Using a screw adjustment, it has a continuous tuning range of 50 nm, a range of continuous control of 0.4 nm, and a linewidth of 60 KHz.

External-cavity tunable lasers have tuning-speed problems, due only partly to the fact that in order for the laser active region to see a new filter setting, several round-trip propagation times must elapse. The more difficult problem concerns the tuning speed of the external filter. As we have seen from Chapter 4, rapidly tunable filters that resolve many channels have proved difficult to find. For example, rapid (10-microsecond) tuning over a 7.5-nm range has been achieved [25] by use of the electrooptic tunable filter of Figure 4-26, but the selectivity of the filter is fairly poor, as was discussed in Chapter 4.

5.10 Two-Section Tunable DBR Laser Diodes

The search for rapid (submicrosecond) tunable lasers led first to modifying the DBR laser of Figure 5-27(E) so that the Bragg mirror could be tuned by a separate current injection. Later, further modifications were introduced.

In the two-section device, separate electrodes carry separate injection currents, one for the active region and the other controlling the index seen by the Bragg mirror. Thus, the active section pump I_a serves to control the radiated power P, and the Bragg section current I_b controls the wavelength λ.

If we had tried to use variable pump current into a DFB laser, some tuning would have occurred, but the power-output variation would render this solution relatively useless.

We shall start with the two-section DBR laser and sketch out historically how more sections were added to this basic structure to improve its performance. The starting premise of all multisection DBR laser diodes is that, by designing the Bragg region to have a large waveguide thickness, and by heavy injection of current, we may shift the Bragg wavelength by as much as 9 nm at 1.5 μ, or 0.6 percent [22]. All the work on tunable DBR lasers having two or three sections has been aimed at achieving continuous tunability across this range. The four-section and phase-matching approaches, to be described later, attempt to exceed even this limit by entirely different approaches.

The two-section device provides a simple example in which the overall tuning range is interestingly large, but the continuous tuning range is not. Figure 5-30(B) shows the situation. The cavity length $L_t = L + L_{dbr}$ is sufficiently long that there are several Fabry-Perot cavity modes v_g/L_t Hz apart within the bandwidth $2\pi/\tau$ Hz of Equation 5.40, where $\tau = L_{dbr}/v_g$. As the current I_b is varied, mode jumping is observed when a new Fabry-Perot mode is nearest the peak of the Bragg mirror's reflectivity as a function of frequency. This is clearly seen as a void in the frequency coverage of Figure 5-30(B) at the point where the resonance jumps to another mode.

For this two-section device, we see that the continuous tuning range, which equals the range of continuous control, is 1 nm. The overall tuning range is 3 nm, far short of the 9-10 nm potentially available.

5.11 Three- and Four-Section Tunable Laser Diodes

Several years ago, it was realized that it would be possible to fill in the gap in the continuous tuning range of the two section DBR laser by adding a third section between the active and Bragg regions. Recall that to tune a Fabry-Perot cavity over an entire free spectral range (spacing between the modes), it is only necessary to change the effective length of the cavity by $\lambda/2$. By introducing the third *phase-shift section*, and then varying its injected current I_p, one can vary the phase of the wave incident on the Bragg mirror section so as to cover the gap that otherwise exists in the two-section arrangement when there is no cavity mode for the narrowband Bragg mirror function of Equation 5.39 to resonate with. The phase-shift section provides that function, and all it has to do is vary the overall effective length by at most $\lambda/2$ to cover any gap.

The addition of the phase-shift section lengthens the cavity and thus places the cavity modes closer together. But the gaps in the overall coverage provided by the Bragg section that would otherwise exist can all be reached by tuning the phase-shift section.

As we get further into the complexities of three- and four-section laser diodes, it is probably worth stopping to unify things in a single diagram to keep track of the various interacting factors. Figure 5-31 is intended to perform that function. We decompose the various elements in the same way we did in Chapter 4, where the overall frequency transfer function of some filter structure was gotten by taking the product of the individual transfer functions. In order to keep the depictions of Figure 5-31 simple, we emphasize what the output power looks like, and ignore the phase.

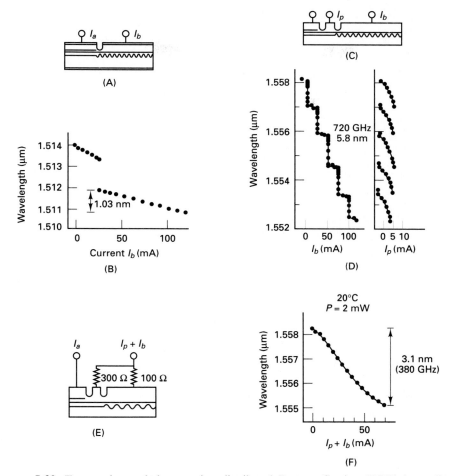

Figure   5-30. Two-section and three-section distributed Bragg reflection (DBR) laser diodes.
(A) Two-section structure.   (B) Its tuning characteristics.   (C) Three-section
structure.   (D) Its tuning characteristics.   (E) Reducing the number of control
parameters from 3 to 2.   (F) Enlarged range of continuous control.   (From [22]
© 1988, IEEE).

The figure starts out with the simple gain curve (A), and proceeds to the simple
bulk or quantum-well Fabry-Perot laser (B), then the DFB laser (C), the tunable DBR
laser (D), the three-section DBR laser (E), and finally the four-section DBR laser (F).
All devices but the last are analyzed in just this way in [22]; the four-cavity device is
described in [26].

The leftmost column lists the devices whose spectra are to be shown.   The
second column diagrams each structure.   The third shows whatever new spectral filtering is introduced that was not there in the diagrams above.   The fourth shows the
resulting spectrum and lists the earlier ones that have combined to produce that spec-

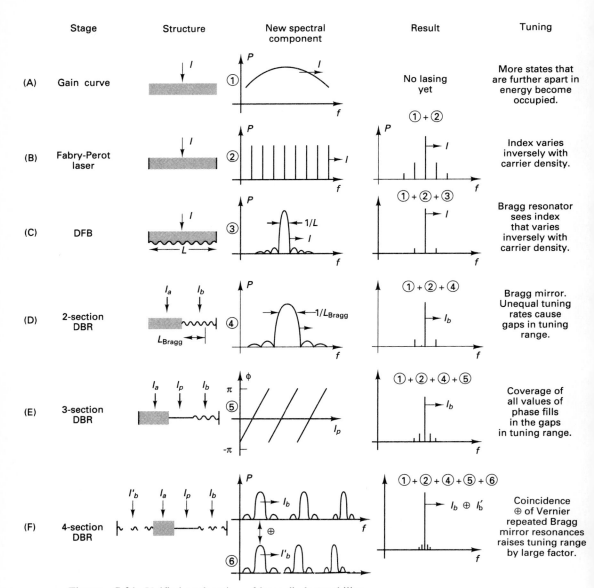

Figure **5-31.** Unified explanation of laser diode tunability.

tral shape. Arrows pointing to the left or right in the spectra show whether an increase of the designated injection current increases or decreases the frequency. The fifth column says in words what is going on.

The gain curve (A) moves in a positive frequency direction with an increase in the current I_a through the active section, and the Fabry-Perot lines in (B) move to the right too (as was described in Section 5.8), but not as fast. Since these motions are in general unsynchronized, as the current I_a increases steadily, the main mode frequency moves to the right for a while and then, when the next higher Fabry-Perot order is closest to the gain-curve peak, it is substituted instead. Thus mode hopping can occur using current tuning of a Fabry-Perot laser. Also, as Figure 5-20 shows, such tuning function is bought at the expense of very large changes in the output power P.

The same mode hopping and sensitivity of P to pump current make pure current tuning of the DFB laser (C) an unattractive choice.

As discussed in the last section, the two-section DBR device of (D), seen earlier in Figure 5-30(A) and (B), is an attempt to isolate the process of light production from that of tuning. As far as index tuning of the Bragg section is concerned, the diagram in the third column shows behavior similar to that of Figure 5-31(C), except that the frequency selectivity of the grating section is not nearly so great, for two reasons: First, the DBR grating does not occupy the entire length L occupied by the DFB grating, and, second, the light penetrates only partway into the grating before being reflected. As discussed before, there are gaps in the coverage where there is no Fabry-Perot mode to resonate with the Bragg section.

This is remedied by the three-section DBR laser of (E), and such devices have proved capable of a continuous tuning range of almost the full 9 nm [22]. Figure 5-30(C) shows the device and (D) the tuning behavior of an early version. The continuous tuning range of 5.8 nm is clearly seen. The developers of this device determined empirically that keeping I_p and I_b proportional using the resistor network of Figure 5-30(E) allowed the Bragg resonance to track the Fabry-Perot mode over more than 3 nm before jumping occurred, as shown at (F).

Figure 5-31(F) summarizes the principles of the four-cavity tunable laser diode due to Coldren [26]. It uses the vernier idea introduced in Section 4.3 to get very large effective free spectral range from a cascade of two Fabry-Perot filters with different individual free spectral ranges. The idea is to cause the Bragg mirror function of Equation 5.40 to occur not just once, centered around a frequency determined by Equation 5.39, but at a number of periodically spaced places on the frequency axis, as shown in the third column of Figure 5-31(F). This serves in place of the original Bragg region, and the position in frequency of this comb of replicas of Equation 5.40 varies with its injection current I_b. The other element in the vernier scheme is a second Bragg mirror, driven by current $I_b{}'$ and placed at the other end of the device.

Figure 5-32 shows how this Bragg-mirror periodicity is effected. At (A) is the effective depth $d_g(z)$ of the original Bragg section as a function of spatial coordinate z. Thinking of this, as before, as mapping over into an impulse response, (B) shows the power transfer function in frequency $T_g(f)$, i.e., the squared magnitude of the Fourier transform of $d_g(t)$. At (C) is the effective depth $d_{sq}(z)$, a (0,1) squarewave function of position z, and at (D) its transfer function $T_{sq}(f)$. Etching out the periodic Bragg grating in this (0, 1) pattern is equivalent to multiplying (A) and (C) in the delay domain. The equivalent in the frequency domain is the convolution of $T_b(f)$ and $T_{sq}(f)$, which is shown at (E). In this way, the Bragg function of (B) is made to repeat

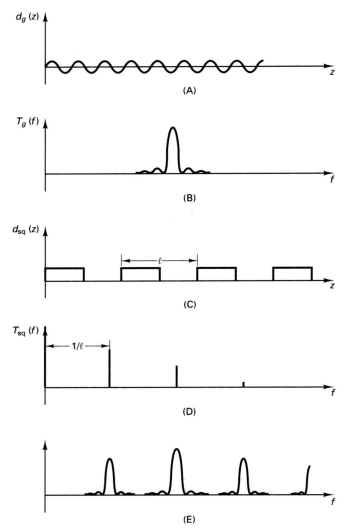

Figure 5-32. Explanation of four-section tunable laser diode. (A) Bragg grating. (B) Its reflectivity as a function of frequency. (C) Periodic interruption of the grating. (D) Frequency transfer function of the interruption pattern. (E) Final result.

itself, the repetitions falling off slowly in power to either side of the central one with a width in frequency proportional to $1/\ell$, where ℓ is the period of the squarewave.

The first and fourth sections of the four-cavity device of 5-31(F) use two different values of ℓ, the squarewave period. As the third-column figure shows, there is only one frequency at which peaks of the two frequency functions coincide, and this coincidence can be tuned (with jumps) over a very wide range, even when each Bragg region can tune over only a much smaller range. It is only necessary that ℓ and the ratio ℓ / ℓ' be properly chosen, and that each Bragg region be able to tune over the range of frequencies between the periodic teeth of the other.

Early versions of the four-section tunable laser have recently achieved a tuning range of 30 nm [27].

5.12   The Phase-Matched Tunable Laser Diode

As we have seen, the fact that index can be tuned over less than one percent by current injection poses a limit on tunability range. This is analogous, in some sense, to the one percent upper limit of piezoelectric dimension change met earlier in Chapter 4 on tunable filters. Analogously to the piezoelectric case, is there a way we can apply the tuning-range multiplication principle to tunable lasers? In other words, is there a scheme by which the emitted wavelength of the device can be made to depend on the *difference* in the indices of two current-injected regions, rather than on an index itself?

The answer is yes; a phase-matching filter can be inserted within the device itself, as shown in Figure 5-33 [28], which depicts a proposed such device. In this case, the filter does phase matching between two regions having different indices n_1 and n_2. Its function is analogous to the acoustooptic tunable filter of Section 4.9 and Figure 4-19(A), which used an acoustically produced grating to couple light from an input to an output if and only if the one-dimensional momentum triangle of Figure 4-19(B) was satisfied. Here again, the collinear version of the momentum triangle must be satisfied by an analogous version of the Equation 4.38 describing the one-dimensional triangle

$$k_1 - k_2 \;=\; K \qquad \text{or} \qquad \lambda = \Lambda(n_1 - n_2) \qquad (5.41)$$

Current I_a controls the light output from the active region, and produces an index n_2 in the region around the gain region. I_b controls the index n_1 seen by the grating of period Λ.

Since mirror facets are necessary on the ends in order that the device lase, mode hopping would result if the phase-shift region were not provided. Phase is controlled by I_p, as with the three- and four-section DBR lasers.

A tuning range of 57 nm has been achieved recently using a phase-matched tuning laser [29], which makes this component, as of this writing, the most attractive possibility so far for optical packet switching.

5.13   Switched Surface-Emitting Arrays

All the devices that have been discussed are edge emitters, as in Figure 5-4(A) through (C). Surface emitters, as depicted at (D), have many interesting properties, of which

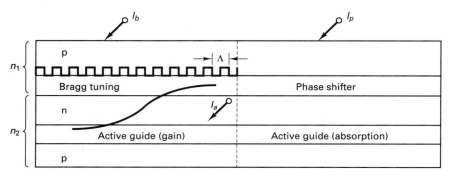

Figure   5-33. Structure of a phase-match tunable laser diode. (From [26]).

two are particularly important in the present context. First, since the cavity is so short (several tens of microns maximum compared to several hundred microns), side-mode problems are virtually absent because the cavity resonances are so far apart. Second, the circular geometry and flexibility in choice of diameter make the problem of matching the numerical aperture of the fiber much easier. Third, their small size laterally and the fact that they emit vertically allows them to be fabricated in dense arrays.

Figure 5-34(A) shows a top view of a 7×11 array of 20-μ-diameter stagger-tuned surface-emitting laser diodes on 354-μ centers fabricated on the same wafer [30]. The measured wavelengths are shown at (B).

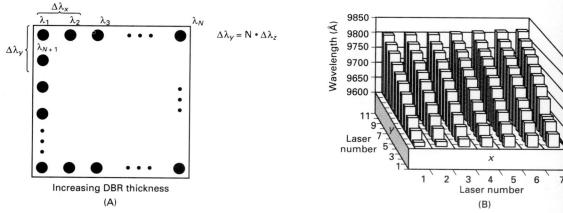

Figure   5-34. Single-chip laser array. (A) Top view, (B) Radiated wavelengths of the various elements of the array. (From [30]).

Such devices may be used to form a tunable laser by activating only one laser at a time. The tuning time is then determined by the speed of electrical switching from one laser to another. Unfortunately, the large physical size of the arrays of this type that have been devised so far makes coupling of the light from all possible active lasers quite difficult.

5.14 LED slicing

Earlier we mentioned that LEDs seem completely unsuitable for any communication application in which narrow source bandwidth was required, since the output spectrum is essentially the gain curve, which is very wide.

There is one circumstance, however, in which an LED can be useful in narrowband applications, provided one can tolerate very low transmitter power levels. By placing a narrowband optical filter after the LED, and tuning the filter to some frequency within the gain curve, a quasi-monochromatic signal can be obtained. As mentioned in connection with Figure 5-2, the output power available from an LED is fairly small, around 100 mW at most, and when only a small slice of this is used, the power transmitted is many dB below that which would be afforded by a DFB laser, for example. However, LEDs are fairly cheap.

In addition to their low output, there is a second problem, namely the random nature of the signal, which exactly a penalty in signal detectability [31] that would not be present for narrow linewidth sources.

Nevertheless, for short-distance WDM applications in which the number of wavelengths required is not large, LED slicing has proved to be a practical alternative in several cases. One of these will be described in Chapter 16.

5.15 What We Have Learned

In this chapter we have learned quantitatively how the injected electrons can be converted into carriers (electron-hole pairs) so that when a suitable highly reflective facet is provided at each end of the active region, a strong coherent-light output is produced by stimulated emission. By adding some further sort of resonance to the basic Fabry-Perot resonant character of the cavity between the facets in order to strongly select out one of the cavity resonances, essentially single-frequency operation can be achieved. It has proved possible even to make this single frequency tunable over useful wavelength ranges. Because of the small dimensions of laser diodes, this tuning can be extremely rapid, but usually has to be managed by adjusting several variables simultaneously and in a discontinuous way.

We learned that on-off keying of a laser diode produces undesirable transient frequency-chirping effects, and that even with CW operation of the device, the presence of some level of undesired spontaneous emission causes the spectral line of stimulated emission to have a nonzero width, more of the randomness being in the phase than in the amplitude, unlike classical narrowband noise sources.

By delving into the subject of density of permitted energy states, we learned that suitably small-scaled dimensioning of the active region or its subdivision in orthogonal directions leads to quantum confinement in quantum wells, wires, or dots, with many advantages, not the least of which is a broadening out of the wavelength region that can be covered from around 30 nm to over 200 nm. As we shall see in the next chapter, this is of considerable system importance, not only for laser sources, but for photonic amplification as well.

Finally, it has become amply clear from this chapter that the laser diode itself is basically just another semiconductor device, capable not only of very small size but of considerable potential for cost improvement, just as with the other semiconductor devices, those for memory and logic, that have so changed the landscape of computing and communicating.

5.16 Problems

5.1 (*Properties of a typical laser diode*) A long-wavelength laser diode has the following parameters:

Wavelength $\lambda = 1.3\ \mu$
Cavity length $L = 150\ \mu$
Cavity depth $d = 0.2\ \mu$
Active region width $w = 2\ \mu$
Optical mode confinement factor $\Gamma = 0.3$
Cavity index of refraction $n = 3.2$
Facet reflectivity $R = 0.35$
Internal loss per unit length $\alpha = 40 \text{cm}^{-1}$
Gain constant $a = 2.5 \times 10^{-16}\ \text{cm}^2$
Carrier density at transparency $N_0 = 1 \times 10^{18} \text{cm}^{-3}$
Carrier lifetime at threshold $\tau_e = 2.2$ nsec
Photon lifetime $= 1$ psec

Calculate (A) The carrier density at threshold (B) The current at threshold (C) The (external) quantum efficiency (D) The slope of the $P - I$ curve (E) The current density in mA/cm^2 for a light output of 1 mW. (F) How close has this diode, operated at 1 mW, approached the ideal of producing one useful photon per injected electron? (G) What would you be most likely to try to improve this number?

5.2 For a certain InGaAsP/InP laser diode intended for use at $1.3\ \mu$, the mole fraction of indium vs. gallium is 0.2. What should be the relative proportion of arsenic to phosphorus in order that the bandgap exactly matches the energy of the desired laser output radiation?

5.3 Why do you think there is a factor of two in the denominator of Equation 5.12?

5.4 (A) What is the physical significance of the curvature of the $\mathcal{E}-k$ curve? (B) We have shown the curvature of the conduction- and valence-band $\mathcal{E}-k$ curve representations as equal, whereas in actuality the valence-band parabola curves more gently. Quantify the expected cause for the upper curve to be one-half as wide as the lower one.

5.5 Explain why all three curves are parabolas: (A) the $\mathcal{E}-k$ curve, (B) the $\mathcal{E}-k_i$ curve, and (C) the density of permitted states as a function of energy, $\rho(\mathcal{E})$.

5.6 For the laser diode of Problem 5.1, what fraction of the main mode power is in one of the first side-modes, when the device runs at 1 mW output optical power? Assume that the spontaneous-emission rate is 5.3×10^{31} photons per cubic centimeter per second.

5.7 For the laser diode of Problem 5-1, what are (A) the relaxation frequency, and (B) the bandwidth of the resonance for $I = 1.5 I_{th}$?

5.8 Consider various kinds of 1.5-μ laser diodes, all having a linewidth of 1 GHz. Compute N_{max}, the maximum number of WDMA channels for the following cases. Assume that, in order to minimize crosstalk and for engineering reasons of wavelength stabilization, one must separate the channels by at least four times their linewidth. Use the values mentioned in the text for typical laser parameters. (A) The limitation is the bandwidth of silica fiber over which the attenuation is no more than twice its minimum value. (B) The limitation is the width of the laser's or laser diode amplifier's gain curve, if the device is a bulk laser diode. (C) A heavily biased multi-quantum-well device is used either as a transmitter laser or a laser diode amplifier. (D) Tuning is done by varying the injection current to change the cavity index. Also, by what multiple does the threshold current I_{th} vary while tuning? (E) Temperature tuning is used. (F) The system uses erbium doped-fiber amplifiers. (G) The system uses the three-section DBR laser of Figures 5-29 and 5-30(E). (H) Name three important factors overlooked in these calculations that are likely to make these values not easily achievable in practice.

5.17 References

1. J. M. Senior, *Optical Fiber Communications*, Prentice Hall, 1985.
2. J. Gowar, *Optical Communication Systems*, Prentice Hall, 1984.
3. J. T. Verdeyen, *Laser Electronics - Second Edition*, Prentice Hall, 1990.
4. G. P. Agrawal and N. K. Dutta, *Long-Wavelength Semiconductor Lasers*, Van Nostrand Reinhold, 1986.
5. T. P. Lee, "Recent advances in long-wavelength semiconductor lasers for optical fiber communication," *Proc. IEEE*, vol. 79, no. 3, pp. 254-276, 1991.
6. W. Shockley, *Electrons and Holes in Semiconductors*, Litton/Van Nostrand, 1950.
7. P. Zory, ed., *Quantum Well Lasers*, Academic Press, 1991.
8. Y. Arakawa and A. Yariv, "Quantum well lasers − Gain, spectra, dynamics," *IEEE Jour. Quantum Elec.*, vol. 22, no. 9, pp. 1887-1899, 1986.
9. B. G. Streetman, *Solid State Electronic Devices - Third Edition*, Prentice Hall, 1990.
10. B. A. Saleh and M. Teich, *Fundamentals of Photonics*, Wiley, 1991.
11. E. Kapon, "Progress in quantum wire lasers," *Conf. Record, Optical Fiber Commun. Conf.*, February, 1991.
12. H. Tabuchi and H. Ishikawa, "External grating tunable MQW laser with wide tuning of 240 nm.," *Elect. Ltrs.*, vol. 26, no. 11, pp. 742-746, May, 1990.
13. D. Marcuse and T. P. Lee, "On approximate analytical solutions of rate equations for studying transient spectra of injection lasers," *IEEE Jour. Quantum Electronics*, vol. 19, no. 9, pp. 1397-1406, 1986.
14. C. H. Henry, "Theory of the linewidth of semiconductor lasers," *IEEE Jour. Quantum Electronics*, vol. 18, no. 2, pp. 259-264, 1982.

15. T. L. Koch and U. Koren, "Semiconductor lasers for coherent optical fiber communications," *IEEE/OSA Jour. Lightwave Tech.*, vol. 8, no. 3, pp. 274-293, March, 1990.

16. A. Mooradian, "Laser linewidth," *Physics Today*, pp. 43-48, May, 1985.

17. W. I. Way, "Subcarrier multiplexed lightwave system design considerations for subscriber loop applications," *IEEE/OSA Jour. Lightwave Tech.*, vol. 7, no. 11, pp. 1806-1818, 1989.

18. S. Kobayashi, Y. Yamamoto, M. Ito, and T. Kimura, "Direct frequency modulation in AlGaAs semiconductor lasers," *IEEE Jour. Quantum Electronics*, vol. 18, no. 4, pp. 582-595, 1982.

19. A. Yariv, *Quantum Electronics - Third Edition*, Wiley, 1989.

20. C. S. Li, F. F. Tong, K. Liu, and D. G. Messerschmitt, "Channel capacity optimization of chirp-limited dense WDM/WDMA systems using OOK modulation and optical filters," *IEEE/OSA Jour. Lightwave Tech., to appear*, vol. 10, 1992.

21. F. R. Chung, J. A. Salehi, and V. K. Wei, "Optical orthogonal codes: Design, analysis and applications," *IEEE Trans. on Info. Theory*, vol. 35, no. 3, pp. 595-604, 1989.

22. K. Kobayashi and I. Mito, "Single frequency and tunable laser diodes," *IEEE/OSA Jour. Lightwave Tech.*, vol. 6, no. 11, 1988.

23. C. H. Henry, "Phase noise in semiconductor lasers," *IEEE/OSA Jour. Lightwave Tech.*, vol. 4, no. 3, pp. 298-311, 1986.

24. J. Mellis, S. A. Al-Chalabi, K. H. Cameron, R. Wyatt, J. C. Regnault, W. J. Devlin, and M. C. Brain, "Miniature packaged external-cavity semiconductor laser with 50 Ghz continuous electrical tuning range," *Elect. Ltrs.*, vol. 24, no. 16, pp. 988-989, 1988.

25. F. Heisman, R. C. Alferness, L. L. Buhl, G. Eisenstein, S. K. Korotky, J. J. Veselka, L. W. Stulz, and C. A. Burrus, "Narrow-linewidth, electrooptically tunable InGaAsP-Ti:LiNbO3 extended cavity laser," *Appl. Phys. Ltrs.*, vol. 51, pp. 164-166, July, 1987.

26. L. Coldren, N. Dagli, J. Bowers, V. J. Jayaraman, and Z. M. Chuang, "Frequency agile lasers with extended wavelength coverage," *Final Report, Year 2 – IBM Shared University Research contract – Univ. of California, Santa Barbara*, February, 1992.

27. V. Jayaraman, D. A. Cohen, and L. A. Coldren, "Extended tuning range in a distributed feedback InGaAsP laser with sampled gratings," *Conf. Record, Optical Fiber Commun. Conf.*, p. 165, February, 1992.

28. Z. M. Chuang and L. A. Coldren, "A novel broadband tunable laser diode," *IEEE LEOS Integrated Photonics Topical Meeting*, April, 1991.

29. R. C. Alferness, U. Koren, L. L. Buhl, B. I. Miller, M. G. Young, T. L. Koch, G. Raybon, and C. A. Burrus, "Widely tunable InGaAsP/InP laser based on a vertical coupler intra-cavity filter," *Conf. Record, Optical Fiber Commun. Conference*, pp. 321-324, February, 1992.

30. C. J. Chang-Hasnain, J. R. Wullert, J. P. Harbison, L. T. Florez, N. G. Stoffel, and M. W. Maeda, "Rastered, uniformly separated wavelengths emitted from a two-dimensional vertical-cavity surface-emitting laser array," *Appl. Phys. Ltrs.*, vol. 58, no. 1, pp. 31-33, January, 1991.

31. K. Liu, "Noise limits of spectral slicing in wavelength-multiplexing applications," *Conf. Record, Optical Fiber Commun. Conference*, p. 174, February, 1992.

CHAPTER 6

Lightwave Amplifiers

6.1 Overview

In the earliest days of what came to be called electronics, ways of amplifying weak electrical signals did not exist. The design of any sort of radio communication link was an exercise in managing the *link budget*, the detailed accounting for every loss of power as the transmitted signal made its way to the receiver over radio or wire propagation paths, antennas, lossy tunable devices, noisy receivers and so forth. Then in 1906 Lee Deforest invented his "Audion," the vacuum-tube triode amplifier, and immediately the link-budget issue began to recede in importance. As long as the system designer never allowed the information-bearing signals to fall to so low a level that the signal-to-noise ratio became too small for the intended error rate, cheap and effective amplifiers were always available to boost the signal level. A transcontinental analog trunking system has literally thousands of cascaded electronic amplifier stages, a fact that everyone seems to accept as unremarkable.

For the last several years, as it became clear to lightwave system designers that it was desirable to keep the signal in photonic form throughout the length of the path between end nodes, there has been an active search for good purely photonic amplifiers. As long as the fiber was carrying only a single bitstream and the rate of this bitstream was less than a few gigabits, we could use a *repeater*, consisting of a detector followed by electronic amplification followed by retransmission in optical form. However, as single-stream bitrates have risen, and the use of one fiber for many wavelength-multiplexed bitstreams has proved attractive, the need for wideband purely photonic amplifiers has become more urgent.

This chapter summarizes the current state of these devices. As the photonic amplifier art progresses even further in the years ahead, it seems a safe prediction that for lightwave-system architects, link-budget questions will begin to recede into the background, just as they did for electronic communication systems.

From the preceding chapter, it should be clear that a purely photonic amplifier can be built from any structure that lases, using some combination of low facet reflectivity to discourage lasing, combined with strong pumping to create as strong a population inversion of the carriers as possible. There are essentially two practical approaches *laser-diode amplifiers* and *doped-fiber amplifiers*. At present, the latter seem particularly attractive because of their simplicity of manufacture and of coupling into the fiber link, their polarization independence, their wide bandwidth, and their comparative freedom from crosstalk effects. The competing laser-diode amplifiers have until recently been in eclipse because of their expense of manufacture and coupling into fiber pigtails, their polarization sensitivity, and the high level of crosstalk.

However, this preference of system designers for doped-fiber amplifiers could be changing. For one thing, as mentioned in the last chapter, heavily injected quantum-well laser devices can be made to exhibit gain-curve widths as great as 240 nm, the entire width of the low fiber attenuation window at 1.5 μ, which suggests that corresponding amplifier bandwidths can be achieved. Second, erbium fiber amplifiers operate only in the 1.5-μ band and no equally-usable solution is known for wideband doped-fiber amplifiers at 1.3 μ, the wavelength band often favored for long links because of its low chromatic dispersion (Section 3.6). Laser-diode amplifiers can handle this wavelength region as easily as the 1.5-μ region.

Figure 6-1 depicts the basic electronic amplifier or photonic amplifier as a three-terminal black box. In both cases, some power from an external supply drives a process in which a weak input signal is transformed into a strong output signal. Internal to the amplifier, some noise is added, and this can be represented as a component additive to the input signal.

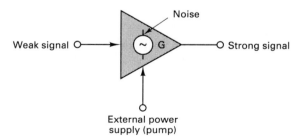

Figure 6-1. Basic amplifier function.

The bandwidth of photonic amplifiers is at least three orders of magnitude greater than anything conceivable with electronics. However, lightwave amplifiers can have at least two undesirable properties. First, although both forms of amplifier will eventually saturate as the input signal level is raised, the level of input signal at which this *gain saturation* occurs is often not far above the upper end of the operating range for photonic amplifiers, whereas, as any hi-fi buff knows, with electronic amplifiers the input levels at which significant nonlinearities due to saturation begin to set in are usually greatly in excess of the usual operating point. The second undesirable property, crosstalk, is a consequence of the first. In laser-diode amplifiers, since the gain is dependent on signal level, when the operating point is near saturation an on-off keyed signal at one wavelength can cause the gain to fluctuate up and down, which affects the output signal at other wavelengths. For erbium doped-fiber amplifiers, this effect goes away above bitrates of around 500 bps; the corresponding number for semiconductor laser-diode amplifiers is several Gb/s. As we shall see, this is because the carrier lifetime for semiconductors is a nanosecond or less, but more than ten milliseconds for the erbium doped-fiber amplifier. Electronic amplifiers usually have so much inertia in the power supply that this effect is not noticeable.

There are basically three roles that amplifiers play in lightwave systems.

- The *power amplifier* is a photonic amplifier placed immediately after the transmitter laser to boost the transmitted power level to a level greater than that available from the laser.

- The *line amplifier* substitutes for a repeater (analog electronic amplifier) or a regenerator (electronic waveform reshaper and amplifier) at one or more places between transmitter and receiver. Its purpose is to buy back the losses incurred along the path, those due to attenuation or splitting or both.

- The *receiver preamplifier*. As we shall see in Chapter 8, the use of an amplifier at the threshold signal levels occurring in photonic receivers allows one to approach receiver detectability limits very close to those achievable with more exotic forms of receivers. For example, if the system is using amplitude-shift keying (ASK), then a suitable receiver consists of a properly designed photonic preamplifier followed by a simple energy-detecting photodetector. This theoretically requires a minimum received-signal level for a given error rate that is essentially equal to that of a heterodyne coherent receiver, but considerably simpler and cheaper.

In this chapter, we shall discuss the various kinds of lightwave amplifiers that have found favor for systems use. We start with the laser-diode amplifier, partly for continuity with the immediately preceding chapter on lasers. We then proceed to doped-fiber amplifiers and conclude with a brief treatment of several forms of amplification that, while no longer strong contenders for system use, are nonetheless interesting and capable of a revival as this active field of research unfolds over the next few years.

We shall be interested in several properties of each type of amplifier: gain as a function of wavelength, output power, power level at which the gain saturates, the character of the noise introduced by the amplifier, and the crosstalk behavior of the amplifier when several high speed on-off keyed signals are applied at the input.

6.2 Laser-Diode Amplifiers

Basics

We start with laser-diode amplifiers. It will turn out that once these have been discussed in some detail, most of the important points will have been covered, and the other amplifier types can be dealt with more briefly.

A laser-diode amplifier [1] can be thought of as a "poor laser," one which uses either such low injection current or low facet reflectivity R or both that the device operates above transparency (Equation 5.10) but below threshold (Equation 5.11). In analyzing amplifiers, we can use all the ideas from Chapter 5 that deal with the subthreshold case.

Two types of laser diode amplifiers can be distinguished, *Fabry-Perot amplifiers* and *travelling-wave amplifiers*, In the former, R is kept high enough for the light to make many bounces within the cavity, just as with the Fabry-Perot filter of Section 4.4. Typical values of power reflectivity R are up to 30 percent, just as with laser diodes. The result is an amplifier with a series of narrow passbands with the gain curve as the envelope. With the travelling-wave amplifier, on the other hand, great care is exerted to make the reflectivity R of both facets as small as possible, typically as low as 10^{-5} to 10^{-4}. If this is done properly, and the gain per unit length g is high enough, the device will amplify an incident signal sufficiently in a single pass through the active region. There will then be little or no cavity-resonance effect visible in the output spectrum. Obviously the travelling-wave form of laser diode amplifier is the one required for use in wavelength-division networks.

We can determine the performance of the travelling-wave amplifier from the analysis of the Fabry-Perot amplifier by letting the facet reflectivity go to zero.

Cavity effect on device gain spectrum

The analysis of laser-diode amplifiers begins at the same place as with lasers, namely Equation 5.10

$$g(N) = a(N - N_o) \tag{6.1}$$

where g is the gain per unit length of the material, N is the carrier density per unit spatial volume, N_o is this density at transparency (gain equals loss), and $a = dg/dN$ is the differential gain. Actually, as we know, g is also a function of optical frequency, the *gain curve* of Figure 5-19. In general g depends not only on optical frequency f, but also on carrier spatial density N, which in turn depends on light intensity P, and on z, the position along the active region. In what follows, for notational brevity, we shall usually only write out that dependency of g that is of interest at that part of the discussion.

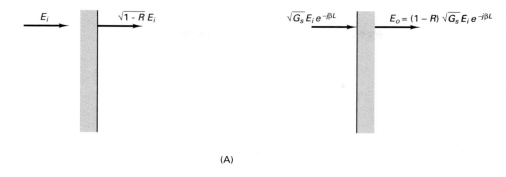

(A)

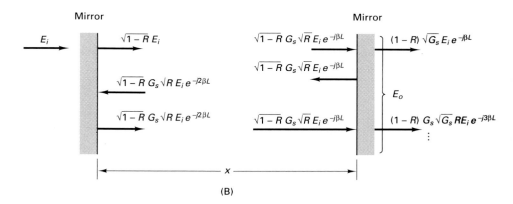

(B)

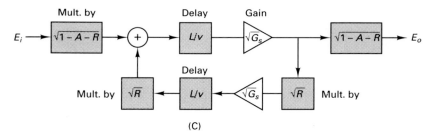

(C)

Figure   6-2. Analysis of two different kinds of laser diode amplifier, (A) Fabry-Perot amplifier, and (B) Travelling-wave amplifier, (C) Equivalent block diagram.

If α is the loss per unit length, then the net excess of gain per unit length over loss per unit length is

$$\Gamma g - \alpha \qquad (6.2)$$

where the confinement factor Γ (typically 0.3 to 1.0) represents the fractional transverse concentration of the light flux within the active region. In an incremental length dz the light power P increases by

$$dP(z) = (\Gamma g - \alpha)P(z)\, dz \qquad (6.3)$$

from which the growth of light intensity inside the active region can be obtained using

$$\int_{P_{in}}^{P_{out}} \frac{dP(z)}{P(z)} = \int_0^L (\Gamma g - \alpha)\, dz = (\Gamma g - \alpha)L \qquad (6.4)$$

for Γ, g, and α independent of z. L is the cavity length, $P_{in} = P(0)$, and $P_{out} = P(L)$. The overall single-pass gain G_s experienced by light making a single traverse within the active region of the device is then

$$G_s = \frac{P_{out}}{P_{in}} = \exp\left[(\Gamma g - \alpha)L\right] \qquad (6.5)$$

Figure 6-2(A) shows a travelling-wave amplifier, one in which the light makes only a single pass. (B) shows the multiple reflections of a Fabry-Perot amplifier, and (C) shows the equivalent circuit. These are analogous to Figures 4-4(B) and (C), respectively, for the passive Fabry-Perot filter. Here, the single-pass gain in power G_s has been added, and for simplicity any attenuation A through each mirror has been assumed negligible. For the moment, the facet power reflectivity is R, assumed identical for both facets.

Starting from this equivalent circuit, we can derive the overall power transfer function as a function of frequency as follows. On each reflection, the reflected field is $\sqrt{R}$ times the incident field. The light of field strength E_i enters the filter, $\sqrt{1-R}\,E_i$ surviving a passage through the first mirror, and $\sqrt{R}$ is reflected to the left from the device and lost. The signal then proceeds into the resonant cavity, where it is reflected repeatedly from one mirror to another. The light first arrives at the right-hand mirror with field $\sqrt{1-R}\,\sqrt{G_s}\,E_i\exp(-j\beta L)$, where β is the propagation constant. A portion $(1-R)\sqrt{G_s}\,E_i\exp(-j\beta L)$ passes to the output, and a portion $\sqrt{1-R}\,\sqrt{G_s}\,\sqrt{R}\,E_i\exp(-j\beta L)$ is reflected back toward the left. This arrives amplified at the left mirror and is reflected back to the right, where it emerges still further amplified as $\sqrt{1-R}\,\sqrt{G_s}\,G_sRE_i\exp(-3j\beta L)$.

Adding up all the successive contributions to the output E_o, and using $\beta L = 2\pi f\tau$, where τ is the one-way propagation time L/v across the cavity, the complex transfer function of the field strength is

$$H(f) = \frac{E_o(f)}{E_i(f)}$$

$$= (1 - R)\sqrt{G_s}\ \exp(-j2\pi f\tau) \sum_{m=0}^{\infty} G_s{}^m R^m\ [\exp(-j4\pi mf\tau)] \qquad (6.6)$$

$$= \frac{(1 - R)\sqrt{G}\ \exp(-j2\pi f\tau)}{1 - RG_s\exp(-j4\pi f\tau)}$$

The gain as a function of frequency is the power transfer function $G(f) = |H(f)|^2$,

$$G(f) = \frac{G_s(1 - R)^2}{1 + G_s{}^2 R^2 - 2G_s R \cos 4\pi f\tau}$$

$$= \frac{G_s(1 - R)^2}{(1 - G_s R)^2 + G_s R(2 \sin 2\pi f\tau)^2} \qquad (6.7)$$

our old friend, the Airy function, which we met before in Section 4.4 as the transfer function of any resonant cavity formed by two mirrors. We have not written out the frequency dependence of g and hence G_s.

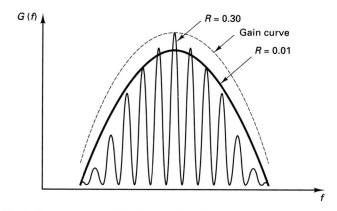

Figure 6-3. Typical gain spectra $G(f)$ for (A) Travelling-wave amplifier $(R = 0.01)$, and (B) Fabry-Perot amplifier $(R = 0.30)$. A parabolic curve of g with frequency was assumed.

Figure 6-3 indicates roughly what $G(f)$ looks like for a fixed value of G_s and for values of R near zero and 0.3, typical of travelling-wave and Fabry-Perot amplifiers, respectively. The transfer function of the travelling-wave amplifier reproduces the gain curve, whereas the Fabry-Perot amplifier shows very large ripple. From Equation 6.7, this ripple, which is analogous to the contrast of a passive filter (Equation 4.15), can be expressed as

$$C = \left(\frac{1 + RG_s}{1 - RG_s} \right)^2 \tag{6.8}$$

By convention, the dividing line between travelling-wave and Fabry-Perot amplifiers is set at the condition that $C = 2$, i.e., 3 dB of ripple.

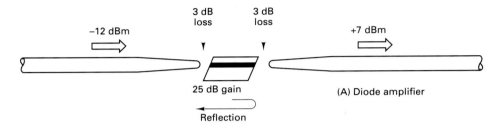

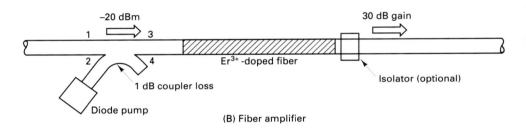

Figure   6-4. Structure of (A) Semiconductor laser diode amplifier, and (B) Erbium doped fiber amplifier.   Typical values are shown for loss in fiber-to-facet coupling using lensed fibers, for facet-to-facet gain, and for shape of the gain spectrum $G(f)$.

Practical considerations

It is clear from this last result, Equation 6.8, that to develop gain from a laser amplifier, while keeping the ripple low, very small facet reflectivities must somehow be achieved. Low facet reflectivity is quite hard come by, and the yields are low for anti-reflection ("AR") coating techniques for values of R below 10^{-3}. To illustrate the problem, suppose we wanted a 25-dB amplifier with only 3 dB of ripple C. According to Equation 6.7 the one-way gain G_s must be 25 dB, and according to Equation 6.8, R must be at most 5.7×10^{-4}.

The expression (6.7) gives only the facet-to-facet performance of the device. The overall gain must include the losses incurred by coupling the laser to the fiber pigtail at each end of the active region, as shown in Figure 6-4(A). Avoidance of pigtailing costs is one of the principal reasons that doped-fiber devices have become so popular. Laser-diode amplifier facet-to-facet gains $G_{\max}(f)$ of as much as 25 dB have been achieved, but by the time light is coupled out of a fiber at the input side and back in at the output side, it has proved difficult to maintain a fiber-to-fiber gain of more than 15 dB or so. Figure 6-4, which compares the structure of laser-diode and doped-fiber amplifiers, also gives some typical values for these coupling losses.

Figure 6-5 shows the facet-to-facet gain $G_{max}(f)$ as a function of pump I for two similar experimental devices [1], one a travelling-wave amplifier and the other a Fabry-Perot amplifier. It is seen that, although similar maximum gains were achieved, the Fabry-Perot device is obviously much more sensitive to operating conditions. It achieves its high facet-to-facet gain only very close to threshold and is therefore very sensitive to fluctuations in bias current, temperature, and signal polarization, the last of these because of the high degree of asymmetry of the active region cross-section dimensions.

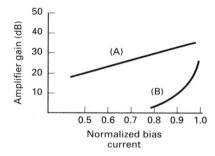

Figure   6-5. Curves of $G_{max}(f)$, the maximum gain in dB, versus normalized pump current I/I_{th} for (A) A typical semiconductor travelling-wave amplifier, and (B) A typical semiconductor Fabry-Perot amplifier. (From [1], © 1988 IEEE).

In spite of the more graceful curve of gain versus pump current shown in the Figure 6-5, travelling-wave amplifiers, have their operational difficulties too. Because their stable operation and low gain ripple depends on keeping the facet reflectivity R very low, they are vulnerable to reflections elsewhere in the system. Reflections from splices, connectors, filter components, and so forth are likely to cause the device to start lasing at some wavelength near the gain peak. Frequently, isolators are required following travelling-wave amplifiers, for example between the receiver preamplifier and a Fabry-Perot tunable filter, because the reflectivity of these filters as a function of wavelength is near unity for all wavelengths except those near the resonance peaks.

Because of the difficulty and uncertainty of the antireflection coating processes with very low R, resort is often made to *angled facets*, as shown in Figure 6-6. A light ray approaching one of the facets along the axis within the active region will suffer a change of $2\theta_{tilt}$ upon reflection. This angle is designed to be greater than the angle that will support internal reflections, θ_1 in Figure 3-15. With θ_{tilt} typically at around 7°, by Snell's Law (Equation 3.46), the output angle is about 25°, because the index of the semiconductor medium is roughly 3.5 times that of air. The practical difficulties and expense of attaching two pigtails at such strange angles, when the in-line attachment process is tricky enough, are obvious.

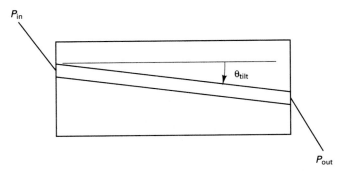

Figure   6-6.   Structure of angled-facet semiconductor travelling-wave amplifier.

6.3 Amplifier Noise

With electronic amplifiers, the noise is a consequence of the random nature of the thermal motion of electrons in the input resistor or of random processes inside the amplifier itself, analogous to thermionic shot noise in a vacuum tube. For the photonic amplifier, the dominant noise is *amplified spontaneous emission (ASE).* Spontaneous noise is generated in all the portions of the active medium, but the part of the spontaneous noise that most degrades system performance is that generated at points near the amplifier input, where the signal is still weak.

The spontaneous noise in the amplifier can be modelled as a stream of random arrivals, each an infinitely short impulse, so that the power spectrum of the noise at the point of generation within the amplifier is flat with frequency. We need to know the power spectrum of the noise at the amplifier output. Then in Chapter 8 we can combine this noise with all the other noises to get a comprehensive picture of signal detectability and probability of bit error.

It has been shown [2, 3] that the amplifier noise output power spectral density $N(f)$ is

$$N(f)df = hf\chi\, n_{sp}\, [G(f) - 1]df \tag{6.9}$$

where $G(f)$ is the gain spectrum of Equation 6.7, and n_{sp} is the spontaneous-emission factor, as before. The quantity n_{sp} is also called the population-inversion factor, capturing the fact that the less complete the population inversion, the more the spontaneous noise power.

The quantity χ is the *excess noise factor,* represents the effect of asymmetry in mirror reflectivity. In the derivation of Equation 6.7, it was assumed that both mirrors had the same reflectivity $R_1 = R_2 = R$. Little would change in Equation 6.7 if we had made $R_1 \neq R_2$, so long as $R = \sqrt{R_1 R_2}$. While an asymmetry in mirror reflectivity is irrelevant for calculating the overall gain, it is not irrelevant in dealing with noise.

Noise due to spontaneous emission is introduced all along the active region, but the part of it that degrades system performance the most is that which occurs near the first facet, before the light has been amplified. The noise at the output will be less if the first mirror allows light to escape, since some of that light will be noise that would otherwise have been reflected back into the resonant cavity at just the point where the signal is the weakest. This effect is captured by the parameter

$$\chi = \frac{(1 + R_1 G_s)(G_s - 1)}{(1 - R_1)G_s} \tag{6.10}$$

It is seen that for the travelling-wave amplifier with $R_1 \approx 0$ and $G_s \gg 1$, $\chi = 1$, whereas with the Fabry-Perot amplifier, where R_1 can be nonzero, χ can be much larger.

6.4 Gain Saturation at a Point

We now consider how the gain saturates at a single point along the amplifier active region for the case of a laser-diode amplifier pumped by current I. The same behavior occurs at a point within a doped-fiber amplifier.

As the input signal power is raised, the gain per unit length g of Equation 6.1 eventually begins to fall off. This is because at the higher levels of light intensity P, the carriers available from the fixed pump current I begin to be used up. The gain at at any value of z along the active region can be expressed as

$$g(\mathrm{P}, z) = \frac{g_o}{1 + \mathrm{P}(z)/\mathrm{P}_{sat}} \tag{6.11}$$

where P is the light intensity in watts per unit area (not to be confused with total power P), g_o is the gain per unit length in the absence of light input, and P_{sat} is the *saturation intensity*, defined as that light intensity at which the gain per unit length has been halved.

We can derive Equation 6.11 and determine g_o and P_{sat} from the rate equation 5.35 and from Equation 6.1 as follows. First of all, we ignore the light produced by spontaneous emission in this calculation since, at any one frequency, the amount of light reaching the output from this source is usually at least two orders of magnitude less than that due to the input signal. (If it were much larger, the amplifier would be unusably noisy).

Under steady state conditions, Equation 5.35 becomes

$$\frac{I}{qLwd} = \frac{N}{\tau_e} + \frac{\Gamma ac}{n}(N - N_o)\,\Pi(z) \tag{6.12}$$

If the photon density $\Pi(z)$ is zero, the corresponding carrier density is $N' = \dfrac{I\tau_e}{qLwd}$, from which we can get, using g_o of Equation 6.1,

$$g_o = g(N') = a\left(\frac{I\tau_e}{qLwd} - N_o\right) \tag{6.13}$$

Now, for any photon density $\Pi(z)$ greater than zero, with constant current I, there are new values of N and consequently of g. From Equations 6.1 and 6.12

$$g = a(N - N_o) = \left(\frac{I}{qLwd} - \frac{N}{\tau_e}\right)\frac{n}{\Gamma c\Pi(z)} \tag{6.14}$$

while from Equation 6.1

$$N = \frac{g}{a} + N_o \tag{6.15}$$

From Equation 5.37, setting facet reflectivity R to zero, the power *density* is

$$P(z) = \frac{hfc}{n}\,\Pi(z) \tag{6.16}$$

Combining these last four equations gives

$$g = g_o\left(\frac{hf}{\Gamma a\tau_e P(z)}\right)\left(1 + \frac{hf}{\Gamma a\tau_e P(z)}\right)^{-1} \tag{6.17}$$

which is the same as Equation 6.11, with

$$P_{sat} = \frac{hf}{\Gamma a\tau_e} \tag{6.18}$$

6.5 Gain Saturation of the Entire Device

Fabry-Perot amplifier

For the Fabry-Perot amplifier, in which the light makes many passes back and forth through the active region, the light intensity is more or less uniform throughout the cavity, so that taking the spatial average over the entire length is permissible. It is sufficiently accurate to replace the optical intensity $P(z)$ at a point by the average value

$\overline{P} = P_{out}/(1 - R_2)$, where R_2 is the power reflectivity of the output mirror, and P_{out} is the output intensity. Using $\overline{P}$ as the appropriate value in Equation 6.11, and using Equation 6.5, the overall gain of the device is then

$$G = \frac{P_{out}}{P_{in}} = \exp\left(\frac{\Gamma g_o}{1 + \overline{P}/P_{sat}} - \alpha \right)L \tag{6.19}$$

This is all we shall have to say about Fabry-Perot amplifiers. For network purposes, the travelling-wave amplifiers are much more interesting because of their flatter overall gain spectrum and the greater ease of controlling their operating points. For travelling-wave amplifiers, the Fabry-Perot amplifier equations continue to be of use only when ripple in the spectrum must be calculated.

Travelling-wave amplifier

From now on, we shall assume that all the gain occurs on the first pass, and will therefore write the overall gain of the device G instead of G_s. We must integrate over z along the entire active-region length L, since the carrier density N at any z depends on the signal level $P(z)$ at that point, and may be different from one value of z to another. The upstream portions (small z) may not have reached saturation at the same time that the downstream portions could have.

The required expression for $G_s = G$ can be gotten by assuming zero mirror reflectivity, and using Equation 6.3 with $\Gamma = 1$ and $\alpha = 0$, and also Equation 6.11.

$$\frac{dP(z)}{dz} = g(z)P(z) = \frac{P(z)\, g_o}{1 + P(z)/P_{sat}} \tag{6.20}$$

from which

$$\int_0^L g_o\, dz = \int_{P(0)}^{P(L)} \frac{1 + P(z)/P_{sat}}{P(z)}\, dP(z) \tag{6.21}$$

$$g_o L = \left[\ln P(z) + \frac{P(z)}{P_{sat}} \right]_{P_{in}}^{P_{out}} \tag{6.22}$$

Defining the single-pass gain in the absence of input light to be $G_o = \exp g_o L$, and $G = P_{out}/P_{in}$, gives this transcendental equation for G:

$$\frac{P_{in}}{P_{sat}} = \frac{\ln (G_o/G)}{G - 1} \qquad \text{or} \qquad G = 1 + \frac{P_{sat}}{P_{in}} \ln\left(\frac{G_o}{G} \right) \tag{6.23}$$

A typical dependency of gain G on input intensity P_{in} is plotted in Figure 6-7. It is seen that, as the input signal power P_{in} is increased, G remains at its weak-signal value G_o for a while, and then begins to drop, eventually levelling off at unity.

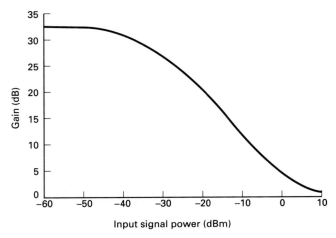

Figure 6-7. Gain saturation effect. Single pass gain G_s versus input optical power P_{in}, assuming saturation power $P_{sat} = -6$ dBm, and small signal single pass gain $G_o = 32.5$ dB. (From [4], © IEEE, by permission).

While this equation is valid for semiconductor laser-diode travelling-wave amplifiers, matters are considerably more complex for the doped-fiber travelling-wave amplifier. The reason is that when the pumping is done sideways by current injection, the density of injected carriers along the medium can be considered constant, while in a fiber amplifier, where the optical pump signal is injected at one end, this pump energy is depleted as it propagates along the amplifying medium.

An exact analysis is very difficult, but the general idea of what is happening as a function of L and input pump power can be gotten from Figure 6-8 [5]. These curves apply to the doped-fiber amplifier, to be discussed shortly, in which the pump is fed into one end, rather than being applied uniformly along the entire length of the device. The top set of curves shows the pump power being attenuated as it propagates. The middle set shows the signal being amplified by an ever-decreasing factor G as the pump attenuates. Eventually, there is so little pump power left that inversion is less and less complete, and for lengths L greater than this, the signal is actually strongly attenuated, for reasons that will be discussed presently. Meanwhile, as the bottom curve shows, the amplified spontaneous emission grows. Clearly, for a given pump power, the optimum length L of the active amplification region depends on how much gain is to be extracted from the device and how large a buildup of ASE can be tolerated.

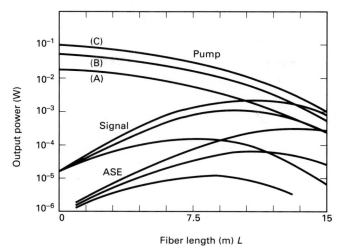

Figure   6-8.   Measured performance of a travelling-wave amplifier as a function of length of the active region. The particular device was an erbium doped fiber amplifier, and the optical pump power at 515 nm was either (A) 20 mW, (B) 55 mW, or (C) 100 mW. (From [5], © 1991, OSA, by permission).

Gain saturation is not the same thing as nonlinearity

The fact that the curve of gain as a function of input light level is not a straight line, but takes the form shown in Figure 6-7, does not mean that the device is nonlinear in the usual sense. This distinction is important because, as we shall see in Chapter 8, one of the most important noise components that must be considered in designing the system is the "signal × spontaneous" amplifier noise, that is, the result of the beat between the signal and the ASE. This does not mean that signal × spontaneous beat noise is generated in the amplifier, it means that when the signal and the spontaneous noise arrive at the photodetector, which is a nonlinear device, this beat component is generated. (Recall from Section 5.2 that the photodiode converts milliwatts of light into milliamperes of current, and is therefore nonlinear).

If the amplifier gain could change within a small fraction of the period of the light signal (some 10^{-15} sec), then the amplifier could be classed as a nonlinear device in the usual sense, but this is impossible. The nanosecond time that a semiconductor laser diode takes for the gain to change is six orders of magnitude less than the optical period. The millisecond time that an erbium doped-fiber amplifier requires is seven more orders of magnitude still less.

In a practical optical network, the only thing that could conceivably be slow enough for the gain of an erbium doped-fiber amplifier to respond to would be *inter-modulation* between wavelength-spaced channels, where the spacing was a few hundred Hz or less. For semiconductor laser-diode amplifiers, this number is one or two GHz, or less than 0.1 Å. Both these numbers are smaller than the expected channel spacing in most wavelength-division systems, so intermodulation [6] should not be an issue.

6.6 Crosstalk in Semiconductor Amplifiers

The carrier-density equation 6.12, from which the effect of gain saturation was calculated, assumed that the time rate of change of carrier density could be neglected. Therefore, the results that followed are appropriate to situations in which none of the parameters are time-dependent. However, the original time-dependent form of this rate equation, namely 5.35, for constant pump current I, is

$$\dot{N}(t) = \frac{I}{qLwd} - \frac{N(t)}{\tau_e} - \frac{\Gamma c}{n} [N(t) - N_o] \, \Pi(t) \tag{6.24}$$

Suppose that, at $t = 0$, the photon density $\Pi(t)$ is suddenly reduced to zero. Then

$$\dot{N}(t) + \frac{N(t)}{\tau_e} = \frac{I}{qLwd} \tag{6.25}$$

or

$$N(t) = \frac{I\tau_e}{qLwd} \, [1 - \exp(-t/\tau_e)] \tag{6.26}$$

so that the carrier density, and thus the gain, are seen to respond with a time constant

$$T_\downarrow = \tau_e \tag{6.27}$$

when the light is reduced to small values, usually as a transition between a "1" bit and "0" bit.

When the transition is in the other direction, the incident light intensity going from essentially zero to a large value, and inspection of Equation 6.24 shows that the time constant is smaller. The time constant is now given by

$$T_\uparrow = \left(\frac{1}{\tau_e} + \frac{\Gamma ac\Pi}{n} \right)^{-1} \tag{6.28}$$

This means that if the bit duration T is much longer than these time constants, then the amplifier gain $G(f)$ will go up and down with the instantaneous input power density P_{in} expressed by Equation 6.23. If the bit duration is much shorter than τ_e, then in Equation 6.23, P_{in} must be replaced by its time-averaged value.

Practical values of τ_e are seven orders of magnitude different for semiconductor laser-diode amplifiers and erbium doped-fiber amplifiers, as we have said. As discussed in Chapter 5, the carrier lifetime for semiconductors is a nanosecond or less. For the erbium transition involved in doped-fiber amplifiers it is over ten milliseconds. Since most lightwave networks are likely to use bitrates from Mb/s rates to around a Gb/s per channel, crosstalk can be a severe problem with laser-diode amplifiers but likely never to be a problem with doped-fiber amplifiers.

An extensive analysis of crosstalk effects in semiconductor laser-diode amplifiers is presented in [4]. A number N of statistically independent on-off keyed (OOK) signals are assumed to arrive at an amplifier with gain G given by Equation 6.23 and Figure 6-7. As a worst-case assumption, the signals are taken to be bit-synchronized with one another. The input power P_{in} can take on $N+1$ values with probabilities that exactly k of them are sending "1," given by the binominal distribution

$$\text{Prob}\,\{P = kP_1 + (N-k)P_o\} = \frac{N!}{k!(N-k)!}\,\frac{1}{2^N} \qquad (6.29)$$

where P_o is the power representing a "0" bit and P_1 is the power representing a "1" bit. If the *extinction ratio* r of the on-off modulation process is

$$r = P_o/P_1 \qquad (6.30)$$

then in terms of the average amplifier input power per channel P_{in}

$$P_{in} = \frac{(P_o + P_1)}{2} \qquad (6.31)$$

$$P_1 = \frac{2P_{in}}{1+r} \qquad (6.32)$$

and

$$P_o = \frac{2rP_{in}}{1+r} \qquad (6.33)$$

The example of Figure 6-9 shows just how badly the gain G can vary when gain saturation can keep up with the bitrate, as is the case with semiconductor laser amplifiers. Ten signals were assumed, each having a power level of -50 dBm and a zero extinction ratio r. The amplifier had a P_{sat} of -6 dB, as in Figure 6-7. There is one case in which only one transmitter sends a 0 and one case where only one sends a 1. These points are almost invisible in the figure, since probabilities are so small; they occur at 550 and 1810. There are many more cases in which some are sending 0 and some 1. The relative probability of the gain taking up various values is plotted as the ordinate in the figure. The gain is seen to vary by a ratio of over three to one over these various cases.

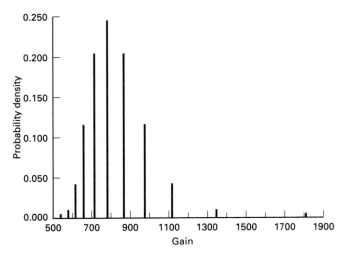

Figure 6-9. Probability density function of the amplifier gain G for a typical laser-diode amplifier driven by many signals concurrently. (From [4], © IEEE 1990, by permission).

This form of multiplicative noise upsets the ability of the detector to distinguish "0" from "1." In Chapter 8 we shall present some examples from the remainder of the analysis from [4] that show quantitatively how bad this crosstalk effect can be.

6.7 Doped-Fiber Amplifiers

Principles

In Chapter 5 it was seen that the outer orbital electrons in an atom can be raised to higher energy levels through pumping. For gas lasers, we spoke of the energy levels of *atoms*, and the pumping was effected by incident optical radiation of an appropriate wavelength shorter than that at which stimulated emission was desired. For semiconductor lasers, it is still the outer orbital electrons whose energy levels are being raised by pumping (using current injection in this case), but it was convenient to introduce a fictitious quasi-particle called the *carrier*, the electron-hole pair, as the physical unit with which to do the energy bookkeeping.

In doped fibers, it is still the outer electrons whose changes of energy underlie the pumping and radiation processes, but the operative jargon is to speak of the *ions* as the physical units for the energy bookkeeping. An ion is an atom that has completely lost one or more of its outer electrons. For example, in erbium doped-fiber amplifiers, it is the Er^{3+} ion that has all the nice properties that we shall now discuss. This is an erbium atom that has lost three of its outer electrons,

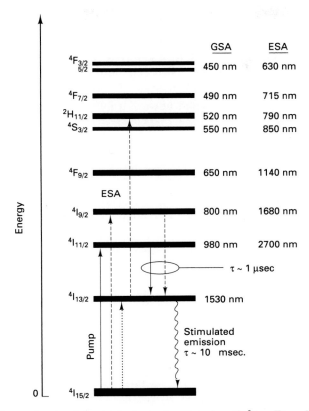

Figure 6-10. Energy levels of the triply ionized erbium ion, Er^{3+}. (From [7], © IEEE 1991, by permission).

Figure 6-10 shows the energy levels of such an atom. The first column of numbers at the right shows the wavelengths equivalent to the energy difference between the various levels and that of the ground state. They are labelled *GSA* for "ground-state absorption." Light of those wavelengths will pump the ions to a higher state than the ground state $^4I_{15/2}$. Once they are pumped to a level higher than the $^4I_{13/2}$ state, many will find their way back to that state and be available for either spontaneous or stimulated emission of light of the desired 1.5-μ wavelength, as was discussed in connection with Figure 5-6. The $^4I_{13/2}$ state is a *metastable* state, meaning that this level's transition to ground state has a very long lifetime compared to those of the states that led to it. The right-hand set of levels are the "excited-state absorption" levels, using the metastable state as a reference. We shall discuss these in connection with desired pump wavelengths.

There are several attractive things about this picture from the optical communication point of view. First, the transition from the metastable state to the ground state is in the wavelength region of low fiber attenuation. Second, it turns out that the lifetime τ_{21} to decay to the ground state is remarkably long, some 11 milliseconds, much slower than the other downward transitions, so that enormous amounts of energy remain stored in the inverted erbium ions at that level. Third, some of the upper levels correspond to wavelengths where commercially available lasers exist to do the pumping. High-power gallium-arsenide lasers or laser arrays have been available around 800-850 nm for some time. Gallium arsenide laser diodes can be made to deliver high power at 980 nm by deliberately introducing strain between layers by a careful mismatching of lattice constants.

As we learned in Chapter 5, the narrow character of each energy level becomes upset when there are other atoms nearby (that is, within the span of the wave function of an electron), and each level becomes smeared out in energy. In the doped-fiber case, the dominant operative mechanism is *Stark splitting*, the splitting that any atomic spectral line will undergo in the presence of a dc electric field. If the doped-fiber material were a regular crystal, each line in Figure 6-10 would split into a finite number of individual lines, but actually the material is an amorphous glass. Viewed in the ensemble sense (in other words, averaging spatially over the ions in the glass), a continuum smearing is seen.

This smearing can be exacerbated by the addition of special co-dopants into the host glass along with the erbium atoms [7]. Co-doping also makes it easier to get higher concentrations of erbium into the glass, which normally resists this intrusion because the erbium atoms are physically so much larger than the atoms of silicon and oxygen in silica. Some co-dopants are themselves much more soluble in silica than erbium, and once in place form interstices of a size that allows more erbium to be dissolved than if the co-dopant were absent. The co-dopant atoms, being physically very close to the erbium atoms, and of different electric field strength than those of the silica host, assist in getting still more line broadening, partly by enhancing the Stark splitting.

A wide variety of combinations have been tried in order to increase erbium solubility and to smear out the transition between metastable and ground states to form a gain curve that is as wide as possible and has as flat a top as possible. Not all of these have led to fiber that is easy to use from the standpoint of fragility, ease of fusion splicing into the normal silica transmission fiber, and so forth. The co-doping with alumina (AlO_3), with the resulting spectrum shown in Figure 6-11, has proved to be the most satisfactory solution. Not counting the peak at 1.532 nm, the overall 3-dB bandwidth is over 35 nm. Using alumina as a co-dopant, erbium concentrations of 1000 parts per million can been achieved routinely, with numbers as high as 2500 parts per million having been reported.

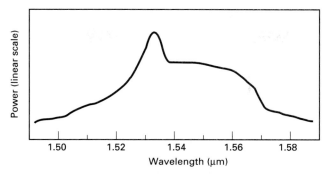

Figure   6-11. Fluorescence (spontaneous emission) spectrum of Er^{3+} in a silica glass host with alumina co-doping. (From [7], © IEEE 1991, by permission).

Physical structure

The doped-fiber amplifier structure depicted in Figure 6-4(B) consists of several parts. First there is the *dichroic coupler*, a 2×2 coupler designed to pass into the doped-fiber section most of the energy at both the signal wavelength and the pump wavelength. As was seen in Section 3.10, unless special tricks are played with Z, the length of the coupling region, and the fiber core diameters, if most of the signal power entering input 1 in Figure 6-4(B) is to go out 3 and into the doped fiber, then most of the pump power at 2 must go into the other coupler output 4. The special tricks take advantage of the oscillatory and wavelength-dependent output power as a function of Z given by Equation 3.80. They allow one to make a dichroic coupler that insures that both the pump power and the signal power go into the doped section, because they are at widely different wavelengths.

For optimum gain and minimum noise output, the doped-fiber section itself will have lengths ranging from several meters to several tens of meters, as illustrated in the examples of Figure 6-8. In order to develop maximum gain from the fiber, it is not sufficient to spread the pump power over the entire core by making it all from doped material. Sufficient pump power density is usually concentrated into a doped material region at the core center only 4 microns in diameter, and of higher index than the rest of the core.

There are several useful figures of merit for doped-fiber amplifiers. A survey of the state of the art of these parameters is given in [7]. In the power-amplifier function, output power is important, and values as large as 500 mW have been achieved. For line-amplifier or receiver-preamplifier service, the total dB gain and the incremental ratio of small-signal gain G_o to pump power are important. Values as high as 46 dB total gain and 10.2 dB increase in gain per milliwatt increase of pump power, respectively, have been reported, using semiconductor pump laser diodes.

Advantages and disadvantages

The difficulties and expense of facet reflectivity control are avoided by the doped-fiber geometry. Due to negligible facet reflectivity, doped-fiber amplifiers have lower Fabry-Perot passband ripple and better noise performance than laser-diode amplifiers, the noise effects being expressed by the factor χ in Equation 6.9. However, the advantage of smaller Fabry-Perot ripple is offset by the fact that the gain curve is not smooth, as Figure 6-11 shows. To reduce the erbium resonance peak at 1532 nm, a notch filter has sometimes been fabricated by using special fiber geometries [8]

The coupling losses are very much lower in a doped-fiber amplifier, because the fiber section is simply spliced into an existing fiber, rather than going through a lensing system. Being rotationally symmetric in cross section, the fiber amplifier is completely polarization-insensitive. Finally, since the fiber amplification process is controlled by discrete atomic lines rather than by continuous energy bands, as with semiconductor lasers and amplifiers, the former have a much smaller sensitivity to temperature (Figure 5-7).

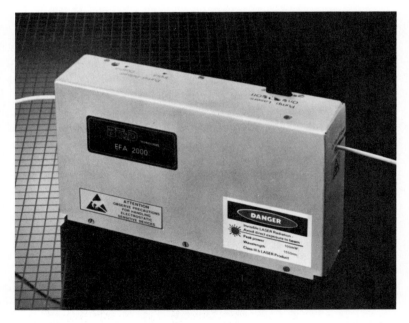

Figure   6-12. Commercial erbium doped fiber amplifier. (Courtesy of BTD Technologies).

Much of this has been known for years, but two things held up wide adoption of doped-fiber amplifiers: the fact that they have to be pumped optically, and the fact that both the available bandwidth and the available erbium concentration were low until alumina co-doping was discovered. The advent of small high-power pump diode lasers of the correct wavelength has meant that a fiber amplifier can be made without physically large and expensive pump sources.

Figure 6-12 shows how far the amplifier state of the art has come today. It shows an erbium doped-fiber amplifier that develops 22 dB gain over 35 nm band-width from a 1480-nm pump laser diode.

The one remaining disadvantage of fiber amplifiers relative to laser-diode ampli-fiers concerns their spectral parameters. A gain-curve width of about 35 nm seems to the most that is easily achievable. In a laser-diode amplifier, the gain curve can be placed at a wide variety of wavelengths (Figure 5-9), is relatively smooth (Figure 5-19), and the multiple-quantum-well laser option is available. MQW lasers can be made with gain curves over 200 nm in width (Figure 5-21). It has proved difficult to build doped-fiber amplifiers at other than 1.5 μ, and, as we saw in Figure 6-11, the gain curve does not have a smooth flat top. Doped-fiber amplifiers at 1.3 μ have been built using the element praesodymium, but require host glasses that are difficult to splice into silica fibers.

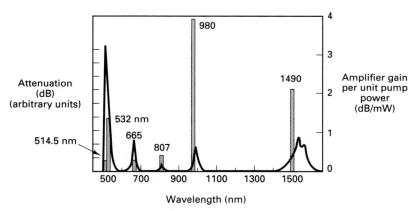

Figure 6-13. Showing the different absorption (and radiation) bands of Er^{3+} (thin lines) and the pump efficiency solid bars), expressed as dB gain per milliwatt of pump power. (From [7], © IEEE, 1991, by permission).

Choice of pump wavelengths

As Figure 6-13 shows, there are several levels above the $^4I_{13/2}$ metastable level to which the erbium ions could be pumped, after which they would eventually decay down to that level, thus providing the desired population inversion. Several factors govern the choice of such pump wavelengths. First of all, to build amplifiers that are practical in real lightwave networks, the pump wavelength must be one that is avail-able with semiconductor-diode lasers. Figure 5-9 shows the limits that can be achieved with III-V alloys. This leaves the 650, 800, 980, and 1490 nm wavelength regions as candidates, of which we can discard 650, since insufficiently high power laser diodes exist at that wavelength. Figure 6-13 shows as vertical bars the efficiency with which population inversion of the ions is produced for the various wavelengths. The curves (narrow lines) in the same figure show the attenuation due to absorption, which is roughly proportional to the strength of stimulated emission as a function of wavelength. Note from Figure 6-8 that when the signal is attenuated, it is attenuated by amounts orders of magnitude larger than those we normally associate with fiber propagation. As the light propagates along the fiber, it is the strength of this absorption, setting in after the pump is no longer able to provide complete inversion, that causes the signal to be so strongly attenuated.

The 800 nm wavelength is not a good candidate because of low efficiency, which is partly caused by *excited-state absorption* (ESA; not to be confused with ASE, amplified spontaneous emission, another laser-amplifier problem we have already discussed). ESA is defined as the absorption of the pump by some undesired transition that just happens to be at or near the right wavelength. In the 800-nm case, there is the unlucky situation that at 790 nm energy offset from the metastable state there is another transition that absorbs some of the 800 nm pump power, because these two lines are actually unresolved, having been smeared, principally by the Stark effect. The right-hand upward-pointing arrow in Figure 6-10 show this. The right-hand column of figures in Figure 6-10 shows the most harmful of these ESA wavelengths, those that take away ions from the metastable state itself.

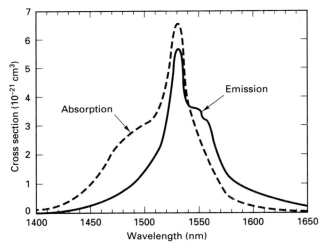

Figure 6-14. Detailed view of the 1.5 μ erbium line in a typical glass host. The line has been spread out by Stark splitting, and differs in absorption and emission. (From [7], © 1991, IEEE, by permission).

This leaves 980 and 1490 nm. It seems counterintuitive that we could pump at 1490 when that figure is so close to the 1525-1560 band to be amplified. The reason this works is that the absorption spectrum is shifted somewhat relative to the spontaneous-emission spectrum, as shown in Figure 6-14, so that we can pump in more energy than is radiated out spontaneously. The nonzero height of the emission spectrum at 1490 causes much light to be lost to both stimulated and spontaneous emission, and that is one reason why the efficiency, shown by the solid bar for 1490 in Figure 6-13, is no taller than it is. Also, the inversion being less complete for 1490 than with 980, the amplifier noise output (Equation 6.9) is higher because n_{sp} is higher. There is still one more thing to be said in favor of 980 over 1490, and that is that, since the former lies at so much greater separation from the 1525-1560 nm range to be amplified, the design of the dichroic coupler is much easier for 980.

Thus, we might be tempted to think that 980 wins hands-down, and indeed it was by using this pump wavelength that the previously mentioned 500 mW output power and 10.2 dB/mW pump efficiency were achieved (compared with 0.6, 0.43, and 5.9 dB/mW for 650, 800, and 1490 nm, respectively as the best recently-published figures [7]). However, 1490 nm is not out of the running because this is the only one of the candidate wavelengths at which the pump can be sent to the amplifier from a remote source over the fiber itself. This constitutes a useful topological system design option that will be explored further in Chapter 11.

6.8 Raman and Brillouin Amplifiers

Long before erbium-doped fibers became practical as the basis of lightwave amplifiers other forms of fiber amplifier were being considered and implemented that did not involve optical pumping of the electrons in an atom to a higher energy state, but instead involved processes at the molecular level. They used one of two such physical phenomena [9], either *stimulated Raman scattering (SRS)* or *stimulated Brillouin scattering (SBS)*. Both these were discussed in Section 3.14.

Both are based on the fact that when light is incident on a solid material, the molecules of the material are set into vibration. With SRS, some of the energy of a given photon goes into setting each molecule into mechanical vibration and some into a reradiated (scattered) photon of a lower energy, and therefore lower frequency. The frequency shift is called the *Stokes frequency*, and the reradiated light is called *Stokes light*.

For SBS, the energy of the incident photon is partly delivered to the reradiated Stokes light and partly to an acoustic wave in the material. As we saw in Chapter 4 in connection with acoustooptic Bragg diffraction, any propagating wave has a dual wave/particle personality, and the particle representation of an acoustic wave is the equivalent *phonon*. So with SBS, the energy of the incident photon is divided between the radiated Stokes photon and the phonon.

When viewed externally, the physical manifestations of the two processes are very different. The SBS process is about two orders of magnitude more efficient than the SRS process. The spatial directivity pattern of SRS is more or less isotropic, whereas that of SBS has a null in the forward direction. For SRS, since the silica fiber is an amorphous glass, there is a smear of molecular vibration frequencies and thus a smear of Stokes frequencies, shown in Figure 6-15. This figure plots the

Raman gain factor, the operative parameter in the design of a Raman fiber amplifier, against the downshift between incident light of 1.0 μ and radiated Raman light.

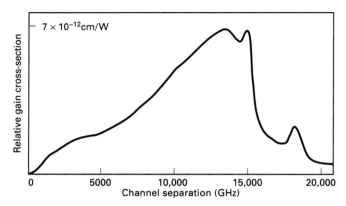

Figure   6-15. Raman gain (per unit length per unit fiber cross-sectional area per watt) as a function of pump frequency minus signal frequency.

With SBS, there is only one value of the sound velocity in the medium (even when it is an amorphous material), and therefore the phonon energy is quantized to a narrow energy level. The frequency of the reradiated light is downshifted from that of the incident light by some 11 GHz (for 1.5-μ light), and the spectrum is extremely narrow in frequency, the order of only 20-30 MHz! By the same phase-matching arguments we used in Sections 4.8 and 5.7, the fractional amount of downshift is found to be proportional to the ratio of sound velocity to light velocity

$$\frac{\Delta f}{f} = 2n\,\frac{V}{c} \tag{6.34}$$

where f is the pump frequency, n is the index, V is the sound velocity, and c is free-space light velocity. As with Equation 5.39, the factor of two takes into account the counterpropagating nature of the two waves. (We can associate $V/\Delta f$ with Λ in Equation 5.39).

Given all this, it is clear that a Raman amplifier can be built along the lines of Figure 6-4(B) with pumping from either end, whereas with a Brillouin amplifier, the pump and the signal must be counterpropagating. Spectrally, the Raman amplifier will have a curve of $G(f)$ that looks like Figure 6-15, whereas the Brillouin amplifier will have a gain spectrum $G(f)$ only a few megahertz wide at a fixed frequency offset from the pump given by Equation 6.34.

Both the SRS and the SBS effects are so much weaker than the gain that stimulated emission provides in a doped-fiber amplifier that Raman and Brillouin amplifiers tend to involve very long lengths of fiber and very high pump powers. Typical values reported in the literature are 10 kilometers and 200 milliwatts for a Raman amplifier of

8.5 dB gain [10] and 30 kilometers and 8 milliwatts for a Brillouin amplifier of 26 dB gain [11].

6.9 What We Have Learned

By revisiting the laser diode and looking at it as an amplifier rather than as a source, we have seen that providing low facet reflectivity, suitable pump input, and high internal differential gain *dg/dN*, an amplifier can be built that will handle signals whose frequencies lie anywhere within the upper portions of the gain curve of the material. For both bulk laser-diode amplifiers and doped-fiber amplifiers, this has meant around 35 or so nm of useful operating wavelength range. For multiple-quantum-well laser-diode amplifiers, the desired 200 nm is achievable.

However, this performance is bought at the expense of two effects common to any amplifier, whether electronic or photonic: gain saturation and noise. The consequences of gain saturation are important in designing the physical topology of the system, as will be seen in Chapter 11, and for semiconductor amplifiers have the additional consequence of potentially introducing large amounts of interchannel crosstalk, a problem that is absent with erbium doped-fiber amplifiers.

To compensate for the crosstalk difficulties of laser-diode amplifiers, as well as their expense, there is the fact that gain-curve broadening by use of quantum confinement, discussed in the last chapter, is an attractive option with laser-diode amplifiers that is not currently available with doped fibers. This is because the latter are basically designed around a set of macroscopic effects that start out as narrow atomic transition lines and can only be broadened a modest amount, rather than being based on the broad energy bands of semiconductor devices.

6.10 Problems

6.1 (*Gain ripple in amplifiers*) (A) As mentioned in the text, the dividing point between a travelling-wave and a Fabry-Perot amplifier has been set by convention as the condition that the Fabry-Perot ripple in the gain spectrum is 3 dB. What value of the product of single-pass gain and facet reflectivity does this correspond to? (B) The device of Problem 5.1 is made into a laser diode amplifier by applying an AR (antireflection) coating that reduces the facet reflectivity to 10^{-3}. Pump current happens to be at a value to produce a carrier density 2.5 times the value at transparency. What is the gain? (C) Suppose you wanted to use this same device as a filter-amplifier that converts FSK (frequency-shift keying) to ASK (amplitude-shift keying) by passing the monochromatic output of some source laser diode through the optical amplifier. You would choose the "0" and "1" frequencies of the source to lie at the peaks and troughs of the amplifier's transfer function, respectively. What gain-reflectivity product would you need if you wanted a 20-dB extinction ratio, defined as the ratio of the minimum power (in the "0" bit) to maximum power (that in the "1" bit)?

6.2 The value of light intensity per unit area at a given point at z along the amplifier that causes enough saturation for the gain $g(z)$ at that point to be halved is P_{sat}, a number that is independent of z. (A) What is the value of input light intensity that causes the gain of the entire travelling-wave device to be halved? (B) Explain why this is or is not the same as P_{sat}.

6.3 A 20-Mb/s ASK bit pattern of incident light is applied to the input laser diode of Problem 5-1, used as an amplifier. The photon density for a binary "1" is 0.65×10^{18} /cm^3 and zero for a binary "0." Sketch the waveform of gain (in arbitrary units) as a function of time for the bit pattern "001011."

6.4 Consider Figure 6-8, which applies to a doped-fiber amplifier. (A) Why does the output power actually start to fall after a certain length L? (B) Would an optical bandpass filter centered at the signal wavelength be expected to improve this picture? (C) At what value of length L is the ratio of signal power to amplified stimulated emission noise power a maximum, and why?

6.5 A coherent optical receiver embodies a strong optical signal generated at the receiver. This signal, called the "local oscillator signal," is applied to the photodiode additively with the arriving optical signal from the transmitter. Since the photodiode is nonlinear, the difference-frequency beat components appear in the output and can be selected out with an electrical narrowband filter. A homodyne coherent receiver has the local oscillator signal exactly at the carrier frequency of the arriving received signal. How would you use Brillouin amplification to build a coherent homodyne receiver in which the local oscillator signal is not generated locally? The arriving OOK signal is at exactly 1.52 μ, the velocity of sound in silica is 5.87 mm/μsec, and the refractive index of the fiber is assumed to be 1.45.

6.11 References

1. M. J. O'Mahoney, "Semiconductor laser optical amplifiers for use in future fiber systems," *IEEE/OSA Jour. Lightwave Tech.*, vol. 6, no. 4, pp. 531-544, 1988.
2. Y. Yamamoto, "Characteristics of AlGaAs Fabry-Perot cavity type laser amplifiers," *IEEE Jour. Quantum. Electr.*, vol. 16, no. 10, pp. 1047-1051, 1980.
3. Y. Yamamoto, "Noise and error rate performance of semiconductor laser amplifiers in PCM-IM optical transmission systems," *IEEE Jour. Quantum. Electr.*, vol. 16, no. 10, pp. 1073-1081, 1980.
4. R. Ramaswami and P. A. Humblet, "Amplifier induced crosstalk in multichannel optical networks," *IEEE/OSA Jour. Lightwave Tech.*, vol. 8, no. 12, pp. 1882-1896, 1990.
5. E. Desurvire, J. R. Simpson, and P. C. Becker, "High-gain erbium-doped traveling-wave amplifier," *Optics Ltrs.*, vol. 12, no. 11, pp. 888-890, 1987.
6. T. G. Hodgkinson and R. P. Webb, "Application of communication theory to analyse carrier density modulation effects in travelling-wave semiconductor laser amplifiers," *Elect. Ltrs.*, vol. 24, no. 25, pp. 1550-1552, 1988.
7. W. J. Miniscalco, "Erbium-doped glasses fro fiber amplifiers at 1500 nm.," *IEEE/OSA Jour. Lightwave Tech.*, vol. 9, no. 2, pp. 234-250, 1991.
8. M. Wilkinson, A. Bebbington, S. A. Cassidy, and P. McKee, "D-fibre filter for erbium gain spectrum flattening," *Elect. Ltrs.*, vol. 28, no. 2, pp. 131-132, January, 1992.

9.   A. R. Chraplyvy, "Limitations on lightwave communications imposed by optical-fiber nonlinearities," *IEEE/OSA Jour. Lightwave Tech.*, vol. 8, no. 10, pp. 1548-1557, 1990.

10.   D. M. Spirit, L. C. Blank, S. T. Davey, and D. L. Williams, "System aspects of Raman fibre amplifiers," *Proc. IEE*, vol. 137, Part J, no. 4, pp. 221-224, August, 1990.

11.   C. G. Atkins, D. Cotter, D. W. Smith, and R. Wyatt, "Application of Brillouin amplification in coherent optical transmission," *Elect. Ltrs.*, vol. 22, no. 10, pp. 556-558, May, 1986.

CHAPTER 7

Modulation

7.1 Overview

In this brief chapter we discuss the various ways in which information-bearing signals may be impressed (*modulated*) onto a photonic energy source, either an LED or a laser diode. In some applications, the information is in analog form, as shown in Figure 7-1(A). In this case, we can use a single sinusoid as the typical information source, knowing that a Fourier composite of such sinusoids can be used to represent any analog input waveform. In the digital case, shown in Figure 7-1(B), we shall assume that the information stream is binary, a string of statistically independent, equally likely "0"s and "1"s.

We shall find that some forms of modulation are extremely cheap and simple, notably direct-modulation ASK (intensity modulation), while others can become quite complex and sensitive to component misadjustment.

There is another important topic that comes most naturally under the heading of modulation techniques. When one WDMA all-optical network passes data to another, if this is to be done completely optically, then the node that does this must be a *wavelength-swapping gateway*. The chapter closes with a few words about what little is known about this class of component. All-optical networks installed in local and metropolitan areas must be linked together and it is desirable to do so without going through purely electronic gateway nodes. Wavelength-swapping technology is required for this, as discussed in Section 11.13.

7.2 Forms of Modulation

What can be modulated

Only a few parameters characterize the light emitted by the source and are therefore candidates for conveying information. They are the amplitude (or the power), frequency, phase, and polarization. The modulation formats involving the first three of these parameters are summarized in Figure 7-1.

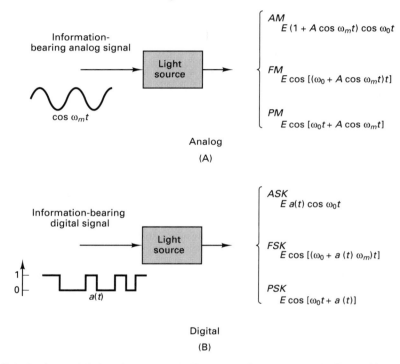

Figure 7-1. Basic modulation formats. (A) For an analog source. (B) For a binary digital source.

For analog information channels, the key word is "modulation," represented by the "M" in AM, FM, PM and PolM. For digital sources, the accepted usage is to add "SK" for "-shift keying," as in ASK, FSK, PSK and PolSK. These represent switching back and forth between values of amplitude, frequency, phase, and polarization, respectively. If the data source is binary, these acronyms are sometimes preceded by "B" – for example BPSK for binary phase-shift keying.

There are also modulation formats that work on the difference in one parameter from one bit time to the next. For example, *differential phase-shift keying (DPSK)* involves encoding a binary "0" as no change in the transmitted phase, and a binary "1" as a 180° change.

Although higher-order digital sources could be used, in principle (for example, *quaternary PSK − QPSK*), these will not be discussed here, since the extension is fairly obvious. In the literature, anything but binary modulation is usually preceded by the modifying letter; if such a letter is absent, a binary source is usually understood. Actually, in optical communication there is not nearly the incentive to use signalling alphabets of any higher order than binary that there has been in communication over traditional transmission media. With voice-grade telephone lines, for example, because the bandwidth is about 3500 Hz, modems for 2400 bps and higher work at only 2400 signalling elements *(bauds)* per second. That is, there are 2400 signalling elements per second for all these bitrates, but each baud can represent more than two signal values. For 2400 *bit-per-second* transmission, (B)PSK is used, involving only two antipodal signal values, either in-phase or antiphase. 4800-bps and 9600-bps modems both use 2400-baud transmission, but with four quadrature (QPSK) values of phase or 16 values in the phasor space, respectively. One speaks of so many "bits per baud.," the number of bits per baud being 1, 2, and 4 for modems at 2400, 4800, and 9600 bps, respectively. The loss in signal discriminability by crowding the signal points close together in phasor space is felt to be a price worth paying in order to limit the bandwidth to a value that the telephone line can handle. With fiber optics there is no such bandwidth limitation. We shall always assume one bit per baud when discussing optical modulation and reception.

ASK is sometimes called OOK for *on-off keying*, but ASK is the better term, because it captures the fact that the light source is usually not turned all the way off during a "0" bit. The ratio of "0"-bit power to "1"-bit power is the *extinction ratio r*, defined as

$$r = \frac{P_0}{P_1} \leq 1 \tag{7.1}$$

ASK is much more easily implemented in optical systems than the other forms of modulation, but has several dB of receiver detectability disadvantage compared to PSK, PolSK, or even FSK, as we shall see in the next chapter.

Direct versus external modulation

As we saw in Section 5.6 and Figure 5-23, laser diodes, while having many positive attributes, do not like to have their injection current I changed violently. Direct modulation does exactly this: the injection current is varied directly with the bitstream waveform.

Small amounts of current change can be counted on to modulate the laser optical frequency, by roughly 1 nm per milliampere, as mentioned in Section 5.8. This provides a simple and effective scheme for performing FM or FSK.

However, in doing AM or ASK the current shift becomes much larger, and chirp enters the picture. Figure 7-2(A) shows the imposition of a squarewave ASK signal on the bias current of a typical laser diode with $r > 0$. The $r = 0$ case is illustrated in Figure 7-2(B), and, from comparing the two cases, it is seen that the price paid for avoiding chirp and keeping the operating regime in the linear region of the $P - I$ curve is a nonzero extinction ratio. In the detectability discussions of the next chapter, when we examine the effect of having r greater than zero we will see that the received power signal-to-noise ratio is multiplied by $(1 - r)$.

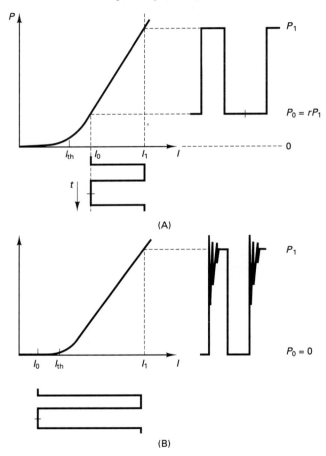

Figure 7-2. Direct ASK modulation. (A) The excursions of the binary modulating current waveform do not go below I_{th}, and therefore the extinction ratio $r > 0$. (B) In an attempt to reduce r to zero, the modulation is caused to make excursions below I_{th}, resulting in chirp.

If one attempts to decrease r by turning the light off during a "0" bit interval, as shown in Figure 7-2(B), the troubles really begin. The large drop in current required results in significant frequency shifts (*adiabatic chirp*). As if this weren't bad enough, recall that if the the light output has been zero for some nanoseconds because the injection current has been near zero, then, when the next "1" comes along, light output will begin to appear only after the random *turn-on delay*, which can be several nanoseconds. What's more, strong relaxation oscillations will ensue, and these will take many nanoseconds to die down, as shown in Figures 5-23 and 5-26. These two effects constitute *transient chirp*.

In a direct detection receiver without optical filtering before photodetection, considerable adiabatic and transient chirp might be tolerated, but in a coherent system or a wavelength-division network using either coherent or incoherent receivers, the fact that the laser may chirp to a frequency outside the receiver's passband constitutes a major system design constraint. If chirp is not minimized, energy is lost from the desired channel, and appears as undesired crosstalk in adjacent channels, so that the chirping is doubly harmful.

Needless to say, these effects must be avoided at all costs in practical systems, and one of the costs that is willingly paid is to avoid turning the laser off completely.

Chirp and turn-on transient effects can be avoided completely if the light generation and modulation processes can be completely separated. The laser can be allowed to run CW, and an *external modulator* is interposed between the laser and the node output, as shown in Figure 7-3. As long as the reflections from the modulator are small enough, they will not increase the laser linewidth beyond the desired amount (Section 5.5). Often the Faraday isolators described in Section 3.13 are required.

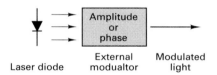

Figure   7-3. External modulation to avoid inducing chirping and other laser transients.

Another ASK scheme that allows the laser to run at almost constant power level is *FSK/ASK* [1], illustrated in Figure 7-4. The transmitter emits an FSK signal with a frequency excursion large enough that at the receiver a narrow optical filter passes only one of the two frequencies to the photodetector. Because of its simplicity, and the fact that essentially zero extinction ratio can be achieved, this scheme is very attractive for use at the transmitter. When used at the receiver, it has the disadvantage in very dense WDM systems that it uses up at least twice the bandwidth that would be needed if external ASK modulation had been used at the transmitter.

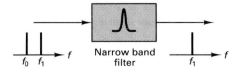

Figure   7-4. FSK/ASK modulation.

External modulators are required in almost all coherent systems in order to maintain a narrow transmitter linewidth. This is because, like a radio receiver, a coherent optical receiver must see the ASK, PSK, FSK, or PolSK modulation imposed on a sinusoid, not on a random narrowband noise. The penalties for nonzero linewidth are surprisingly stiff, as will be seen in Section 8.13.

7.3 External Phase and Amplitude Modulators

The acoustooptic deflector

Several mechanisms are candidates for use in external modulators. Early on, a reasonable candidate seemed to be the acoustooptic deflector (Section 4.8) because it had proved so successful with laser printers. In this form of printer, introduced in the mid-1970s, a beam of light from a CW helium-neon gas laser was acoustooptically deflected to hit or to miss a knife-edge obstruction, depending on whether the printer was to print white or black, respectively. This was done either by modulating the RF drive on and off or by using FSK. The extinction ratio was thus essentially zero, very important for printing. It was this application that propelled the acoustooptic deflector to the point that it became a commodity component (Figure 4-22(A)). The advent of the high-power laser diode eventually displaced the acoustooptic deflector in all laser printers, and in the process made low-cost consumer laser printers possible.

The acoustooptic deflector is a reasonable candidate for an ASK modulator only if the bitrate is low enough, less than around 10 Mb/s. What is a fast access time for printing (tens of nanoseconds) is prohibitively slow for external modulators in even subgigabit nodes. For this reason, the search for fast external modulators for optical communications led to electrooptic approaches.

Quantum-well modulators

We learned in Section 5.4 that the curve of device gain with wavelength has a much sharper drop at the long-wavelength side for all quantum-confinement laser types than for bulk lasers. This effect allows one to ASK-modulate some fixed wavelength by current injection in a quantum-well external modulator. To get a very wide choice of wavelengths to be modulated, different active-region compositions are required.

First- and second-order electrooptic effects

The most useful external modulators are based on a voltage-dependent phase retardation in some material [2, 3], either semiconductor, insulating crystal, or organic polymer. Figure 7-5 shows a commercial ASK (intensity-) modulator using lithium niobate insulating crystal.

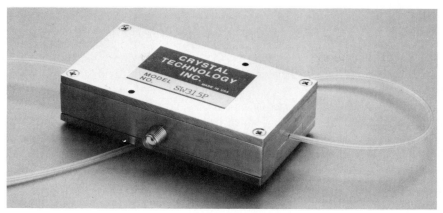

Figure   7-5. A commercial external Mach-Zehnder ASK modulator (Courtesy of Crystal Technology, Inc.).

In the semiconductor, the change in phase retardation is effected by changing the index via the injection current. With crystals or anisotropic polymers, the electrooptic effect is used, i.e., the voltage dependence of index. The modulator structure is usually implemented as a waveguide lithographed in or on a substrate material. The semiconductor approach has the potential advantage that the laser diode and the modulator could, in principle, be fashioned from the same material in a single monolithic structure.

The effect of an applied field intensity $\mathbf{E}$ on the change of index seen by incident light of a particular SOP is

$$n = \alpha E + \beta E^2 \qquad (7.2)$$

E is the magnitude of the vector electric field $\mathbf{E}$, and α and β are tensors, expressing the fact that the electrooptic effects are very anisotropic, depending critically on the orientation of the input light relative to the axes of the material. For certain orientations, the β term is identically zero, and we have the *linear electrooptic* or *Pockels effect*. If, on the other hand, an axis is chosen such that α is zero, then the *quadratic electrooptic* or *Kerr* effect is observed. It is the linear Pockels effect that is usually employed, and in this case the difference between the index when a field E is applied, and that when $E = 0$, is

$$\delta n = \frac{n_o^3}{2} r_{ij} E_j \qquad (7.3)$$

where n_o is the index for $E = 0$, and r_{ij} is the *linear electrooptic coefficient*, an element of a tensor, the i and j being the axes of the appropriate coordinate system of the anisotropic material.

From this,

$$\Delta\phi \;=\; \frac{2\pi}{\lambda}\,\delta n_i\, Z \;=\; \frac{\pi Z}{\lambda}\, n_o^3\, r_{ij}\, E_j \tag{7.4}$$

the desired linear dependence of phase on electric field strength. Z is the interaction length.

To date, most external modulators have been implemented as separate devices in lithium niobate (LiNbO$_3$), which has a very high electrooptic coefficient along certain axes. However, one of the problems with LiNbO$_3$ is that it has to be grown as a crystal. Quite recently, polymers have become even more attractive, since they can be deposited over large areas fairly inexpensively via spin coating. The desired optical anisotropy of the polymer is "frozen in" by applying a voltage across the material during the solidification process. In this way an anisotropic material composed of long molecules of the desired properties can be created.

External phase modulators

Using the simple idea embodied in Equation (7.3), an external phase modulator can be built in the form shown in Figure 7-6. The two electrodes apply the desired electric field along the distance Z.

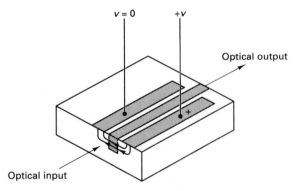

Figure 7-6. Structure of an external phase modulator made by forming titanium waveguide regions in a lithium niobate crystal.

External intensity modulators

There are two widely implemented forms of external intensity modulators, as shown in Figure 7-7(A) and (B).

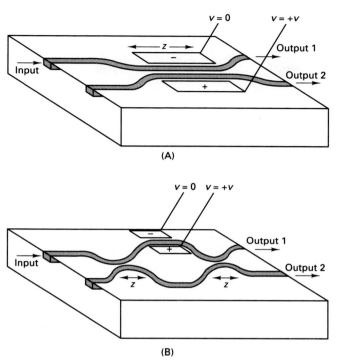

Figure 7-7. Two forms of external ASK modulators in integrated optic configuration. (A) Directional coupler geometry, and (B) Mach-Zehnder configuration.

The *directional coupler* modulator is simply the fused biconical tapered coupler of Section 3.10 and Figure 3-22 reimplemented (but without the taper) in, say, lithium niobate, with two electrodes suitably positioned, as in Figure 7-7(A), so as to cause a voltage-dependent phase shift in the part of both waveguides that lies within the coupling region. Recall that that fraction of the input light that is delivered to, say, output 1, is a function of the effective length Z of this coupling region, and that this function is a raised sinusoid. That is, given a value of Z, an applied voltage can be found that delivers all the light to output 2, and another that delivers it all to output 1. By varying the modulating voltage between these two values, ASK modulation is obtained.

The *Mach-Zehnder amplitude modulator* of Figure 7-7(B) is nothing more than one section of the MZI chain discussed in Section 4-6. Ideally, when the path-length difference for the two arms is zero or a multiple of 2π, full light output appears at output 1 and none to output 2, but when the difference changes by π, output 1 carries zero light, all of it going to the output 2.

The device in Figure 7-5 is a Mach-Zehnder external ASK modulator for 1.5 μ. It requires an applied voltage of 11 volts, achieves an extinction ratio of 0.006, and can support bandwidths up to 3 GHz with an insertion loss of about 6 dB.

Generally speaking, Mach-Zehnder intensity modulators tend to have a better extinction ratio than directional coupler modulators because the phase cancellation, on which they both depend, can be more precisely controlled in the former case.

Polarization sensitivity

The devices just described depend on the incident light having a very particular SOP, usually plane-polarized in a specified plane. Since the device is usually situated immediately adjacent to the light source, random SOP drift, which can be an issue at the receiver, is absent. However, the light that does enter the modulator must not contain very much light of the orthogonal polarization, because this light will not be modulated. It will appear at the output and will degrade the extinction coefficient.

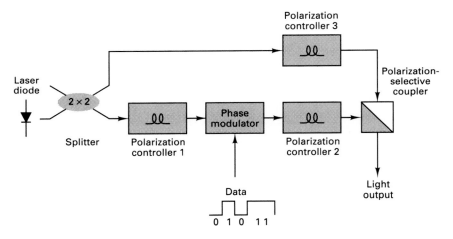

Figure   7-8. Polarization-shift keying (PolSK) external modulator.   (From [4], by permission IEE).

7.4 Polarization Modulators and Scramblers

In the next chapter, one of the more curious forms of receiver we shall discuss will be one that operates from changes of received signal state of polarization (SOP) between two orthogonal conditions. There are many ways to switch the polarization between two orthogonal states. One that has been used experimentally for lightwave transmission [4] is shown in Figure 7-8. The incoming linearly polarized light is split and fed to the two inputs of a bulk optic polarization-selective coupler, which has the property that when the phase modulator imposes a 0° phase shift, right-circular polarization results, due to setting the polarization controller to produce a 90° rotation of the plane of linear polarization of its input.   Then, in order to produce the orthogonal left-circular state, the phase modulator is made to introduce a 180° phase shift.

If they can be modulated sufficiently rapidly, the same devices that are usable as polarization modulators can be used as polarization scramblers. These devices simply dither the polarization state back and forth between orthogonal states more than once per bit time in order that a receiver set to respond with 100 percent detectability at some arbitrary SOP will always have a partial detectability.   In fact, assuming that during one bit time the light spends equal time in each of the two orthogonal states, if one considers the Poincaré sphere of Section 3.13 and Figure 3-29, and recalls that any two orthogonal states are antipodal on the sphere, it is clear that no matter where the receiver SOP point is, it will be no worse than 3 dB away from one of the two signal SOP points. (Remember that a 90° great-circle distance on the Poincaré sphere means the two radiations are 50 percent correlated). For the gigabit and subgigabit data rates of interest for lightwave networks, such devices as the lithium niobate phase modulator described in the previous section are capable of the speed needed to do the scrambling at least twice per bit time T.

Another clever polarization scrambling scheme is the one due to Maeda [5]. Just before arriving at a polarization-sensitive receiver element, the light is passed into a short section of multimode fiber that has been mechanically distorted (for example, by being tightly twisted), so that a large number of modes are excited, each of whose SOP evolves differently over the short length of multimode fiber. By the time the signal arrives at the receiver (over the multimode fiber section), the receiver will see a mixture of modes having different SOPs.

7.5 ASK Remodulators for All-Optical Gateways

Although the bandwidth available in silica fiber is copious, it is not infinite, and as an individual all-optical LAN or MAN grows in the number of nodes N that it supports, sooner or later there will not be enough wavelengths available to handle the entire address space of all users.   For example, while $N = 1000$ might be an adequate number for an individual LAN or MAN, at least 100,000 is a more realistic estimate for an interconnected set of such systems, for example the total network of a large corporation. Public networks will be even larger.

This need for *wavelength reuse* is exactly the analogue of a problem met in traditional nonoptical networks, the exhaustion of the address space.   If there are not enough bits in the address field to reach all users everywhere, the same set must be reused concurrently in all the subnetworks, and there must be a gateway function between them that maps an address in one subnetwork to another address (chosen from the same address space) in the next subnetwork.

To do this in a WDMA network requires the capability that the information-bearing modulation reaching a gateway at a particular wavelength be reimposed on light of a different wavelength when it is forwarded into the second network.

This is easiest to do when the modulation format is ASK. Figure 7-9 shows a scheme [6] developed at Nippon Electric Research Laboratories.

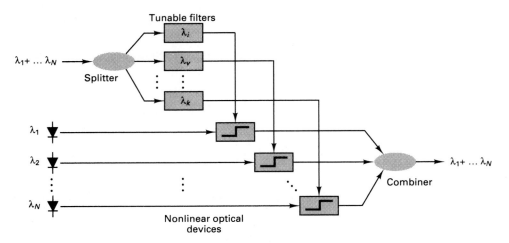

Figure 7-9. Wavelength-swapping remodulator for use in all-optical inter-network gateways. (From [6], © 1990, IEEE).

Suppose that the modulation entering the input at λ_i is to be remodulated onto light at λ_j emanating from the output. A tunable filter applied to the input and tuned to λ_i drives the control input of a bistable optical device that passes light from a laser running CW at λ_j or not, depending on whether the light intensity reaching the control input is high or low, respectively. The desired bitstream of the λ_i signal into input 1 will be reimposed on the CW light at λ_j at the output, as desired. Note that the system serves not only as a wavelength swapper, but also as a power regenerator.

An experimental gateway structure [6] was built that could handle up to 8 channels over a range of 6 Å, provided 13 dB of remodulation gain, and could switch channels in about 7 nsec. The fast switching time was obtained by using a three-section tunable laser diode of Section 5.11 as a tunable filter by biasing it below threshold.

7.6 What We Have Learned

Modulating binary ASK information onto a laser diode seems a trivial thing to do until one looks at the consequences of letting the injection current stay below threshold very long. The resultant transient effects produce so much chirp that it becomes necessary either to sacrifice extinction ratio by using direct ASK or to use external modulation.

FM by direct modulation of the injection current is feasible, simple, and attractive, and when used with a filter-based receiver in the FSK/ASK mode requires only that there be twice as much bandwidth available as with, say, PSK or PolSK.

External modulation is required for both PSK and PolSK. In these categories, it is also possible to work from the difference between successive bits (DPSK or DPolSK), where a "no-change" condition signals a "0" and a reversal signals a "1."

All-optical signal remodulation is feasible, but complex, and one may expect to see heightened activity in improving the state of this art, as the need for all-optical interconnection of all-optical subnetworks becomes acute.

7.7 Problems

7.1 What transformation would you make on a squarewave bit pattern so that when you apply it to the input of a PSK external modulator, you get binary FSK out of the modulated device? Draw the waveform for a 001011 bit pattern.

7.2 Referring to Figure 5-26, estimate (A) The extinction ratio r, (B) The adiabatic chirp in GHz, and the peak transient chirp in GHz.

7.3 A 1.55-μ laser diode is operated at 100-percent quantum efficiency and a power output of 1.0 milliwatt. Assuming very low modulation frequencies, how wide an FSK swing can be gotten before the optical power output change that accompanies this frequency shift exceeds 10 percent?

7.4 If the ASK direct-modulation currents of the laser diode of Problem 5-1 are 6 mA and 30 mA for "0" and "1," respectively, (A) What is the extinction ratio r? (B) What is the frequency shift between the two states?

7.5 You are given the choice of a number of electrooptic materials, all with an index of around 2.0 along all crystal axes, and asked to design an external modulator for use at 1.55 μ with the electric field applied longitudinally. Because of space limitations, you are required to constrain the interaction length to 1.0 cm, and to keep the RC time constant small, you must use an electrode spacing no greater than 1 μ. You are also given a driver-amplifier that produces 0 or 20 volts, depending on whether its input is a binary "0" or "1." (A) What are the units of r_{ij} generally? (B) What value of r_{ij} would you need in order to make a fairly linear PSK phase modulator that swings the phase over $\pm 45°$? (C) What value of r_{ij} would you need to produce a Mach-Zehnder ASK intensity modulator?

7.6 You would like to put a 1-centimeter length of $LiNbO_3$ waveguide along one of the arms of a Mach-Zehnder tunable filter stage (Section 4.6 and Figure 4-14) in order to tune it. The device is to be used at 1.3 μ. Assume that the electric field is applied across the waveguide between parallel 1-centimeter electrodes 5 μ apart. What is the applied voltage necessary for achieving the required 180° phase shift? The appropriate electrooptic coefficient for $LiNbO_3$ is $r_{33} = 31 \times 10^{-12}$ meters per volt and the index is 2.1.

7.8 References

1. I. P. Kaminow, "FSK with direct detection in optical multi-access FDM networks,," *IEEE Jour. Sel. Areas in Commun.*, vol. 8, no. 6, pp. 1005-1014, August, 1990.

2. E. Hecht, *Optics - Second Edition*, Addison-Wesley, 1987.

3. R. Alferness, "Waveguide electrooptic modulators," *IEEE Trans. on Microwave Theory and Tech.*, vol. 30, no. 8, pp. 1121-1137, 1982.

4. T. G. Hodgkinson, R. A. Harmon, and D. W. Smith, "Polarization-insensitive heterodyne detection using polarization scrambling," *Elect. Ltrs.*, vol. 23, no. 10, pp. 513-514, May, 1987.

5. M. W. Maeda and D. A. Smith, "New polarization-insensitive detection scheme based on fibre polarization scrambling," *Elect. Ltrs.*, vol. 27, no. 1, pp. 10-12, January, 1991.

6. S. Suzuki, M. Nishio, T. Numai, M. Fujiwara, M. Itoh, S. Murata, and N. Shimosaka, "A photonic wavelength-division switching system using tunable laser diode filters," *IEEE/OSA Jour. Lightwave Tech.*, vol. 8, no. 5, pp. 660-666, 1990.

CHAPTER 8

Detection and Demodulation of Optical Signals

8.1 Overview

In other chapters, we have discussed some of the lightwave system building blocks–the sources of light, the modulation of binary information onto the light, and the various processes involved in the fiber, couplers, splitters, amplifiers, and filters that intervene before the light finally impinges on the optical receiver and is converted back to binary information in electrical form. Before we can go further in understanding the architecture of the total network and considering specific designs, we must understand the receiver: what it takes to distinguish between a "1" bit and a "0" bit reliably using one of the available modulation formats and receiver structures.

Issues to be settled include the proper figure of merit for receiver performance, basic limits on that performance, which type of photodetector component to use, what other elements in addition to the photodetector constitute the receiver, and whether this receiver is expected simply to register presence or absence of light (direct detection) or to permit determination of "1" or "0" by exploiting knowledge of optical frequency or phase (coherent detection).

This chapter plays a pivotal role in the present volume. On the one hand, it looks forward, in that it serves to complete most of the background needed to discuss the architecture of the network as a whole. On the other hand, it looks backward, in that it integrates the various factors that interact at the lightwave receiver, matters that arose in Chapters 3 through 7.

Several helpful references deal with the theory of optical detection and demodulation at length, in particular the classic 1980 study by Yamamoto [1] and several tutorial surveys, [2, 3]. The physics of photodetection is treated in depth in References [4-7].

Types of receivers

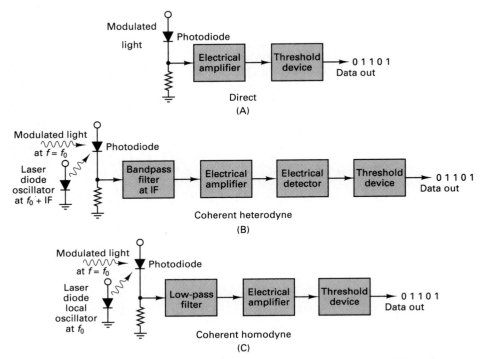

Figure 8-1. Three basic receiver types. (A) Direct detection with a PIN or avalanche photodiode, (B) Coherent heterodyne reception, and (C) Coherent homodyne reception.

Figure 8-1 depicts, in a very rudimentary form, the three different types of receiver. In the *direct-detection* receiver of Figure 8-1(A), some form of photodetector serves to convert the stream of incident photons into a stream of electrons, this current is amplified, and a determination is made of whether the current is mostly above or below a certain threshold.

The direct-detection form of receiver, as used with amplitude-shift keying (ASK), has been called, with some justification, "inelegant, if not downright tacky" [9]. It is about as sophisticated in principle as the blinking lights used in the classical Roman civilization to communicate between mountaintops [10]. All the receiver is doing is telling whether "there is or is not light" (Genesis 1:4). Or, we might say that the principle of operation of a direct-detection receiver is merely that of the earliest crystal radio receivers.

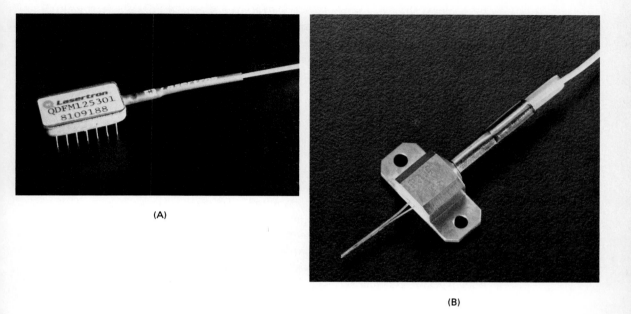

(A)

(B)

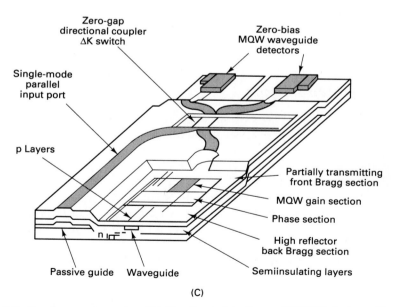

Zero-gap
directional coupler
ΔK switch

Zero-bias
MQW waveguide
detectors

Single-mode
parallel
input port

p Layers

Partially transmitting
front Bragg section

MQW gain section

Phase section

High reflector
back Bragg section

Semiinsulating layers

Passive guide   Waveguide

(C)

Figure 8-2. Receiver components. (A) For direct detection, a PIN/FET device. (Courtesy of Lasertron, Inc.). (B) For direct detection, an avalanche photodetector packaged with its high-voltage supply. (Courtesy of ATandT.). (C) An experimental coherent receiver chip. (From [8], © 1990 IEEE).

By analogy, the two forms of *coherent receiver* shown in Figure 8-1(B) and (C) are patterned after the later superheterodyne radio receivers. A monochromatic CW light source forms the *local oscillator*. Light from this source falls on the photodetector, along with the much weaker arriving information-bearing optical signal, and, due to the nonlinear nature of the photodetector, a strong output beat signal is developed at the difference frequency. This *intermediate frequency (IF)* is chosen to be at the band center of a fixed-tuned *IF filter*. Amplification of the weak signal at the IF takes place electrically. Then, after the signal is sufficiently strong, an appropriate demodulation and thresholding takes place. If the transmitter uses ASK, simple energy detection precedes the thresholding operation; if phase or frequency modulation is imposed at the transmitter, the receiver is somewhat more complex.

The difference between the heterodyne optical receiver of Figure 8-1(B) and the homodyne version of Figure 8-1(C) is that in the latter the IF is exactly zero. That is, the optical frequency of the incoming information-bearing signal and that of the local oscillator are identical. As we shall see, even though this identity is somewhat difficult to achieve, it has often proved worth trying, because less received signal power is required for a given BER than with a heterodyne coherent receiver.

Both types of coherent receiver require a smaller received SNR for a given bit error rate than does the direct-detection receiver. However, the latter option has consistently been the one to be implemented in practical lightwave systems because of its simplicity. It has recently gained additional respectability by the advent of convenient photonic amplification (Chapter 6) with the result that when these amplifiers are incorporated into a direct-detection system, the performance is almost as good as that of a coherent system.

Types of photodetectors

There are many ways of converting incident light into a current or voltage, and the general class of devices that do so are called *photodetectors*. There are bolometers, photomultiplier tubes, phototransistors, photodiodes, and others. Of these, only the semiconductor *photodiode* subclass has proved to have the right combination of sensitivity, low noise, small size, low cost, and high speed of response to serve satisfactorily in a fiber optic communication system. The internal processes by which they convert a photon flux P into a current I are exactly the reverse of the processes outlined in Chapter 5 that function within a light-producing semiconductor LED or laser diode.

There are three types of semiconductor photodiodes, and these will be discussed in some detail later in this chapter: PN diodes, PIN diodes, and avalanche photodiodes (APDs). As in Chapter 5, the P and the N refer to positively- and negatively-doped extrinsic semiconductor materials, respectively. The I refers to intrinsic (undoped) material. PN and PIN diodes act to convert a large fraction of the incoming photons, one-by-one, to electrons that flow into an electrical amplifier circuit, whereas the APD acts to generate many electrons from an individual incoming photon, and thus provides an internal form of gain. Because the current generated by a PIN diode is so small and therefore so vulnerable to noise, it is common to package the amplifier, often one made from field-effect transistors (FETs), into a single *PIN/FET* package.

Figure 8-2(A) shows such a commercial PIN/FET component, and 8-2(B) an APD. At (C) is shown an experimental single-chip coherent heterodyne receiver developed at Bell Laboratories [11].

Ways of combatting thermal noise

It will turn out that, under most operating conditions met in practice, the dominant noise component that must be dealt with is likely to be thermal noise developed at the input to the electrical amplifier. There are three approaches to overcoming this limitation by amplifying the signal before it leaves the photodetector. The first is to use an APD. The second is to precede the photodetector with one of the photonic amplifiers described in Chapter 6. The third is to employ coherent detection with such a high local-oscillator power level that the thermal noise is much smaller than the useful beat products between local oscillator and incoming signal.

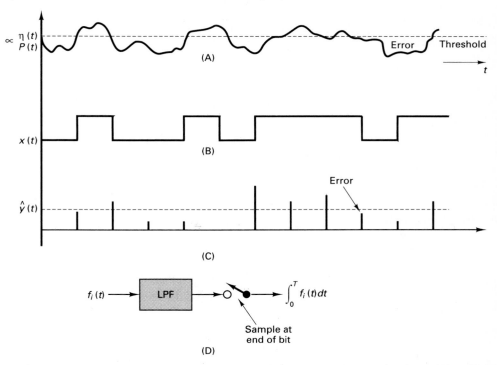

Figure 8-3. Showing (A) that the optical receiver sees $\eta(t)$, an impure signal consisting of a mixture of noise with the clean signal $x(t)$, shown at (B). (C) The receiver uses $y(t)$, which is derived from $\eta(t)$, to determine whether the observed mixture was mostly "1" or mostly "0" and uses a threshold (dashed line) to decide. Occasionally an error is committed (arrow). (D) Ideal matched filter, represented in the *integrate-and-dump* form.

Bit error rate as the key system parameter

Figure 8-3(A) shows a typical plot of optical power as a function of time for ASK, along with the corresponding hypothetical version (B) that would have existed if all sources of noise and distortion had been absent. If one could be sure that the random fluctuations during a "1" bit and a "0" bit were exactly equal and that the probabilities that the transmitter sent a "0" and a "1" were equal, then one would set some sort of measurement threshold exactly in the center, as indicated by the dashed line. If during a given bit time the waveform seemed to be predominantly above the threshold, a "1" would be declared, otherwise a "0," as in Figure 8-3(C).

Clearly the unpredictable noise in the system will cause this decision process to commit an occasional error, as shown by the arrow. In Figure 8-3(D) is shown the *integrate-and-dump* filter receiver (to be discussed later) which uses bit-clock information to sample each bit at its very end where all the energy in the bit has arrived. As we shall see, this form of receiver is the one we want because it makes errors less frequently than others.

If one assumes that there is no statistical correlation between successive bits at the transmitter and no correlation between the noise waveforms in successive bit periods, and that "0" and "1" are equally likely, then everything we need to know about the transmitter-receiver path is contained in the *bit error rate* (BER), the probability that a "0" will be mistaken for a "1" or vice versa. From the BER, one can deduce the probability of occurrence of all sorts of higher-level protocol events, such as the probability that a long block of bits will contain one or more errors, that a request for a connection will never arrive properly, that the address on a packet will be wrongly received, and so forth.

In this book we shall make these assumptions of statistically independent transmission of successive bits, independence of noise waveforms from bit to bit, and equiprobable "0"s and "1"s.

The link budget

When it comes to designing the system, no issue is more important than accounting for all the things that can happen to the power of the signal on its way from the laser diode or LED to the photodiode and thence to the threshold decision point, and also all the ways that noise and other things can corrupt this signal. At the receiving end, one finally calculates the *received signal-to-noise ratio (SNR)*. This is often defined as the ratio of signal power to noise power at the threshold point.

It is very important, when discussing SNR, to be careful to distinguish electrical SNR and optical SNR. Since the power generated in the photodiode load resistance is proportional to the square of the incident optical power, the same number of dB of SNR means different things when dealing with the optical and the electrical environment. An ideal noiseless photodetector that saw an incident mixture of signal and noise with D dB of optical SNR would produce at its output a mixture having $2D$ dB of electrical SNR. Similarly an SNR loss of D dB at the optical level will be twice as

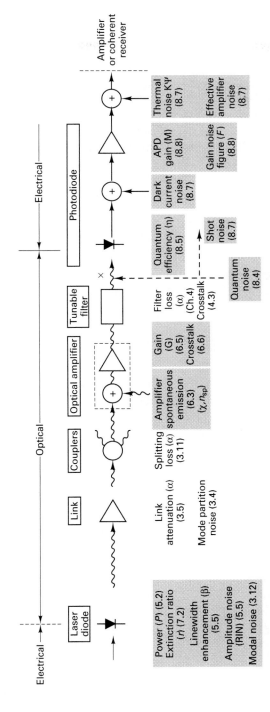

Figure 8-4. Elements of the link budget, listing the symbol for each appropriate parameter and the section of this book where each is discussed. Shading indicates those included in the detectability discussions of this chapter.

harmful to the final bit error rate performance as the same number of dB occurring in the electrical part of the receiver.

In order to achieve a given bit error rate, the minimum acceptable received SNR will be somewhat different for different modulation formats and forms of receiver [2, 3]. This is because, when one compares different schemes experiencing the same optical SNR at the receiver input, there will be differences in the SNR as seen at the point where the threshold decision is made. It will be one of our major tasks in this chapter to explore the relative advantages and disadvantages of these different modulation and receiver options in just these SNR terms. We shall do so from the standpoint of the physical principles involved and their detection-theory consequences, leaving details of practical devices and circuit designs for several excellent reference books on the subject [4, 5, 12].

The link-budget exercise could be viewed as designing the power levels in the system to lie between two extremes, an upper limit imposed by consideration of nonlinearity effects (Section 3.13) or eye safety (Section 3.16), and a lower limit set by the SNR required at the receiver for a given bit error rate.

Photons per bit as a metric

The received SNR will depend partly on the product of received power and duration of a bit—the received signal energy per bit. In preference to speaking of the received signal energy in joules or ergs, it will prove very useful to use the number of photons per bit. Among other things, this helps account for two things in one quantity, the energy and the degree to which the light has become so weak that its discrete nature as individual photon arrivals must be taken into account. Using Planck's equation $\mathcal{E} = hf$ gives the following handy formula, applicable to a wavelength of 1.5 μ

$$\text{One milliwatt} = 7.5 \times 10^{15} \text{ photons per second} \tag{8.1}$$

from which the number of photons per bit for a given bitrate and received power is easily gotten.

8.2 Components of the Link Budget

Figure 8-4 shows a block diagram of the complete path from laser to receiver, broken down according to the various elements that enter into a link-budget calculation. Those that it will be necessary to discuss in this chapter are shown shaded.

Signal link-budget items

Among the variables that determine the received signal power at the point where the decision is made are:

- **Laser output power** actually delivered into the fiber pigtail. This number will be dictated by the choice of components available, or by power limits set by either nonlinearities (Section 3.14), or eye-safety considerations (Section 3.16).

- **Fiber attenuation**

- **Splitting losses**: in couplers, combiners, and splitters, together with the excess losses in these components,

- **Excess losses** in connectors and splices,

- **Extinction ratio** r for an ASK system. Recall that r is the ratio of optical power for binary "0" to that for binary "1."

- **Optical amplifier gain** G, and

- The **quantum efficiency** with which the photodetector converts photons to carriers, generally less than 100 percent.

In this chapter, we shall deal explicitly only with the last three, assuming that the combined effect of the others is to deliver an optical power P for binary "1" and rP for binary "0" to the receiver input point. In Figure 8-4, $\times$ marks the spot.

Noise link budget items

We have discussed several kinds of additive noise and several kinds of multiplicative noise. Additive noise remains when the signal disappears, but multiplicative noise either is inherent randomness within the signal itself or is produced in some device only when the signal is present. Among the sources of additive noise that must be considered are:

- **Dark current noise** within the photodiode, caused by thermal processes,

- **Thermal noise** or Johnson noise in the resistive part of the input impedance of any electrical amplifier that follows the photodetector,

- **Amplified spontaneous emission**, generated in any optical amplifier that appears on the path between laser and photodetector,

- **Electronic amplifier noise** arising within the amplification stages following the photodetector, and

- **Crosstalk** from adjacent channels or from gain variations in laser-diode amplifiers,

Sources of multiplicative noise include:

- **Mode partition noise** in the case of a nonmonochromatic laser sources driving single-mode fiber,

- **Modal noise**, present in multimode fibers only, and due to the randomly varying relative excitation of a small number of propagation modes by an insufficiently wideband source,

- **Laser phase noise**, the process that causes a laser to exhibit a finite nonzero spectral linewidth. This is a factor not only in the transmitter laser, but in a laser that might be employed as a local oscillator in a coherent receiver.

- **Laser amplitude noise** (expressed by the *RIN* parameter), present in both transmitter and local oscillator lasers.

- **Gain-factor noise**, present in APDs, which have the advantage that they provide gain but the disadvantage that the gain is randomly time-varying,

and last but not least

- **Quantum noise**, present when the light signal gets so weak that its quantum nature becomes apparent by the individual photon arrivals' becoming evident. At higher levels, the photon arrivals are so frequent that they are smoothed out. Quantum noise may be modelled as a series of arriving impulses of equal height.

- **Shot noise**, the quantum-noise waveform as shaped by the finite bandwidth of the optical receiver. The two terms "quantum noise" and "shot noise" are often misused in a way that does not distinguish them from each other. Also, sometimes "shot noise" is (confusingly) used to refer to any noise component that is so weak that its quantum nature is observable. In this book, we shall use the term "shot noise" to refer only to the intrinsic quantum noise in the signal itself, *after* it has been smoothed by the finite-bandwidth elements of the photodiode and the circuitry that follows.

We shall omit further discussion of the first two of this long list of possible multiplicative noise sources, but will deal with the last five.

In all interesting cases, the various noise waveforms may be considered statistically independent and usually (but not always) gaussian. As long as they are independent, their powers add, so that one may reckon the relevant SNR by adding in the denominator all the noise power components being considered at that point in the discussion.

As this chapter proceeds, we shall use different versions of the block diagram of Figure 8-4 to highlight the problem at hand. For example, the ideal photon-counting receiver omits almost all the elements in the figure and focusses only on the level of the signal appearing at the receiver input. For simple direct-detection systems, neither laser phase noise nor amplitude noise is usually an issue, so they will never appear in the block diagrams appropriate to that discussion. Coherent systems never use tunable filters and seldom use APDs, so optical filter crosstalk and random fluctuations of APD gain do not appear in those discussions.

8.3 Basic Detection Theory

The maximum a posteriori probability (MAP) receiver

In its idealized form, a digital communication system can be represented in the form shown in Figure 8-5 [13]. The bits actually transmitted arise from an *information source* that generates a stream of bits, each of which is assumed to be independent of what went before. In other words, the probability

$$P \, (x \text{ is sent}) \equiv P(x) \tag{8.2}$$

(where $x = 0$ or 1), does not depend on what was sent during any of the previous bit times. $P(x)$ is called the *a priori* probability. $P(x)$ can be thought of as a probability density, a function such that the probability that x lies between some x and $x + dx$ is $P(x)dx$. In our binary communication situation, $P(x)$ consists of two impulses, one at $x = 0$ and the other at $x = 1$.

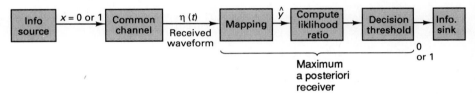

Figure 8-5. Idealized binary digital communication system.

Let $\eta(t)$ be the entire received signal waveform and let y be some single parameter derived from it at the end of each bit time (Figure 8-3). Thus, for each possible $\eta(t)$ there is a single value of y. In general, the communication channel is noisy, and so there will be a distribution or smear of possible y values for any given x, which could only have been either 0 or 1. When the received signal waveform $\eta(t)$ is mapped onto the observed value $\hat{y}$ of the variable y, then

$$P(y \leq \hat{y} \leq y + dy) \; = \; P(y) \, dy \tag{8.3}$$

where $P(y)$ is called the *a posteriori* probability density.

Now the *joint probability density* $P(x, y)$ that x was sent **and** y was received can be written in two ways

$$P(x, \, y) = P(x)P(y\,|\,x) = P(y)P(x\,|\,y) \tag{8.4}$$

where $P(y|x)$ and $P(x|y)$ are both *conditional* probability densities, the first being the probability that y was received **if** x was transmitted, and the second the probability that x was transmitted **if** y was received.

What we really want to know is the probability that if a given y was received, then a particular x was transmitted, and we can get this from rewriting the right-hand equality as

$$P(x|y) = P(y|x)\,\frac{P(x)}{P(y)} \tag{8.5}$$

known as *Bayes' rule*. From this we can form the ratio

$$\frac{P(1|y)}{P(0|y)} = \frac{P(1)}{P(0)}\,\frac{P(y|1)}{P(y|0)} \tag{8.6}$$

where the second factor on the right is called the *likelihood ratio*, R. We have assumed that "0" and "1" are *a priori* equally likely, and so the first factor is unity.

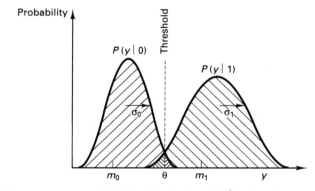

Figure 8-6. The *a posteriori* probability distributions $P(y|x)$ for the two cases in which the transmitted signal x is either "0" or "1". y is the receiver derived signal variable.

It can be shown that the optimum receiver, one that has the minimum probability of making an error, is a *MAP receiver* ("maximum a posteriori probability"), one that, having observed a received signal waveform $\eta(t)$ and having computed $\hat{y}$ from it, goes on to declare its choice to be that alternative for x for which the likelihood ratio is the maximum. In other words, it implicitly computes $P(1|y)$ and $P(0|y)$, then takes the ratio R of the two, and always declares "0" if R is less than unity and "1" if the ratio is greater. The bit error rate, or probability of error, is

$$BER = \frac{1}{2}\,\text{Prob}\,(R\rangle 1\,|\,0) + \frac{1}{2}\,\text{Prob}\,(R < 1\,|\,1) \tag{8.7}$$

In reality, of course, there is no real "computing" going on in the receiver, unless you want to call mapping $\eta(t)$ onto y a computation. As Figure 8-6 shows, knowing the probability density functions $P(y|0)$ and $P(y|1)$, a *decision threshold* is set at the point at which the area under the two tails of the $P(y|x)$ curves that lie on either side of the threshold are equal, as shown by the dashed line. If, at the end of each bit time, the observed value $\hat{y}$ of the variable y falls to the left, "0" is declared (passed on to the information sink in Figure 8-5), and if to the right, "1" is declared.

Given the a priori probabilities $P(x)$ that describe the data source, and $P[\eta(t)|x]$ describing the channel, the optimum (minimum BER) MAP receiver will be the one that performs the most effective mapping of η into y and then sets the threshold as just described. By "most effective" is meant the particular mapping of η onto y that serves to differentiate the values of y most completely. This in turn means that the receiver should be designed so that the areas under the tails beyond the threshold are as small as possible.

Setting the threshold. The Q-function

In practice it is usually true the variable y can either be proved to have gaussian first-order statistics or to have first-order statistics that can be well approximated as gaussian. In this case, the probability density function $P(y)$ is of the form

$$P(y|x) = \frac{1}{\sqrt{2\pi\sigma}} \exp\left[-\left(\frac{y-m}{\sqrt{2}\,\sigma}\right)^2\right] \tag{8.8}$$

where σ is the standard deviation (square root of the variance), and m is the mean.

In general, both the mean and standard deviation will depend on whether $x = 0$ or $x = 1$, as the subscripts in Figure 8-6 show.

For a gaussian distribution, the area under a curve $P(\xi)$ to one side of a point $\xi = \alpha$ (e.g., the decision threshold θ) is given by the well-known Q-function

$$Q(\alpha) = \frac{1}{\sqrt{2\pi}} \int_\alpha^\infty \exp(-\xi^2/2)d\xi \approx \frac{1}{\sqrt{2\pi}\,\alpha} \exp\left(-\frac{1}{\alpha^2}\right) \tag{8.9}$$

The approximation is accurate to one percent for $\alpha \approx 3.0$ and gets even more accurate as α increases.

In the binary communication case of Figure 8-6, we want to set the threshold (dashed line) for minimum probability of error. Assuming that the "0"s and "1"s are equally probable, and that mistaking a "0" for a "1" is no more nor less harmful than the converse, it is clear that the threshold value θ should be set for equality of the areas under the tails of the curves (shaded areas).

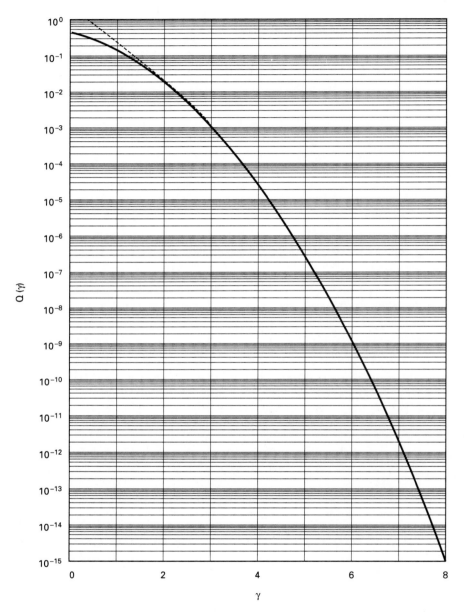

Figure 8-7. The Q-function for determining bit error rate when the noise is gaussianly distributed in amplitude.

In general, this does not mean setting θ where the curves are of equal height, but it can be shown [3] that the following approximation gives answers that are quite close to the optimum:

$$Q\left(\frac{\theta - m_0}{\sigma_0}\right) = Q\left(\frac{m_1 - \theta}{\sigma_1}\right) \tag{8.10}$$

Equating arguments,

$$\theta = \frac{m_0\sigma_1 + m_1\sigma_0}{\sigma_0 + \sigma_1} \tag{8.11}$$

so that the bit error rate is

$$BER = Q\left(\frac{m_1 - m_0}{\sigma_0 + \sigma_1}\right) = Q(\gamma) \tag{8.12}$$

where we can identify γ^2 as roughly the *power* signal-to-noise ratio (SNR). Note that the argument of the Q-function, as given in the left-hand equality, takes into account that the noise power appearing when a "0" is being received can be different from that for a "1." This possibility, which does not usually occur in electronic communication systems, will be the case for many of the optical receivers to be discussed.

The Q-function is widely useful and is therefore given here to large scale in Figure 8-7. The approximation of Equation 8.7 is shown by the dashed line. Note how rapidly the probability of error decreases with a small increase in the argument (roughly the square root of the power SNR); a change from 6.0 to 8.0 (2.5 dB) in the argument produces six orders of magnitude improvement in BER! Thus we have a handy rule of thumb that, for the low error rate targets of fiber-optic systems, there is something over two orders of magnitude improvement in error rate for every dB improvement in SNR.

This sensitivity of *BER* to electrical *SNR*, plus the fact that an improvement in optical link budget pays off in twice as much improvement in electrical link budget, encourages us to believe that improving the error rate from 10^{-9} to 10^{-15} may not be the heroic undertaking it might at first seem to be.

8.4 The Ideal Direct-Detection Receiver and Its Quantum Limit

In designing a communication system, it is always nice to know when to stop trying to improve things. Fundamental bounds on what the laws of physics allow one to achieve are always useful in this way. This was one of the things that Shannon's information theory did for communication engineers. The importance of his theory lies not only in his constructive proofs about ways of doing source coding to achieve bandwidth compression or doing channel coding in order to minimize corrected bit error rate. Probably more important, he gave engineers ways to calculate limits in both cases, the information rate of a source and the channel capacity of a channel, respectively. The first bound tells the engineer when to quit trying to develop better bandwidth compression schemes. Since the channel capacity is the rate below which completely error-free communication is theoretically possible, the second bound tells the engineer when to stop trying to invent methods of improving the bitrate that could be supported by a given communication link.

In a similar spirit, the *quantum limit* is a bound that gives the lightwave network architect a bound on what is ultimately achievable from a particular form of optical communication link. It is usually stated in terms of the minimum number of photons per bit that will allow one to achieve a given bit error rate with the specified modulation format and type of receiver.

We can quickly deduce this quantum limit for an idealized ASK direct-detection receiver. This is often called *IMDD (intensity modulation, direct-detection)*. Intensity modulation is the same as ASK (also known as on-off keying, OOK). IMDD is the most prevalent form of optical communication link by far, because of its structural simplicity, and, as we shall see as this chapter progresses, for this reason it has proven surprisingly resilient to competition from other modulation-detection combinations, for example, those involving coherent detection.

Planck's law tells us that photons arrive at an average rate per second

$$r = P/hf \tag{8.13}$$

but it does not tell us when they will arrive. The arrivals are completely random, and all one can say is that, on average, the number arriving grows linearly with the duration of the observation at a rate r. Barring the perfection of bizarre schemes for "squeezing" light so as to regularize the photon time of arrival [14], this randomness in arrival times is unavoidable. For a given average rate r photons per second, the Poisson distribution

$$P(N) = \frac{(rT)^N \exp(-rT)}{N!} \tag{8.14}$$

gives the probability that exactly N photons will arrive during the bit interval T seconds long.

The Poisson distribution has the interesting property that the variance and the mean are equal

$$\sum_{N=0}^{\infty} NP(N) = \sum_{N=0}^{\infty} (N - rT)^2 P(N) = rT \tag{8.15}$$

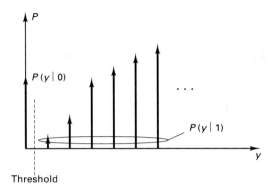

Figure   8-8. The same probabilities as in Figure 8-6 for the special case of the ideal photon counting receiver.

a fact that will prove useful later on. Many of the noise components we shall be discussing later have their physical basis as Poisson arrival processes, and this equality between mean and variance will be relevant.

We can build such an idealized ASK communication system by sending light during a bit time T to signify each "1" bit, and absolutely no light to signify a "0" bit (i.e., $r = 0$). Given a certain target probability of error, what is the lower limit on average number of photons per bit that will guarantee this error performance? We can use the detection-theory approach of the previous section to find out.

All we can know about the received signal is the measured number of photon arrivals in the interval T, and this will be our received variable y. The optimum receiver will be that which sets the decision threshold where the areas under the probability density tails are equal, as in Figure 8-6. Figure 8-8 is a plot of the probability density $P(y|x)$ for the idealized optical ASK situation. Note that the probability distributions for "0" and "1" are definitely nongaussian, so the Q-function definitely does not apply.

From this figure it is clear that, since the probability density is actually a set of impulses, this amounts to setting the threshold anywhere between $y = 0$ and $y = 1$. In other words, the receiver is a *photon-counting receiver*, an ideal device that can only count up to 1 but does so completely reliably (introduces no noise itself). If it sees so much as a single photon, it knows with certainty that a "1" bit is sent. But what if it counts no photon arrivals? This could have been either a *no-error* situation in which the transmitter intended "0" or an *error* situation in which the transmitter sent "1," but the light was so weak that no photons arrived during T.

Therefore, the probability of error of this optimum receiver is (from Equation 8.7)

$$BER = \frac{1}{2}(0) + \frac{1}{2} \text{Prob}\left[\, 0 \text{ photons arrived, but a 1 bit was sent}\,\right]$$

$$= \frac{1}{2}\exp(-rT)$$

(8.16)

For example, for $BER = 10^{-9}$, the average required number of photons rT per "1" bit is 20, and for $BER = 10^{-15}$, it is 34 photons per "1" bit. These numbers constitute the ASK quantum limit for these two BER targets.

Figure 8-9 plots required receiver power in dBm (dB below one milliwatt) as a function of bitrate $1/T$ for the ideal photon-counting ASK receiver and for other important combinations of modulation format and receiver that we shall discuss. The same two values of target bit error rate are assumed, 10^{-9}, and 10^{-15}. The ideal ASK quantum limits of 20 and 34 photons per bit, respectively, are shown at the bottom.

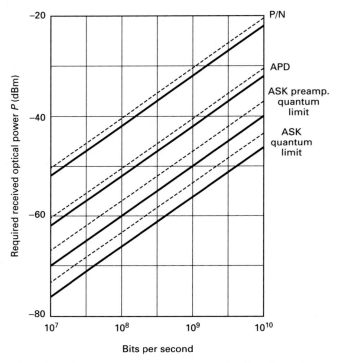

Figure 8-9. Curves of required minimum received power in dBm for various optical receivers and for two values of bit error rate, 10^{-9} (solid lines), and 10^{-15} (dashed lines). For PIN and APD, 100-percent quantum efficiency is assumed, temperature $Y = 300°K$, and $C = 0.2$ pF. $\langle M \rangle = 10$ for the APD. For the preamplifier, $\chi = n_{sp} = 1$.

One must be careful to distinguish between average power and peak power. Since laser diodes are peak-power-limited rather than average-power-limited devices, the transmitter power will be the same for on-off keying as with some constant-power form of modulation, for example frequency-shift keying (FSK). Note that if we had been talking about average power, the quantum limits mentioned earlier would have been half as large: 10 and 17 photons per bit for 10^{-9} BER. Sometimes these average-power numbers appear in the literature as representing the quantum limits, but they are not really applicable, because the limitation on real components is one of peak power, not average power.

In practice, the ideal, noiseless, photon-counting receiver only be built at a temperature of absolute zero, because the counting process itself cannot otherwise be made to be completely reliable (noiseless). Nevertheless, the ideal ASK quantum limit is a useful bound against which to compare what is possible with real receivers.

Table 8-1.   Quantum Limits for Various Modulation/Detection Options

Receiver form	BER equation	Probability densities	Quantum limit (ph/bit) 10^{-9}	Quantum limit (ph/bit) 10^{-15}
Incoherent				
ASK photon counter	$0.5 \exp(-rT)$		20	34
ASK w. optical preampl.	$Q(\sqrt{rT/2})$		76	130
Coherent				
PSK homodyne	$Q(\sqrt{4rT})$		9	16
ASK homodyne	$Q(\sqrt{rT})$		36	64
PSK heterodyne	$Q(\sqrt{2rT})$		18	32
ASK heterodyne	$Q(\sqrt{rT/2})$		72	128
FSK heterodyne	$Q(\sqrt{rT})$		36	64

Table 8-1 gives the quantum limit for each of a number of choices of modulation format and receiver type. Several of these are also plotted in Figure 8-9, The second column in the table gives the appropriate formula for BER, equivalent to Equation 8.12, the noise being assumed to have gaussian statistics, so that BER is expressed in terms of the Q-function. For example, for BER = 10^{-15}, the argument of the Q-function must be 8.0, a useful number to remember. The sketches in the third column relate the means and standard deviations of the probability density distributions of the different cases. The standard deviations are compared in multiples of some quantity a, and the separation of the means in terms of a quantity b.

8.5 PN and PIN Photodiodes

At one time PN photodiodes were the most prevalent detection device used in lightwave systems, but these have now been superseded by PIN devices. We now discuss these two in that order.

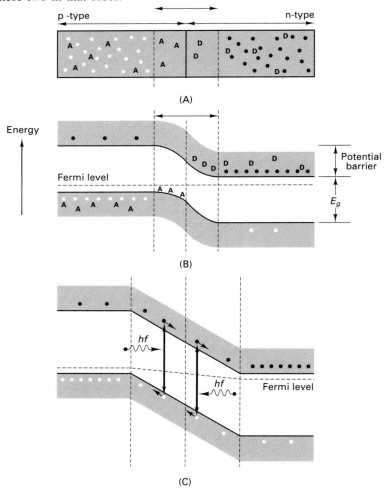

Figure 8-10. PN photodiode. (A) Open-circuit PN junction, (B) Corresponding energy versus position *y*. (C) PN junction energy versus position when reverse bias is applied.

Basic principles

Figure 8-10 shows the structure of a *PN photodiode*, a device consisting simply of a PN homojunction upon which light impinges, as described in Section 5.2. The dia-

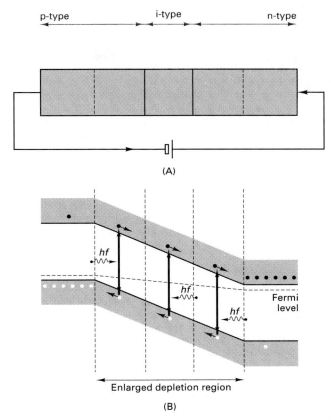

Figure   8-11. PIN photodiode with the depletion region enlarged by the introduction of an intrinsic region in the middle. (A) Structure. (B) Corresponding energy versus position y.

grams of Figure 8-10(A), (B), and (C) are identical to the earlier Figures 5-11(A), (B) and (F), respectively. If the device is open-circuited, the Fermi levels of the two regions, p and n, will line up, as shown by the dashed line in (B), a potential will develop across the junction, and the majority carrier electrons in the n-region will all run downhill to the right and the holes of the p-region will all run uphill to the left. As a result, the depletion layer in the middle will have almost no carriers remaining. When the reverse bias current is applied as with (C), the Fermi levels are displaced from one another, the depletion layer widens, and the potential barrier becomes even higher than in the open-circuit case.

Figure 8-10(C) shows a photon incident from the left impinging on the depletion layer by passing through the p-layer. Any incident photon that has an energy larger than the bandgap may give up this energy by exciting an orbital electron from the valence band into the conduction band, a process that we have viewed before as amounting to the creation of an electron-hole pair consisting of an electron in the conduction band and a hole in the valence band. The electrons will flow out of the photodiode and into the external electrical circuit to produce a useful photodetected signal. We refer to the hole-electron pairs generated by incident light as *carriers*.

Desired properties

Given all this, we would like the photodiode geometry to have several properties. First of all, we would like the depletion region to be as thick as possible, so that as many of the incident photons as possible will be converted to carriers. If the depletion region is too thin, the photons may pass through it unproductively. (It doesn't do much good for the photons to be converted to carriers in the two other regions outside the depletion region, because there is no voltage gradient there to propel them into the external circuit; they will stay there until they either combine via nonproductive processes or diffuse slowly into the intrinsic region). Widening of the depletion region is done, of course, by increasing the voltage (compare Figures 8-10(B) and (C)), but can also be greatly assisted further by the important step of introducing a third region, composed of intrinsic material between the p- and the n-regions of extrinsic (doped) material. The structure is shown in Figure 8-11(A) and the effect on the diagram of energy-versus-y (position) is shown at (B). Now, instead of a *PN photodiode*, we have a *PIN photodiode*.

The geometry of an actual such device is shown in Figure 8-12. Note that the active area of the photodiode is large, so that the problems of coupling to the fiber that were such an issue with laser diodes (Chapter 5) are largely absent.

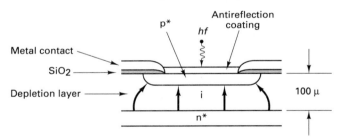

Figure   8-12. Geometry of a typical PIN photodiode.

The second desirable property is that the p-region through which the light travels before reaching the depletion region should be as thin as possible and as transparent as possible, so that most of the light arrives at the depletion region. The device of Figure 8-12 exhibits this thinning. Another helpful trick is to make the layer through which the light passes on its way to the i-region out of a material having a wider bandgap (Figure 5-9). In other words, the p- and i-regions form a heterojunction.

Lastly, the material of the intrinsic region must have as high an absorption coefficient as possible at the wavelength of the incident signal. This will not only maximize the fraction of photons that are converted to carriers, but also will allow the intrinsic region to be shortened so that the device will respond as rapidly as possible.

Profiles of charge, field intensity and voltage

Let us use Figure 8-13 to examine the inside of the depletion layer in a bit more detail [15] than we did when we first discussed it back in Section 5.2. This will not only aid in understanding what is going on inside ordinary PIN photodiodes, but will be particularly useful in understanding avalanche photodiodes. For the moment we treat the PN case and then the PIN case.

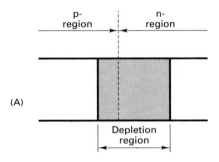

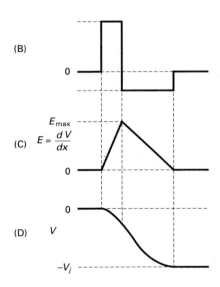

Figure   8-13. Profiles of various parameters for PN diode. (A) Structure, (B) Charge density, (C) Electric field intensity, and (D) Voltage. (From [15], © Holt, Rhinehart and Winston).

As Figure 8-13(A) shows, on the n-side of the junction there is an excess of free electrons in the conduction band, causing the net charge density there to be negative. On the p-side there is an excess of holes in the valence band and the net charge

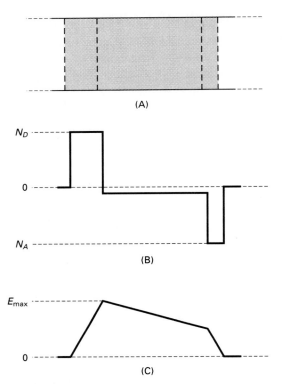

Figure 8-14. Profiles for PIN diode. (A) Structure, (B) Charge density, (C) Electric field intensity, and (D) Voltage. (From [15], © Holt, Rinehart and Winston).

density there is positive. These charge densities are plotted as a function of y in Figure 8-13(B). Now, according to Poisson's equation in one dimension relating charge density N to electric potential (voltage) V,

$$\nabla^2 V = \frac{d^2V}{dy^2} = -\frac{dE}{dy} = -\frac{N}{\epsilon} \tag{8.17}$$

This is actually the one-dimensional version of one of the Maxwell equations, Equation 3.5, generalized to include free charge density N. Figure 8-13(C) shows the resulting electric-field intensity $E(y)$ from Equation 8.17, and (D) shows the voltage $V(y)$. Figures 8-14(A) through (C) do the same thing for the PIN case. For the open-circuit case discussed earlier in connection with Figure 5-11(A), the total voltage V is simply the contact potential. For the back-biased photodiode, it is the applied reverse voltage, which is considerably larger.

Since PN diodes have been superseded by PIN diodes, we shall not discuss PN diodes further.

Responsivity and quantum efficiency

Ideally, every incident photon will produce one electron. The degree to which this ideal is approached in practice is given by the photodiode's *responsivity*

$$\mathcal{R} = \frac{I}{P} = \eta \, \frac{q}{hf} \tag{8.18}$$

The quantity q/hf (which $= 1.28$ amp/watt at 1.5 μ), represents the conversion factor between photons and electrons, and therefore has nothing to do with the device itself, whereas η, called the *quantum efficiency*, characterizes the device. 1.28 amp/watt is a handy number about optical receivers that is worth remembering.

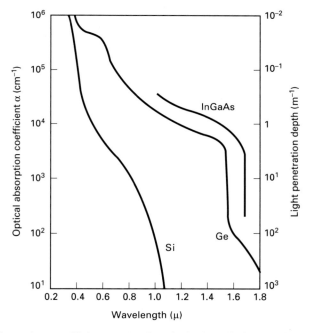

Figure 8-15. Absorption coefficient $\alpha(\lambda)$ of typical photodiode materials as a function of wavelength. (From [16] © 1987 IEEE).

Let us deduce an expression for η as a function of the wavelength λ and the device parameters. If the power reflectivity of the facet through which the light must pass on entering the device is R, the depth of the p-region is δ, and the depletion region has depth d, then the quantum efficiency, the fraction of the total incident optical power that is turned into electric current, is

$$\eta = (1 - R) \, \exp(-\alpha(\lambda)\delta) \, [1 - \exp(-\alpha(\lambda)d)] \tag{8.19}$$

The absorption $\exp[-\alpha(\lambda)]dy$ is defined as the fraction of the incident photons that are converted to carriers in traversing a distance dy.

The first factor gives the amount of energy lost by facet reflection. The second factor, $\exp(-\alpha(\lambda)\delta)$, accounts for unproductive absorption due to carrier generation in the p-region. The last factor represents the productive absorption of photons in the total depletion region. Obviously, we want both the depletion region width d and the absorption α to be as large as possible, and δ to be as small as possible.

Choice of materials

The wavelength dependence of α depends on the materials chosen. Figure 8-15 shows $\alpha(\lambda)$ for several of the most important photodiode materials. Figure 8-16 indicates as dashed lines the responsivity $\mathcal{R}$ as a function of wavelength for 100-percent quantum efficiency η, and as solid lines the actual responsivity achievable with these materials. As would be expected from a consideration of bandgaps, there is a long-wavelength cutoff describing the situation in which the photon no longer has enough energy to make it across the bandgap.

As these figures indicate, the most practical photodiode material for the 0.85-μ band is silicon, whereas both germanium and InGaAs are candidates for the 1.3- and 1.5-μ wavelength bands [17].

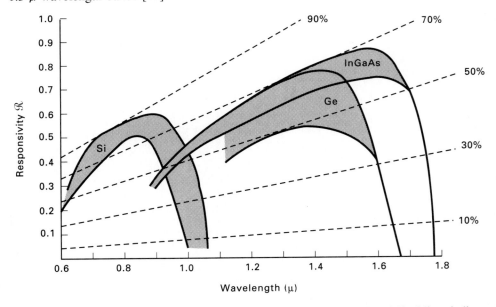

Figure   8-16. Responsivity $\mathcal{R}$ of the three materials of Figure 8-15. The dashed lines indicate various values of the quantum efficiency η. (From [17]).

From Figure 8-16 alone, the reader begins to see why it is not so easy to achieve the number of photons per bit that the quantum limits specify. This departure from the

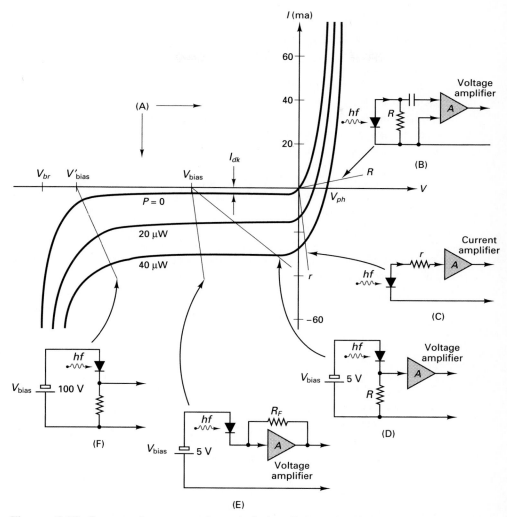

Figure 8-17. Current-voltage curve of a typical photodiode, and corresponding detector possibilities. (A) *I–V* curves. Forward bias is to the right and reverse bias to the left. (B) Operation in open-circuit (photovoltaic) mode. (C) Operation in short-circuit mode. (D) Operation in reverse-bias mode into a voltage amplifier. (E) Operation into a transimpedance amplifier. (F) Operation as an avalanche photodiode.

quantum limit is really nothing compared to the even worse news to come when we take noise effects into account.

Electrical operating characteristics

Figure 8-17(A) shows the family of *I*-vs.-*V* curves of a typical photodiode [17]. Each curve gives the current through the device as a function of the voltage appearing across it, for a particular value of light intensity P. Quite a lot of information can be gleaned from studying such a set of curves.

When V, the voltage applied across the device, is positive, current flows in the "forward-biased" direction, and the behavior is as described in Chapter 5. The current levels shown are much less than that needed to achieve transparency, much less stimulated emission. When negative voltage is applied, the current is almost flat with voltage for a given light level until the *breakdown voltage* V_{br} is approached. As the reverse voltage across the device approaches V_{br}, a kind of avalanche process called *impact ionization* begins to set in. In Section 8.8, we shall see how this avalanche effect can actually be put to good use in an APD to produce much more current for a given light level P than is possible with the PIN photodiode.

The curve for zero light level ($P = 0$) is the familiar semiconductor-diode curve in which the resistance in the forward direction is small and that in the backward direction is large but not infinite. The small current that flows in the back-biased condition, even for zero light level, is called the *dark current I_{dk}*, and represents a source of unwanted noise. In an ordinary semiconductor diode, this is called the *reverse current*. The dark current is due to carriers being generated thermally. That is, the temperature is high enough that here and there in the intrinsic region a few electrons leave their parent atoms and become part of a small current flow, the dark-current noise. Values of dark current in typical commercial photodetectors are 0.1 to 1 $\times 10^{-9}$ amperes for silicon and 1 to 5 $\times 10^{-9}$ amperes for InGaAs at room temperature. From this fact and Figure 8-16 it is clear that silicon is a particularly good photodiode material; unfortunately it is not effective at 1.3 and 1.5 μ.

Conceivably, it would be possible to operate the photodiode device in any one of the three quadrants covered by the curves. One can employ the *photovoltaic* mode by connecting the photodiode across the high-impedance of a voltage amplifier, so as to amplify V_{ph}, as shown in Figure 8-17(B). The value of load resistor for this mode might be many megohms. Since the curves are very close together, it is clear that this is not a very effective way of using the device.

Another possibility might be to operate the photodiode in *short-circuit mode* and use a current amplifier, maintaining the voltage across the device near zero so that the current amplifier sees a light-dependent current at its input. This possibility is shown in Figure 8-17(C). There are several problems with this approach. For one thing, according to Equations 8.18 and 8.19, since the depletion region will be very shallow, the quantum efficiency will suffer, and for another, a strong potential across the depletion region is required in order to minimize the response time of the device by accelerating the carriers, as we shall see.

Therefore, photodiodes for fiber optic communication use neither of these modes, but instead use the lower left quadrant, i.e., reverse biasing, as shown in Figure 8-17(D), (E), and (F). (D) represents the PIN diode operating into a *voltage amplifier*, (E) shows the same thing, but using a *transimpedance amplifier*, and (F) shows operation as an APD. The diagonal lines in the figure are the *load lines* representing the voltage appearing across the appropriate load impedance seen by the diode. To see how this works, consider the load line for the case of the voltage amplifier (D). When the current *I* is zero, the entire reverse bias voltage appears across the device. As the current increases, due to an increase in light level *P*, less and less of the supply voltage appears across the diode because more and more is appearing across the load resistor.

Response time considerations

In the next section, we shall discuss the electrical amplifier circuit constants that the photodiode sees. In many practical cases, it will be these that limit the bandwidth of the receiver, but at very high bitrates (over 1 Gb/s) the finite dimensions of the photodetector come into the picture. Imagine a squarewave of light power being applied to the reverse-biased PIN photodiode of Figure 8-12. In the absence of circuit constraints, the time constant with which the current can rise and fall will, in principle, depend on the *RC* time constant and on several parameters having to do with carrier motion [7, 18], namely, the *transit time* of the carriers across the total depletion region, and the much longer *diffusion time* into the depletion region of those carriers that were generated outside the depletion region.

The device internal capacitance will be inversely proportional to the depletion-region thickness *d*. On the other hand, we have seen that it is desirable to make the depletion region as wide as possible in order to maximize the quantum efficiency (Equation 8.19), but this increases the transit time of the carriers across this region. As long as the carriers in the depletion region see an electric field strength of several kilovolts per centimeter (which is usually the case), the velocity of travel is at its maximum value of around 8×10^6 cm/sec for electrons and about half that for holes, respectively, in all three materials whose parameters are given in Figures 8-15 and 8-16. Thus, for a typical PIN diode with a depletion region of 20 μ, the contribution to the time constant of finite carrier mobility will be about 0.2 nsec.

For maximum speed of response, the thickness *d* is usually adjusted to be several times $1/\alpha(\lambda)$, representing the best compromise between capacitance and quantum efficiency on one hand and transit time on the other [7].

8.6 Electrical Amplifiers for PIN Photodiodes

Wide receiver bandwidth versus clocking

Figure 8-18 shows the various optical receiver elements, of which the photodiode is just a part. Other than the photodiode itself, the key component that defines the receiver performance is the preamplifier. It is at the preamplifier input that the detected signal is the weakest and therefore most prone to contamination by noise.

Two basic approaches are possible. In the first, the bandwidth is allowed to be quite wide, so that the squarewave on-off modulation, illustrated in Figure 8-3(B), is

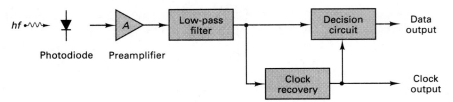

Figure 8-18. Block diagram of the electrical elements of an optical receiver.

most visible in order that a simple slicing circuit can be used to decide on binary "1" or "0." This approach is almost never used, since too much noise is passed by the wide preamplifier bandwidth required. In the second approach, shown in Figure 8-18, a *bit-clock* signal is derived and used to make the decision at exactly the right instant of time in each bit. For example, for the squarewave modulation of Figure 8-3, it can be shown [13] that the optimum MAP receiver is the *integrate-and-dump* receiver, discussed earlier in connection with Figure 8-3(D), one in which a low-pass filter is followed by sampling the output exactly at the end of each bit time and declaring "1" or "0," depending on which side of a threshold the sampled value lies. The integration over the bit interval will be recognized as exactly equivalent to the *matched-filter* condition between incoming waveform (square pulse) and effective receiver impulse response (an identical square pulse). In practice, the integrator is almost any form of lowpass filter.

Two preamplifier types

There are two basic types of preamplifier in wide use, the conventional voltage amplifier, depicted in Figure 8-19(A), and the transimpedance amplifier of (B). A principal design objective in both cases is to make the value R of the resistive component seen by the photodiode as large as possible to reduce thermal noise, but small enough to achieve adequate bandwidth. The tradeoff can be seen from the equations for noise and for bandwidth.

The *thermal noise* (or "Johnson noise") appearing with any resistance is a randomly varying current with variance

$$\langle I_{th}^2 \rangle \;=\; 4KYB/R \quad \text{amp}^2 \tag{8.20}$$

where K is Boltzmann's constant (1.38×10^{-23} joules per $^\circ K$), Y is temperature in $^\circ K$, and B is the bandwidth in Hz.. We use the symbol Y for temperature and T for bit duration, to avoid confusion. This fluctuating current adds to the useful signal current generated in the photodiode, and is one of several sources of noise that will have to be considered when we assess the performance of practical direct-detection systems. Assuming the bandwidth B Hz of the receiver to be limited by the RC time constant of the amplifier input circuits,

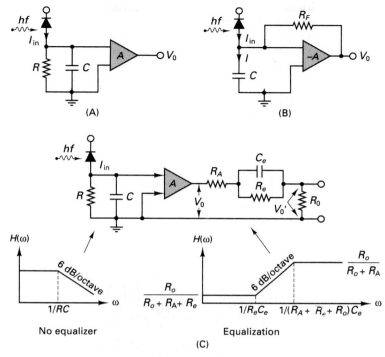

Figure 8-19. Preamplifier circuits. (A) Voltage amplifier, (B) transimpedance amplifier, (C) Equalization circuit for (A).

$$B = \frac{1}{2\pi RC} \tag{8.21}$$

where C is the total parallel combination of the device, parasitic, and amplifier input capacitances. Equations 8.20 and 8.21 place in evidence the conflicting requirements on the resistance R.

The voltage amplifier and the transimpedance amplifier achieve the two objectives of low noise and high bandwidth in different ways.

The equalized voltage amplifier

The voltage amplifier uses a large R, and then improves the bandwidth by a postamplification equalization, as shown in Figure 8-19(C). Let us consider the amplifier and the equalizer separately.

For the unequalized situation of Figure 8-19(A), the ratio of amplifier output voltage to input current, the *transimpedance*, is

$$\frac{V_o(\omega)}{I_{in}(\omega)} = H(\omega) = \frac{AR}{1 + j\omega RC} \tag{8.22}$$

where A is the amplifier voltage gain. A simplified version of the frequency function $H(\omega)$ is shown as the left spectrum in Figure 8-18(C). It has a rolloff at the "corner frequency" of $\omega = 1/RC$, followed by a drop of 6 dB per octave.

The equalizer attenuates the low frequencies and boosts the high frequencies. Let R_A be the effective amplifier output resistance, R_o the output terminating resistance, and let the equalization section be composed of a parallel combination of R_e and C_e. By inspection, the equalizer transfer function is

$$H_e(\omega) = \frac{V_o(\omega)'}{V_o(\omega)} = \frac{R_o}{R_A + R_o + R_e / (1 + j\omega R_e C_e)} \tag{8.23}$$

The equalizer exhibits a 6-dB-per-octave rise whose corner is at $\omega = 1/R_e C_e$ and then a flattening off at a second corner frequency $\omega = 1/(R_A + R_e + R_o)C$. This is shown in the right-hand spectrum of Figure 8-19(C). By placing the left-hand equalizer corner frequency at $1/RC$, one can extend the bandwidth of the photodetector of Figure 8-19(A) by the ratio $(R_e + R_A + R_e)/R_e$. In commercial PIN/FET receivers, this ratio is limited to no more than about 5.

The equalization approach has several disadvantages when pushed too far: (i) Any high-frequency preamplifier noise introduced after the input circuit will be exacerbated by the equalizer network. (ii) In a PIN/FET package it may be necessary to optimize the combination of preamplifier and equalizer component parameters at different values for different operating bitrates, and for non-identical production photodiodes. (iii) Severe dynamic range requirements are imposed on the preamplifier. This last effect occurs because, in order for the cascade of preamplifier and equalizer to end up with a flat response, the preamplifier must handle very strong low frequencies and weak high frequencies. One thing (but not the only thing) that can present very strong low frequencies is a long succession of "1" bits or "0" bits. This can be prevented by suitable line coding, as was mentioned in connection with thermal effects in lasers (Section 5.6). This line coding is usually already provided in most digital links.

It should be mentioned at this point that large receiver dynamic range can become much more important in any kind of lightwave network than it is for a point-to-point lightwave link. In this context, receiver dynamic range is defined as the difference in power levels of the largest and the smallest signals that the receiver can handle.

Receiver dynamic range can become important because of the *near-far* problem, the fact that a given receiver, in order to support connections to a variety of other transmitters, may find itself experiencing a wide variety of transmitter-receiver distances for successive connections, so that the received power level can vary considerably.

The transimpedance amplifier

This form of preamplifier achieves the wideband low-noise result in a different way, one that has great dynamic range advantages over the unequalized-amplifier approach of Figure 8-19(A). In many applications, it has superseded the equalized-voltage amplifier.

The transimpedance amplifier of Figure 8-19(B) can best be considered as a device that performs a direct conversion of the photodiode output current I_{in} to an output voltage V_o, using the feedback resistance R_F around an inverting amplifier. (The circuit diagram is perhaps more familiar to some readers as the "inverting opamp").

Let us again compute the transimpedance, the ratio V_o/I_{in} as a function of frequency. Again, A represents a voltage-amplification factor.

$$V_o(\omega) = -A \frac{I(\omega)}{j\omega C} \tag{8.24}$$

$$I(\omega) = I_{in}(\omega) + \frac{V_o(\omega) - I(\omega)/j\omega C}{R_F} \tag{8.25}$$

from which

$$V_o(\omega) = \frac{-A}{j\omega C} \left(\frac{I_{in}(\omega)R_F + V_o(\omega)}{R_F + 1/j\omega C} \right) \tag{8.26}$$

so that

$$H(\omega) = \frac{V_o(\omega)}{I_{in}(\omega)} = \frac{-A}{(A+1)} \frac{R_F}{1 + j\omega CR_F/(A+1)} \tag{8.27}$$

Several interesting things are clear from this equation. First of all, in the limit of large gain A, the transimpedance (current-to-voltage ratio) is given by the value of the feedback resistor, and not primarily by the amplifier gain. This resistance can be made very large. Second, the effective input RC time constant is reduced by $(A+1)$. The very low effective impedance seen by the photodiode is reflected in the almost vertical load line indicated in Figure 8-17(A) for the transimpedance-amplifier case (E).

A third advantage is that the dynamic range of the amplifier stage is not stressed by the need for equalization. An even more important contributor to large dynamic range is the fact, visible from the load line in Figure 8-17(A), that as the light level P changes, there is little change in the voltage across the amplifier input. That this is the case is clear from the inspection of the circuit diagram of Figure 8-19(B), where it is seen that a small change in input voltage is immediately counteracted by the large negative feedback provided by the amplifier gain and the feedback resistor.

8.7 Direct Detection Using PIN Photodiodes

Let us now examine how close to the quantum limit a real PIN diode is likely to come. As Figure 8-4 told us, there are four receiver noise components: thermal noise, dark-current noise, noise generated within the transistor amplifier, and finally the shot noise that is the result of smoothing the intrinsic quantum noise of the incoming light by the finite-bandwidth photodiode and its accompanying amplifier circuit.

Figure 8-20 shows the conversion of the optical signal into the electrical signal, along with the various noises that contaminate them. At (A), the physical processes are indicated, and at (B) the equivalent electrical circuit at the amplifier input. The dashed wiggly line indicates that the shot noise is caused by the quantum nature of the incoming signal; this form of noise arrives as an intrinsic part of the signal.

Components of the SNR equation

Given an arriving ASK optical signal power P for a "1" bit, and zero for a "0" bit (assuming that the extinction ratio $r = 0$), our task will be to use Equation 8.12 properly to get the bit error rate. We examine the "0" bit, determine its mean m_0 and standard deviation σ_0, and then do the same thing for a "1" bit to get m_1 and σ_1. If none of the noise standard deviations depended on whether a "0" or a "1" is being discussed, then we could take the photocurrent $I_{\text{sig}} = \eta q P / h f$ to be the signal, put this in the numerator of an equation for electrical-power signal-to-noise ratio observed at the amplifier input, then add up all the noise components and put them in the denominator.

$$SNR = \frac{I_{\text{sig}}^2}{\langle I_{noise}^2 \rangle} = \frac{I_{\text{sig}}^2}{\langle I_{sh}^2 \rangle + \langle I_{th}^2 \rangle + \langle I_{dk}^2 \rangle + \langle I_{ampl}^2 \rangle} \tag{8.28}$$

where the notation $\langle \, \rangle$ indicates the ensemble average. In actuality, we shall have to proceed somewhat more carefully in deducing σ_0 and σ_1, since they are not the same for some forms of noise.

Importance of the thermal noise component

The system architect has no control over the shot noise, but does have control over the other three components. The signal and noise components are as follows:

From Equation 8.18,

$$I_{\text{sig}} = \frac{\eta q P}{h f} = \mathcal{R} P \tag{8.29}$$

The thermal-noise variance is given by

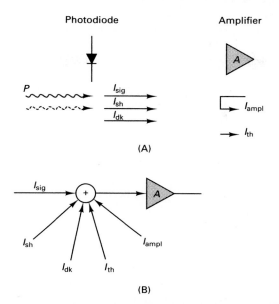

Figure 8-20. Signal and noises in the direct photodetection process. (A) Optical signals and electrical currents, (B) Electrical equivalent circuit at the amplifier input.

$$\langle I_{th}^2 \rangle = \sigma_{th}^2 = 4KYB/R \qquad (8.30)$$

where it is understood that an integration over the bandwidth from 0 to B Hz of a spectral density $4KY/R$ amps^2/Hz has taken place. This is permissible, since the thermal noise, and all the other noise components, may be considered flat over the bandwidth of interest.

We can write $B = 1/2T$, where T is the bit duration, as long as the received signal is integrated and then sampled in the low-pass version of the integrate-and-dump filter of Figure 8-3(D).

Assuming, as was done earlier, that the bandwidth of the receiver is limited by the RC time constant of the amplifier input circuits,

$$B = \frac{1}{2\pi RC} \qquad (8.31)$$

where C is the total parallel combination of the device, parasitic, and amplifier input capacitances, so that

$$\langle I_{th}^2 \rangle = \sigma_{th}^2 = 8\pi KYB^2C \qquad (8.32)$$

In other words, once the system designer has picked the upper frequency limit B of photodetector operation, the remaining variable under his control in minimizing thermal noise is the capacitance. This fact helps explain why it is so important to have the photodetector and the amplifier front end very close together, preferably in the same package.

As for the shot-noise and dark-current noise terms, they are both Poisson arrival processes. In Equation 8.15, we saw that with such processes, the variance is equal to the mean. For real noises specifically, it can be shown [3] that such components have variances

$$\langle I^2 \rangle = 2qIB \tag{8.33}$$

so that the component of shot noise intrinsic to the signal is

$$\langle I_{sh}^2 \rangle = \sigma_{sh}^2 = 2qI_{sig}B \tag{8.34}$$

If we had been talking about the bandpass case, where two-sided spectra are appropriate, the factor of two would be missing from equations such as 8.33, but for the low-pass case considered here, the factor of two appears.

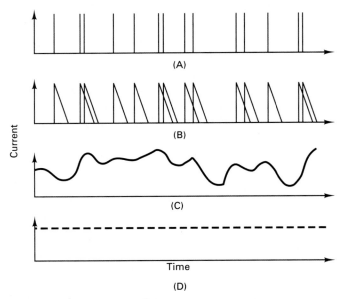

Figure 8-21. How shot noise gets smoothed by the diode and amplifier bandwidth limitations. (A) Poisson impulses, (B) Smoothed impulses, (C) Resulting noise waveform, (D) Average (useful signal).

Figure 8-21 shows how the finite response time of the combination of the photodiode and the amplifier circuits smooths each impulse into a random waveform having a power spectral density that is the squared magnitude of the Fourier transform

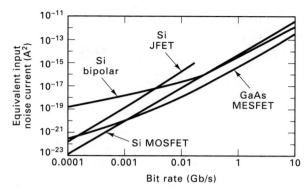

Figure   8-22. Amplifier equivalent input noise power for several forms of amplifier transistor. (From [19], Academic Press, by permission).

of the receiver impulse response. It can be shown [13] that as the power level of a filtered Poisson process becomes larger, the smearing between impulses soon causes what was originally a Poisson process to become a Gaussian process.

The amplifier-noise component is usually generated inside the FET or other semiconductor device doing the amplification. We deal with the amplifier noise as referred back to the amplifier input, i.e., the value that a fictitious equivalent Poission noise source placed at the input would have in order to produce the actual noise power at the actual place in the amplifier where it is generated.

$$\langle I_{ampl}^2 \rangle \; = \; \sigma_{ampl}^2 \; = \; 2qI_{ampl}B \tag{8.35}$$

We shall ignore here the actual frequency-dependences that some components of this noise have.

The dark-current noise is

$$\langle I_{dk}^2 \rangle \; = \; \sigma_{dk}^2 \; = \; 2qI_{dk}B \tag{8.36}$$

We can demonstrate that, for practical PN or PIN photodiodes, the thermal noise predominates. Assume, for example, that the amplifier and succeeding stages can only handle signal and noise components up to 1 GHz. Using a typical value of C of 0.2 pF, representative of the best commercial GaAs MESFET components paralleled with the best commercial photodetectors, Equation 8.32 gives a figure of 2.1×10^{-14} amp^2 for the thermal-noise component. Commercially available PIN photodiodes typically have a dark current I_{dk} of about 10 nanoamperes, and substituting the electronic charge $q = 1.6 \times 10^{-19}$ coulombs, one obtains $\langle I_{dk}^2 \rangle = 6.4 \times 10^{-18}$, about four orders of magnitude less than the thermal noise.

The calculation of the noise introduced by amplifiers of various types is extensive and complex. A comprehensive treatment of the subject can be found in [19], where the data of Figure 8-22 are developed. It is seen that for GaAs MESFETs (metal-semiconductor FETs), $\langle I^2_{ampl} \rangle$ ranges from about 10^{-17} amp^2 at bitrates of 0.5 Gb/s to about 10^{-13} amp^2 at 5 Gb/s. The 1-GHz case of our example corresponds roughly to the 0.5 Gb/s situation, and it is seen that the amplifier noise is roughly three orders of magnitude lower than the thermal noise.

To compare the magnitudes of the shot-noise and thermal-noise components, let us see how large the input optical power P would need to be in order to generate a noise component having the same variance as the thermal noise of our FET amplifier example. Again assume a bandwidth of 1 GHz, room temperature ($Y = 300°K$), a 0.2 pF capacitance, and also assume operation at a wavelength of 1.5 μ (2×10^{14} Hz). Equating $\langle I^2_{th} \rangle$ and $\langle I^2_{sh} \rangle$, and using Equation 8.18 for optical power P and 100-percent quantum efficiency, gives a value of $P = 0.026$ mW (-16 dBm) for the optical power above which the shot noise exceeds the thermal noise. Clearly this is an unreasonably large number for received optical power, because we already know from Figure 8-9 that in actuality the thermal noise only poses a limitation on achieving bit error rates of 10^{-9} to 10^{-15} for much lower values of received optical power P than -16 dBm, as we shall now see in detail.

Probability of error expressions

The probability distributions of the current at the photodetector input will be different for a binary "0" and binary "1," as was illustrated in Figure 8-6. The arrows denote the standard deviations, each of which is the square root of the corresponding variance. Let us neglect the dark-current noise and concentrate on the shot noise and the thermal noise.

When a binary "0" is being received, the mean m_0 is zero and the variance is entirely due to thermal noise, and is therefore $\sigma_0 = \sigma_{th}$, which is given by Equation 8.32. When a binary "1" is being received, the mean is $m_1 = I_{sig}$, as given by Equation 8.29, and the variance is the sum of the shot- and thermal-noise variances, since the two physical processes are statistically independent zero-mean processes. (We are assuming that there is enough light for the shot noise to be considered a gaussian process, rather than a Poisson process). Because of the shot noise, the variance is slightly larger for a "1" than for a "0," a condition that is shown exaggerated in the figure.

$$\sigma_1 = \sqrt{\sigma^2_0 + \sigma^2_{sh}} \qquad (8.37)$$

The decision-threshold element is actually several electrical stages past the amplifier input, but, without loss of generality, we can consider it to be transplanted to the input point and analyze what happens.

From Equation 8.12,

$$BER = Q\left(\frac{I_{\text{sig}}}{\sigma_{th} + \sqrt{\sigma_{th}^2 + \sigma_{sh}^2}}\right) \tag{8.38}$$

Equating the bandwidth B with $1/2T$, and using Equations 8.29, 8.30, and 8.34, we have

$$BER = Q\left(\frac{\mathscr{R}P\sqrt{2T}}{\sqrt{4KY/R} + \sqrt{4KY/R + 2q\mathscr{R}P}}\right) \tag{8.39}$$

If we neglect thermal noise, the answer for *BER* comes out $Q(\sqrt{rT})$, which is wrong and unusable. The reason is the incorrectness of the gaussian approximation at low light levels. (A correct analysis is given in [20]). The case of thermal noise only is much more interesting and is a closer approximation to real life. For that case

$$BER = Q\left(\frac{\mathscr{R}P}{2}\sqrt{\frac{T}{2KY/R}}\right) \tag{8.40}$$

This situation is plotted in Figure 8-9, assuming room temperature ($Y = 300°K$), 100 percent quantum efficiency ($\mathscr{R} = q/hf$), and a total input capacitance of 0.2 pF, as before. Even for this assumption, which represents the very best photodiodes, the achievable result that includes thermal noise only is 24.3 dB short of the quantum limit.

8.8 Avalanche Photodiodes

We have just seen that in the case of even the most efficient PIN photodiodes, the shot noise is usually dominated by the thermal noise of the preamplifier input. The performance limits imposed by this noise component naturally lead one to question whether the signal can somehow be amplified before reaching the electrical amplifier input. There are three approaches to doing this. One could amplify the signal with one of the optical amplifiers described in Chapter 6, one could use coherent detection and swamp the thermal component with a strong local-oscillator input, or one could use the avalanche-breakdown effect indicated by the operating curve for high reverse bias shown in Figure 8-17(F). In this section we discuss the avalanche-photodiode (APD) approach.

Impact ionization

The idea behind the APD is exactly the same as that involved in the old photomultiplier vacuum tubes which used the secondary emission effect. In these

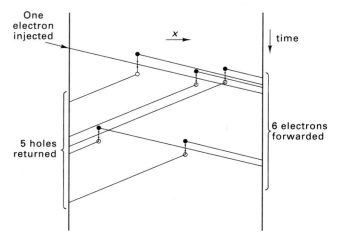

Figure   8-23. Timing diagram showing how the impact ionization effect produces an avalanche of additional 5 hole-electron pairs (carriers) from a single incident photon, giving amplification $M = 6$.   (From [15], © Holt, Rhinehart and Winston).

devices the impact of one incident electron of sufficient kinetic energy knocked several electrons loose from some emitting surface.

Figure 8-23 shows how the avalanche effect works in an APD. The applied reverse-bias voltage is quite high, typically tens to hundreds of volts (Figure 8-17(F)), compared to a few volts for the PIN diode operating modes shown in Figure 8-17(D) and (E). The consequence is that many of the electrons generated by the incident photons reach such a high kinetic energy that, upon colliding with an atom whose orbital electron is in the valence band, they will knock this electron into the conduction band. Now one has two carriers instead of the original one. Both the old and new carriers can be further accelerated to produce still other impact carriers, as illustrated in the figure. If the applied voltage is too high, this process can reach a runaway state (avalanche breakdown) such that the current grows to very large values independently of whether incident light is maintained. To avoid this and thus keep the device responsive to light input level, the applied voltage is set 10 to 15 percent below that which will produce the breakdown condition. The effect of impact ionization and the complete breakdown condition are seen in the lower left-hand portion of the photodiode characteristic curves of Figure 8-17(A).

At any given instant of time, the number of carriers produced by a single incident photon is the *APD gain* $M(t)$, a randomly fluctuating quantity whose ensemble average is $\langle M \rangle$, a quantity typically between 30 and 100.

The fluctuations constitute the *gain noise* of the APD. The instantaneous resulting photocurrent for the APD can be written

$$I_M(t) = \mathscr{R}PM(t) \qquad\qquad (8.41)$$

where $\mathscr{R}$ is the device's responsivity for unity avalanche gain.

The average signal power is increased by the factor $\langle M \rangle$.

$$I_{\text{sig}} = \mathscr{R}\langle M \rangle P \tag{8.42}$$

As for the randomly time-varying gain that multiplies the shot and dark-current noises, Figure 8-24 illustrates how the uniformly high but randomly arriving photon pulses of the shot noise get multiplied by the random gain. The result is to additionally enhance the shot noise by the equivalent multiplier F, greater than unity, called the *noise factor*.

$$\langle I_{sh}^2 \rangle = 2qI_{\text{sig}}B\langle M \rangle F \tag{8.43}$$

The same thing happens with the dark-current noise.

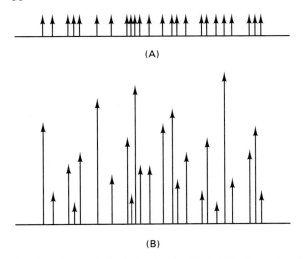

(A)

(B)

Figure   8-24. Showing how the random shot noise is affected by the randomly- varying gain of an APD. (A) Before, (B) After.

The general definition of the noise factor of a device is "the factor by which the noise-to-signal ratio at the output exceeds the noise-to-signal ratio at the input." If all sources of noise other than shot noise were absent from an APD, it is seen that F expresses the entire noise factor of the device, since the SNR at its output is F^{-1} times the SNR at its input.

Creating the high field intensity

Figure 8-25 shows the layer structure [7] and profiles of charge density, field strength, and voltage [15] as a function of x for one form of APD. The various regions shown in (A) are doped in such a way that the charge density (B) has a very large positive value over a certain narrow range, while adjacent to this is a narrow region of negative

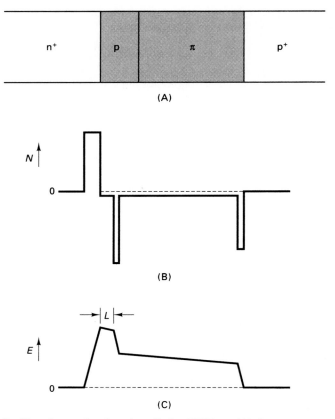

Figure 8-25. Profiles for avalanche photodiode (APD). (A) Layer structure, (B) Charge density, (C) Field intensity. (From [15], © Holt, Rhinehart and Winston).

charge density. This combination results in a strong positive peak in the field-intensity profile $E(x)$, as shown in Figure 8-25(C).

The entire depleted region, consisting of the i-region plus the region near the junction that would have been there without being artificially lengthened by the i-region, is almost completely depleted, because of the high applied voltage. Any incident photon has a good chance of generating a hole-electron pair, and these are accelerated in opposite directions. When the electron enters the high-field region, an impact ionization may take place, each resultant electron is also accelerated, with the possibility that it, too, may cause an impact ionization, and so forth.

Controlling the gain noise

To keep the noise factor F low requires either that the initiation of the impact ionization process be dominated either by electrons, the holes playing a minor role [21], or vice versa. Let L be the length of the region in Figure 8-25 over which the impact ionization takes place. The gain is

$$M = \exp(\alpha' L) \tag{8.44}$$

where α' is the absorption coefficient, analogous to the α that was defined before in connection with Equation 8.19 for the PN or PIN photodiode. However, this time α' expresses the efficiency of impact ionization, rather than the efficiency of conversion of incident photons into carriers. The absorption coefficient α' is, in general, very different for holes and electrons. The ratio between the two is

$$k = \frac{\alpha'_e}{\alpha'_h} \tag{8.45}$$

The avalanche process can be initiated by either electrons or holes. In the illustrative example of Figure 8-23, electrons were the initiators, but a device with a reversed geometry could conceivably have been built that used holes for this purpose.

The value for noise factor F depends on which type of carrier initiates the process.

$$F = \langle M \rangle \left[1 - (1 - \kappa) \frac{(\langle M \rangle - 1)^2}{\langle M \rangle^2} \right] \tag{8.46}$$

where $\kappa = k$ when electron-initiated, and $\kappa = 1/k$ when hole-initiated. For the electron-initiated case of Figures 8-23 and 8-25, Equation 8.46 is plotted in Figure 8-25. It is seen that a small value of k is desired, and in the limit as $k = 0$, the noise factor is 2, i.e., 3 dB.

So we conclude that with an ideal APD we can avoid the thermal-noise component and get away with simple low-cost electrical amplification, paying a penalty of only 3 dB in link margin, if we can find a material in which the absorption per unit length of incident holes and electrons is widely different. Silicon has a particularly favorable value of $\kappa = 1/k$ of about 0.2. For both germanium and InGaAs, $\kappa = k = 0.5$.

Advantages and disadvantages

APDs offer greater sensitivity than PIN diodes, since the multiplication-factor noise is smaller than the electrical-amplifier thermal noise. Noise factors of about 3 dB are achievable. Gains of several tens to 100 (15-20 dB) can be gotten before the noise-factor advantage is lost (Figure 8-26).

However, these benefits are bought at the expense of providing a high-voltage power supply that must be carefully controlled to avoid avalanche breakdown. APDs are considerably more expensive than PIN diodes and even packaged PIN/FETs, because of the complexities of their fabrication. Another disadvantage of APDs is their inability to handle extremely high bandwidths (more than about one Gb/s). As

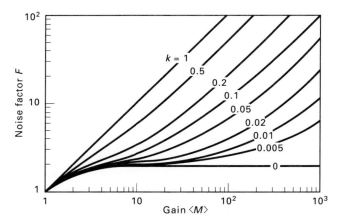

Figure 8-26. APD noise factor F as a function of gain for various values of ionization rate ratio k. (From [22]).

the timing diagram of Figure 8-23 shows, it takes time for the successive impact ionizations to occur, since each electron must have time to accelerate to the required velocity. It is not that the device is larger than a PIN diode, it is that the time taken for the avalanche process to complete is many times greater than the time occupied by a photon-to-carrier conversion.

8.9 Direct Detection Using APDs

Since the gain takes place before the electrical amplification, the thermal-noise and amplifier-noise sources have smaller effects with APDs than with PIN detectors. Analogously to Equations 8.34 and 8.36, one has at the input of the amplifier

$$\langle I_{sh}^2 \rangle = 2qI_{\text{sig}}BF \tag{8.47}$$

and

$$\langle I_{dk}^2 \rangle = 2qI_{dk}B\langle M \rangle F \tag{8.48}$$

and the thermal and amplifier noises remain the same, as given in Equations 8.30 and 8.35, respectively. Again, there is a noise present when a "1" is being received that is not there for a "0," and again two components of it are the shot and dark-current noises, but this time much amplified.

For zero dark current, the BER expression is, using Equations 8.30, 8.42, and 8.47,

$$BER = Q\left(\frac{(\mathscr{R}P\langle M\rangle)\sqrt{2T}}{\sqrt{4KY/R} + \sqrt{4KY/R + 2q\mathscr{R}P\langle M\rangle F}} \right) \tag{8.49}$$

For thermal noise only, this becomes

$$BER = Q\left(\frac{\mathscr{R}P\langle M\rangle}{2} \sqrt{\frac{T}{2KY/R}} \right) \tag{8.50}$$

This result is plotted in Figure 8-9 for $\langle M\rangle = 10$, $Y = 300°$ K, and $C = 0.2$ pF.

It is seen that direct detection using APDs is likely to come closer to the ideal quantum limit than receivers using PIN diodes, but is still 14.3 dB short of the ideal.

8.10   Coherent Detection

Adding the local oscillator to the incoming signal

So far in this book, we have adopted one of two viewpoints about what a monochromatic light wave consists of. Such a signal would be, for example, the coherent radiation emitted by an ideally monochromatic source, such as an ideal semiconductor-diode laser. For the better part of a century it has been known that the two earlier classical views of such radiation, either as a stream of photons (the particle view), or as a sinusoid having an amplitude, frequency, and phase (the wave view), were really two incomplete aspects of a unified picture (the quantum view). At the level of this book, we more or less casually hop back and forth from one view to the other. A completely accurate view, covering all levels of light intensity, would take us into the realm of quantum theory, but for the system-design understanding that is the objective here, we need not do that. We will be able to get away with thinking of very weak narrowband radiation as consisting of a stream of photons of energy hf, with Poisson statistics, and, as the radiation gets stronger, we can think of the light becoming a narrowband sinusoid of frequency f, with gaussian statistics.

Coherent detection exploits this sinusoidal nature of the light by using receiver structures that are in many ways identical to radio receivers. This extensive topic is treated at length in [23, 24]. Both the arriving photonic signal and light from a local oscillator (LO), usually another laser diode, impinge on an ordinary PIN diode. The difference-frequency output of the photodiode is amplified, filtered electrically to a bandwidth $B = 1/2T$, and then the detection threshold generates the output bit stream by distinguishing between two values of integrated intensity (for ASK), frequency (for FSK), phase (for PSK or DPSK), or polarization (for PolSK). The analyses in the following paragraphs leave out considerable relevant detail, much of which can be found in several good tutorials on the subject [2, 3].

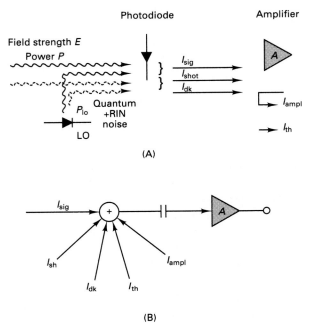

(A)

(B)

Figure 8-27. Signal and noises in the coherent photodetection process. (A) Optical signals and electrical currents, (B) Electrical equivalent circuit at the amplifier input.

The physical picture is shown in Figure 8-27(A), where the photodiode experiences two optical inputs, an arriving signal and a local oscillator signal, both of which we shall consider for the moment to be purely sinusoidal. For illustration, we assume the LO to be of longer wavelength than the signal, but it could be the other way around. The arriving-signal vector electric field is

$$\mathbf{E}_{sig} = \mathbf{u}_{sig}\sqrt{P}\ \exp(-j\omega_o t + \phi) \tag{8.51}$$

and the local-oscillator electric field is

$$\mathbf{E}_{lo} = \mathbf{u}_{lo}\sqrt{P_{lo}}\ \exp(-j\omega_{lo}t) \tag{8.52}$$

where, as in Section 3.13, ϕ is the relative phase. We assume that the states of polarization of the two fields are such that the great-circle distance between the two SOP (state-of-polarization) points on the Poincaré sphere of Section 3.13 is γ. Without loss of generality, we can represent the two electric fields as being linearly polarized in the directions $\mathbf{u}_{sig}$ and $\mathbf{u}_{lo}$, respectively,

$$\mathbf{u}_{sig} \cdot \mathbf{u}_{lo} = \cos\frac{\gamma}{2} \tag{8.53}$$

The summation vector field that impinges on the photodetector is

$$\mathbf{E} = \mathbf{E}_{sig} + \mathbf{E}_{lo} \tag{8.54}$$

so that

$$|\mathbf{E}|^2 = P + P_{lo} + 2\sqrt{PP_{lo}} \, \cos\frac{\gamma}{2} \cos[(\omega_o - \omega_{lo})t + \phi] \tag{8.55}$$

plus an irrelevant double-frequency component.

It is seen that the field intensity $|\mathbf{E}|^2$ that falls on the photodetector has a strong component at the beat (IF) frequency $(\omega_o - \omega_{lo})$, and that the intensity of this component depends on γ, which expresses the mismatch between the two SOPs. The need to match SOPs is one of the most striking differences between optical coherent receivers and conventional RF heterodyne detection, and constitutes a major engineering problem in coherent reception, since, as we saw in Section 3.13, the arriving signal has an unpredictable, slowly-varying SOP that must be somehow compensated if the $\cos \gamma/2$ term is not to exact a significant detectability penalty. We shall deal with mechanisms for controlling the loss due to polarization mismatch in Section 8.12.

Returning to Figure 8-27, it is seen that light of intensity $|\mathbf{E}|^2$ falls on the photodiode, whereupon there are the same signal, shot-noise, dark-current noise, thermal-noise, and amplifier noise components as before (Figure 8-20), except that they are quantitatively different from the previous cases of PIN/FET or APD receivers. Assuming that the SOPs are matched $(\gamma = 0)$,

$$I_{sig} = \mathcal{R}P + \mathcal{R}P_{lo} + 2\mathcal{R}\sqrt{PP_{lo}} \, \cos[(\omega_o - \omega_{lo})t + \phi] \tag{8.56}$$

By making the optical field from the local oscillator much stronger than the arriving optical signal, and by filtering out the *DC* terms (as indicated by the coupling capacitor in Figure 8-27(B)), we have

$$I_{sig} = 2\mathcal{R}\sqrt{PP_{lo}} \, \cos[(\omega_o - \omega_{lo})t + \phi] \tag{8.57}$$

The thermal, amplifier, and dark-current noises are indicated in Figure 8-27, as before with direct detection (Figure 8-20), and are quantified in Equations 8.30, 8.35 and 8.36, respectively.

The shot-noise variance is

$$\langle I_{sh}^2 \rangle = 2q\mathcal{R}(P + P_{lo})B \geq 2q\mathcal{R}P_{lo}B \tag{8.58}$$

the last inequality tending toward an equality as P_{lo} gets large.

Since both the signal power and the shot-noise power are proportional to the local-oscillator optical power P_{lo}, it is always possible to suppress the thermal, dark-current, and amplifier noises to any desired amount just by increasing P_{lo}. We shall neglect all three of them, for this reason, and also for a better reason, namely that it is generally the transmitter or LO laser linewidth that causes more of a problem than these components. The linewidth effects will be treated in Section 8.13.

Ideal homodyne detection

Suppose that the IF frequency is identically zero ($\omega_0 = \omega_{lo}$), and also that the phase difference $\phi = 0$. Suppose further, that at the transmitter, *binary phase-shift keying* (BPSK) is used, in which the transmitted phase is zero for a "1" and π for a "0." In other words the "0" and "1" signals are equal and opposite sinusoids. The decision threshold is set at zero. The means and standard deviations that enter into the BER equation 8.12 are

$$m_1 = 2\mathscr{R}\sqrt{PP_{lo}} \tag{8.59}$$

$$m_0 = -2\mathscr{R}\sqrt{PP_{lo}} \tag{8.60}$$

$$\sigma_1 = \sigma_0 = \sqrt{2q\mathscr{R}P_{lo}B} \tag{8.61}$$

These quantities are depicted in the third column of Table 8-1 for this and other detection options.

From Equation 8.12, and using the matched-filter assumption that $B = 1/2T$,

$$BER = Q\left(\frac{m_1 - m_0}{\sigma_1 + \sigma_0}\right) = Q\left(\frac{4\mathscr{R}\sqrt{PP_{lo}T}}{2\sqrt{q\mathscr{R}P_{lo}}}\right) = Q\left(2\sqrt{\frac{\mathscr{R}PT}{q}}\right) \tag{8.62}$$

It is interesting to compare this expression with Equation 8.39 for direct detection. One striking feature is that the argument of the Q-function goes as P for the incoherent case and $\sqrt{P}$ for the coherent case. This means that, once the designed power level P is established, a given small improvement in the link budget will be twice as effective (in dB) in an incoherent (direct detection) system as it would be with coherent detection.

To see what the quantum limit is for homodyne BPSK, we set the quantum efficiency $\eta = 1$, whereupon, according to Equations 8.13 and 8.18,

$$BER = Q\left(\sqrt{4rT}\right) \tag{8.63}$$

where r is the average photon arrival rate.

If the argument of the Q-function must be 6.0 and 8.0 for $BER = 10^{-9}$ and 10^{-15}, respectively, then the number of photons per "1" bit rT is seen to be 9 and 16, respectively, which is an even lower quantum limit (by about 3.5 dB) than the ASK quantum limit of Section 8.4 and the bottom curves of Figure 8-9.

If ASK is used with homodyne detection, then $(m_1 - m_0)$ decreases by a factor of two, the argument of the Q-function goes down by a factor of two, the required number of photons per "1" bit goes up by a factor of four, and the new quantum limits are 36 and 64 photons per "1" bit, respectively. All these numbers are given in Table 8-1.

Heterodyne detection

If the IF is nonzero, then immediately I_{sig} goes down by 3 dB, because the power in a sinusoid of a given amplitude is one-half that in a DC signal at that amplitude value. For this reason, the argument of the Q-function in the BER equation goes down by $\sqrt{2}$, giving the quantum limits of $Q(\sqrt{2rT})$ and $Q(\sqrt{rT/2})$ for heterodyne PSK and ASK, respectively, as listed in Table 8-1.

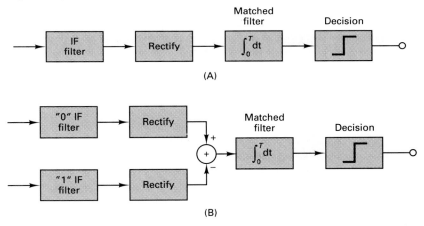

Figure   8-28. Showing how twice as much noise is let through to the threshold decision point for FSK (B) than for OOK or PSK (A).

With FSK, one gains 6 dB over ASK by doubling the spread between m_1 and m_0, but loses 3 dB of it back again. This latter is due to the fact that there are, in effect, two IF filters that let noise through, one at the "0" frequency and one at the "1" frequency. The difference of outputs is taken and the threshold is set at zero. The essential distinction between this and the PSK and ASK receivers is shown in Figure 8-28. For FSK, the decision is no longer made on the difference between two values of the same noisy output (A) as was the case for PSK and ASK, but rather on the difference between two noisy outputs, as shown at (B). There is twice as much noise power as if there had been only one IF channel leading to the decision point.

Practical problems

Given the detectability limits imposed on direct detection by dark-current noise, amplifier noise, and above all, by thermal noise, coherent reception has proved a tempting alternative. All these sources of noise succumb to the tactic of using large local-oscillator power level. While it is difficult to stabilize two laser frequencies so that their difference is maintained at the center of the IF filter passband, it can be done using standard, but carefully engineered, phase-lock loop techniques. Controlling the heterodyne images (aliases) in the frequency domain is also a significant issue [25]. It has even proved possible to stabilize the LO phase against the transmitter laser phase so as to do homodyne detection. This large amount of heroic experimental work has been summarized in [23, 24].

However, this rosy picture is greatly transformed when one tries to use coherent-detection approaches in building systems that must simple, inexpensive, and stable enough to compete in commercial network products. Shortly, we shall discuss two of the important difficulties, those caused by drift of state of polarization, and by the departures from nonzero linewidths of the transmitting and LO lasers.

8.11 Direct Detection Using Optical Preamplifiers

We have now covered two of the three competing methods of overcoming the thermal noise at the receiver amplifier input. We have discussed amplifying inside an APD device by using the impact-ionization effect, and we have discussed swamping the thermal noise with a large local-oscillator signal in a coherent receiver. The trouble with these two approaches is that the first does not buy very much because of randomness of the avalanche gain, while, as we shall see shortly, the second entails considerable system complexity that usually causes performance to fall many dB short of the relevant quantum limits of Table 8-1.

An optical amplifier (Chapter 6), especially of the travelling-wave type, when placed immediately before the photodetector, has neither of these difficulties. However, it has its own problem, namely random spontaneous-emission events that occur within the amplifier and appear at the output as ASE (amplified spontaneous emission).

So each of these three silver linings has its own particular cloud.

However, we shall find in this section that the best solution of the three seems to be photonic preamplification, particularly in the case of the very convenient ASK modulation format. It will turn out that the detectability limit for such an arrangement, which is physically simple and is capable of being greatly cost-reduced, is close to that of the shot-noise imposed quantum limit of a coherent heterodyne ASK system. The systems that outperform direct detection using preamplification will be seen to be those that use coherent heterodyne FSK or to go to the very exacting homodyne class of coherent solutions.

Figure 8-29 shows an experimental optical receiver consisting of an erbium-doped fiber amplifier followed by a tunable filter (Fiber Fabry Perot of Section 4-4) and then a PIN/FET device.

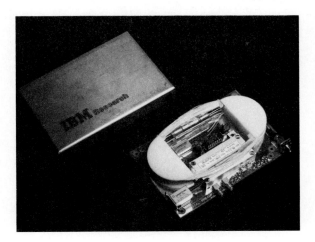

Figure 8-29. Experimental integrated optical receiver consisting of an erbium-doped fiber amplifier, a tunable filter and a PIN/FET detector. (Courtesy K. Liu and D. G. Steinberg, IBM T. J. Watson Research Center).

In Section 6.3, it was determined that the power spectral density of ASE at the amplifier output is

$$N(f) = h f \chi \, n_{sp} \, [G(f) - 1] \tag{8.64}$$

where χ is the excess noise factor (≥ 1) due to nonzero mirror reflectivities, as in Equation 6.10, and n_{sp} is the spontaneous-emission factor.

In calculating the limit of best possible performance, a somewhat artificial assumption is now made, that one-half the spontaneous-emission noise has been eliminated before the photodetector by interposing a *polarization filter*, one that passes only the SOP of the signal and rejects light with an SOP that is orthogonal ($\gamma = \pi$). In practice, this can be eliminated if one can accept the consequent 3-dB increase in $N(f)$. To get the quantum limit, we shall assume the inclusion of the polarization filter, even though this element is omitted in practical doped-fiber amplifiers, such as those in Figures 6-12 and 8-29.

Let us analyze the situation for ASK modulation in which the amplifier gain $G(f)$ is flat across the modulation bandwidth, and in which that portion of the ASE lying within the modulation bandwidth is sufficient to swamp out the thermal noise. Suppose a single ASK signal is passed through the amplifier and is then filtered in a prefilter having an irrelevant bandwidth B_o that is a few times the modulation bandwidth, as shown in Figure 8-30(A). The amplifier output then impinges on the photodiode before entering the usual matched filter and threshold device.

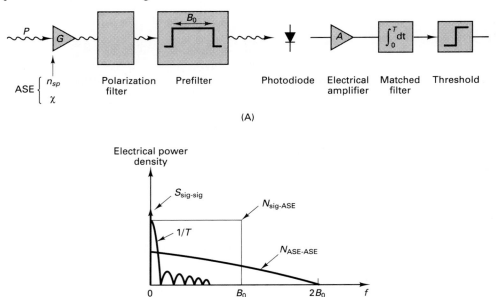

(A)

(B)

Figure 8-30. Direct detection receiver using optical preamplification, as idealized to include a polarization filter. (A) Block diagram, (B) Power density spectrum at the photodiode output.

Assume, for the moment, that the input optical signal of power P is a steady sinusoid. Let us examine signal and noise at the electrical amplifier input. Since the power spectrum of the electrical current out of the photodiode is the result of a squaring operation upon the input optical power, the power spectrum of the photodiode output will contain not only a spike at DC, the *signal-signal* cross-product term, which is the squared optical signal, but will also contain two noise cross-product continuum spectra, *signal-ASE*, and *ASE-ASE*. The latter has a spectral density much smaller than that of the *signal-ASE* component. The spectra that the photodiode amplifier sees are illustrated in Figure 8-30(B), where B_o is the bandwidth of the prefilter, and $B = 1/2T$ is the bandwidth of the electrical filter.

This situation has been analyzed at length [26, 27]. Assuming zero pigtailing coupling losses at the amplifier, no loss of light between amplifier and photodiode, and unity photodiode quantum efficiency, the powers in the useful *signal-signal* term and the harmful *signal-ASE* terms are, respectively

$$S_{\text{sig-sig}} = \left(G\frac{q}{hf}P\right)^2 \tag{8.65}$$

and

$$N_{\text{sig-ase}} = 4\,\frac{q^2}{hf}\,P\,\chi n_{sp}G(G-1)B \tag{8.66}$$

If we can make the assumption that all the noises are gaussianly distributed, and if we replace the steady-state optical signal with one that is squarewave-modulated with bit period T seconds, and assume $B = 1/2T$ is the bandwidth of the matched electrical filter, we obtain the probability of bit errors committed at the threshold decision point

$$BER = Q\left(\sqrt{\frac{2G^2PT}{4hfG(G-1)\chi n_{sp}}}\right) \le Q\left(\sqrt{\frac{PT}{2hf\chi n_{sp}}}\right) \tag{8.67}$$

the inequality becoming an equality for large gain. For travelling-wave amplifiers, $\chi = 1$, so for such devices, we can get the number of photons rT per "1" bit required from

$$BER = Q\left(\sqrt{\frac{rT}{2n_{sp}}}\right) \tag{8.68}$$

For fully population-inverted material, $n_{sp} = 1$, and we have

$$BER = Q\left(\sqrt{rT/2}\,\right) \tag{8.69}$$

This is exactly the bit error rate expression of coherent heterodyne ASK, and gives a sensitivity limit of 72 and 128 photons per "1" bit for 10^{-9} and 10^{-15}, respectively.

Actually, by employing the correct noise statistics rather than the gaussian approximation, the more exact limit has been found to be 76 and 130 bits [28]. This number, which is about 5.8 dB short of the ASK photon-counting quantum limit, is given in Table 8-1 and Figure 8-9. A value of 92 photons per "1" bit has been achieved experimentally [29] using photonic amplification preceding direct detection.

8.12 Dealing with Polarization

Five approaches

It should be clear at this point that there are many circumstances in which the slow random drift of the received signal's SOP (state of polarization) can be ignored. Photodiodes will respond to light of arbitrary polarization, and most tunable filters are either intrinsically polarization-insensitive (e.g., Fabry-Perot etalons) or can be made so (by the double polarization beamsplitter arrangement, for example). However, it was seen that all coherent systems are polarization-sensitive, and so are some forms of filters and modulators.

It is therefore not acceptable to ignore the polarization problem, particularly in a coherent receiver, since detectability in such receivers will be hopelessly ruined during long intervals during which the SOPs of signal and local oscillator are nearly orthogonal.

There are five approaches to dealing with polarization, in roughly increasing order of effectiveness:

- Using **polarization diversity**, so that no matter where the received SOP wanders on the Poincaré sphere, the signal will be received.

- **Scrambling**: Insuring that, during one bit time, several widely spaced SOPs are visited, thus insuring that the signal SOP and the receiver SOP will not stay orthogonal for an entire bit. We discussed this in Section 7.4.

- Installing special **polarization-maintaining fiber**, which forces a condition of fixed received SOP.

- **Polarization tracking**, consisting of measuring the received SOP, and then adjusting the incoming and receiver SOPs into coincidence.

- **Polarization-shift keying**. The argument here is that "if you can't fix it, feature it." The transmitter sends two orthogonal SOPs, as described in Section 7.4, one for a binary "0" and the other for binary "1." The same devices that one needs to measure the SOP to provide inputs to the polarization-correcting operation in a polarization tracker can be used to distinguish between "0" and "1."

Polarization diversity receivers

Polarization diversity [23] is illustrated in Figure 8-31. At (A) is shown a nondiversity receiver, in which the incoming signal has SOP_{sig} and the receiver wants to see SOP_{rec}. The two can become completely orthogonal at times. (B) shows the idea behind a polarization-diversity receiver, and (C) the closely related phase-diversity receiver for binary PSK. The latter might be required when the receiver wants to see a specific value of phase, that is, with a homodyne coherent system. It is possible to cascade polarization diversity with phase diversity in a coherent receiver, but the problem is clearer if we keep them separate.

All such designs are based on the simple "diversity principle": if *two separate receivers* are employed, each responding to a state orthogonal to the other, and if their outputs are combined after photodetection but before the threshold decision point, then a usable signal will always be received. In polarization diversity, the two receiver SOP points are antipodal on the sphere, and in phase diversity, the two receivers are arranged to be 90° out of phase. The "polarization-sensitive components" in Figure 8-31(A) could be, for example, polarization-sensitive tunable-filter direct-detection receivers or they could be coherent receivers. Many forms of polarization splitter are available [12, 30], including some intended specifically for fiber optic usage and therefore pigtailed without lenses and having very low excess loss.

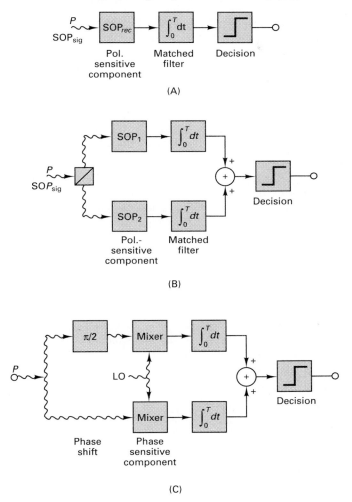

Figure   8-31. Diversity receivers. (A) Reception with polarization-sensitive component. (B) Polarization diversity, (C) Phase diversity receiver for BPSK.

While polarization diversity, described in Section 7.4, seems to be the currently favored fix for polarization drift in propagation, it requires a doubling of receiver complexity. The detection penalty paid is 3 dB of optical power, equivalent to the assumption that the received signal spends half its time in the SOP orthogonal to that which the receiver responds to. Also, as discussed in connection with Figure 8-28, the extra branch lets through twice as much noise as a single-threaded receiver chain, so that even though the received SOP will always produce a useful information-bearing detected signal at the decision point, the combined electrical output at that point will have 3 dB more noise added to it. There is one of the structures of Figures 8-31(B) and (C) for the "0" bit and another for the "1" bit, each feeding into the decision circuit on the right.

Polarization-maintaining fibers

In Chapter 3, it was pointed out that each mode that is supported in a fiber actually consist of several *degenerate submodes*, usually two in number. In particular, the HE_{11} mode of Figure 3-19, the only one that normally propagates in a single-mode fiber, really consists of two degenerate states at right angles to one another transversely.

Suppose HE_{11} light is launched with pure vertical polarization. As mentioned in Section 3.13, the combined effect of small birefringence and slight inhomogeneities in the fiber will couple some of the light into the horizontal state, this coupling will be a random function of distance along the fiber, and the two submodes will propagate at slightly different velocities. In one *beat length*, a complete 360° of phase will have accumulated. The net result, after kilometers or even meters of propagation, is total unpredictability of the received SOP.

However, it turns out [31] that the coupling coefficient between the two degenerate states is stronger the more closely the beat length matches the axial length of the inhomogeneities, and it is also fortunately true that the relative probability of occurrence of an inhomogeneity of a given length ℓ drops off very rapidly with decreasing ℓ. This means that if some way could be contrived to artificially enhance the difference in propagation velocities of the two submodes, the beat length would decrease to the point where few inhomogeneities of that scale size would be seen, the coupling between modes would be greatly reduced, and conversion would be suppressed.

This is the principle behind polarization-maintaining fibers, several examples of which are shown in Figure 8-32. The fiber is fabricated to contain a strong, deliberately introduced, circumferential asymmetry, and the resulting strong birefringence assures that the beat length is reduced from meters down to one or two millimeters. In this way, more than 20 dB suppression of coupling between submodes has been obtained for links several kilometers long [4].

There are two prices to be paid in applying this approach, one major and the other more or less trivial. First, the entire system must be populated with this special fiber (and preinstalled fiber replaced), which has largely discouraged the widespread use of polarization-maintaining fiber. Second, the transmitted SOP must be adjusted to match that of one of the two degenerate submodes.

Polarization tracking

Polarization tracking consists of the two stages, SOP measurement and then SOP correction. Standard techniques from the field of optical laboratory instrumentation are used to measure the SOP of the incoming light, usually by determining the four Stokes

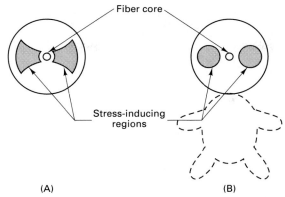

Figure 8-32. Several patterns of refractive index alteration across the cross-section of a silica fiber. (A) Bow-tie, (B) PANDA (The name is not really an acronym for "Polarization-maintaining AND Absorption-reducing", but is instead explained by the dashed lines).

parameters. From these measurements, control signals are derived that are fed to polarization-control devices that move either the signal SOP or the receiver SOP around on the sphere until they correspond. A comprehensive tutorial on polarization tracking is contained in [32].

Polarization shift modulation

We saw in Section 3.13 that when a monochromatic radiation having a certain polarization has imposed on it a change of SOP to the orthogonal SOP (antipodal position on the Poincaré sphere), this orthogonality survives the propagation process, and the received states of polarization remain orthogonal. As described in Section 7.4, we can exploit this fact to do *polarization-shift keying (PolSK)*, or even *differential polarization-shift keying (DPolSK)* [33]. In binary PolSK, one SOP represents a "0" and the orthogonal SOP a "1." Each arriving SOP will be completely different from the state in which it is transmitted, but the two will be orthogonal to one another upon arrival at the receiver. This is analogous to the much more obvious physical fact that even though one may not recover the transmitted phase from the received phase, PSK or DPSK between two phases will work because the shift between two orthogonal phases in phasor space is preserved.

Neither PolSK nor DPolSK has been applied practically, but they are not uninteresting. One intriguing point is that, since the Stokes parameters can be measured without knowing the phase, it is possible to build an incoherent polarization-modulation system. This is an attractive substitute for ASK for the same reason that PSK and FSK are attractive; one avoids the chirping of the transmitter laser caused by large changes in the injection current.

8.13 Effect of Laser Linewidth on Coherent Detection

Two effects of laser noise

As we saw in Section 5.5, the output from a laser-diode source is not purely monochromatic (sinusoidal). On the contrary, the output resembles narrowband gaussian noise in that both phase and amplitude are random functions of time, varying at a rate much slower than the center frequency. However, there is a difference between laser noise and classical filtered gaussian noise in that, for laser noise, the phase part of the randomness is exaggerated by a factor of up to 30-40, the *linewidth-enhancement factor β* of Equation 5.32. Phase noise has a serious effect on coherent systems, particularly on those using the modulation formats that depend on phase: PSK and DPSK.

Coherent systems are twice at the mercy of these fluctuations, once as the signal is transmitted, and a second time as the local oscillator introduces its own random fluctuations into the detection process.

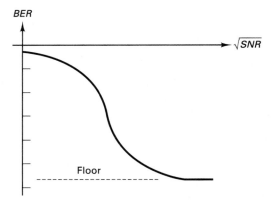

Figure 8-33. An error rate floor caused by some effect that only shows up at received high signal to noise ratios, such as finite laser linewidth.

Effects such as noise due to nonzero linewidth show up as *error-rate floors* of the sort illustrated in Figure 8-33. The normal *Q*-function behavior breaks down as the argument increases, and the error rate is greater than it should be. Error-rate floors are the evidence that points to some hitherto ignored effect that was not significant at lower SNRs. They have been known to appear unexpectedly as the state of the art progresses toward lower and lower BERs. Thus, it may not always be straightforward to make a system designed to work at $BER = 10^{-9}$, for example, to achieve 10^{-15} simply by cranking up the SNR by the expected amount. Often, an error rate floor is found, indicating that some hitherto-ineffective system parameter, among the many listed in Figure 8-4, will have to be revisited.

Amplitude fluctuations and the balanced receiver

The amplitude fluctuations are considerably less harmful than the phase fluctuations, and not only because they are several orders of magnitude smaller than with filtered noise. That part of the amplitude fluctuations occurring in the local oscillator can be cancelled almost completely by the clever artifice of using two back-to-back photodiodes in the *balanced-receiver* format [34] of Figure 8-34, where (A) shows a single photodiode, and (B) shows the balanced arrangement.

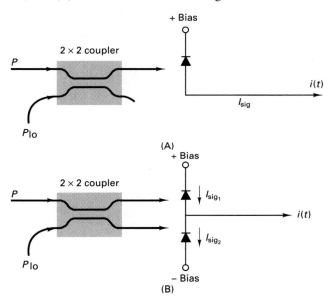

Figure 8-34. The balanced detector. (A) Power loss when a single photodiode is used, (B) The power is preserved in the balanced detector.

The balanced receiver has become widely used in all coherent systems, not only because it balances out the amplitude noise, but more importantly because it removes a 3-dB loss in the optical power reaching the photodiode that could otherwise occur. We can see all this from the following analysis.

In building a coherent receiver, when it comes to combining the received and LO radiations, the most practical method is to use a 2×2 coupler, as shown in the figure. Analysis of this device (Section 3.11) showed that at output 1 the received signal appears with a 90° phase shift, while the LO appears unshifted. At output 2, the received signal appears shifted and the LO unshifted. Note that in the conventional single-ended receiver of Figure 8-34(A) half the optical power is wasted (unavoidably, as we saw).

If we return to Equation 8.57, and reexamine what happens to the phase, we have

$$I_{\text{sig1}} = 2\mathcal{R}\sqrt{PP_{\text{lo}}} \, \sin[(\omega_o - \omega_{\text{lo}})t + \phi] \qquad (8.70)$$

and

$$I_{\text{sig2}} = 2\mathcal{R}\sqrt{PP_{\text{lo}}} \, \sin[(\omega_{\text{lo}} - \omega_o)t - \phi] \qquad (8.71)$$

so that the two sinusoidal difference-frequency photocurrents are equal and opposite in sign. These currents add in the back-to-back configuration of Figure 8-34(B).

A demonstration that a balanced receiver cancels out the local-oscillator amplitude fluctuations is straightforward, if tedious [35]. We can call the complex input-signal electric field strength S, and that of the local oscillator $L + N$, where L is the steady component and N the fluctuating component. Normalizing out the responsivity $\mathcal{R}$ and ignoring shot noise, the currents out of the photodetectors are

$$I_1 = \left| S + j(L + N) \right|^2 \qquad (8.72)$$

$$I_2 = \left| jS + (L + N) \right|^2 \qquad (8.73)$$

The two noise components are the results of subtracting out the steady state S and L fields

$$I_{1n} = I_1 - \left| S + jL \right|^2 \qquad (8.74)$$

$$I_{2n} = I_2 - \left| jS + L \right|^2 \qquad (8.75)$$

After length expansion, where S, L, and N have been split into their real and imaginary parts (e.g., $S = S_R + jS_I$), one has, using the identity $\left| a + jb \right|^2 = a^2 + b^2$,

$$I_{1n} = N_R^2 + N_I^2 + \text{cross products of } N_R \text{ or } N_I \text{ with } S_R, \; S_I, \; L_R, \; \text{or } L_I \qquad (8.76)$$

$$I_{1n} = N_R^2 + N_I^2 + \text{cross products of } N_R \text{ or } N_I \text{ with } S_R, \; S_I, \; L_R, \; \text{or } L_I \qquad (8.77)$$

At the amplifier input, the noise that is left is

$$I_n = I_{1n} - I_{2n} = 4S_I N_R - 4S_R N_I \qquad (8.78)$$

All that is left are two cross-products of the LO noise N against the signal S, which is much smaller than would have been any products of N against the LO field L, since the LO power is orders of magnitude larger than the signal power. Thus, the desired low-noise performance has been achieved.

It should be pointed out that this solves only half the amplitude-fluctuation problem; the fluctuations from the transmitter laser are still present.

Phase noise effects

It was mentioned in Section 5.5 that random phase noise is a Wiener process. Such a process is one in which the variance increases with time, as was illustrated in Figure 5-24. That this makes sense can be seen [2, 15] by imagining what phase one measures as time elapses in a long sample of narrowband random noise. The variance of the phase is given by

$$\langle \phi^2(t) \rangle = 2\pi \Delta f_{\text{tot}} t \tag{8.79}$$

where the total effective linewidth Δf_{tot} is the sum of the signal and LO linewidths

$$\Delta f_{\text{tot}} = \Delta f_{\text{sig}} \beta_{\text{sig}} + \Delta f_{\text{lo}} \beta_{\text{lo}} \tag{8.80}$$

Each unenhanced linewidth Δf is given by Equation 5.30 as augmented using the linewidth enhancement factor β of Equation 5.32. Typical values of β are around 30-40, unless special linewidth-reduction means are employed, such as long external cavities (Section 5.9).

Equation 8.79 displays the linear growth of variance of the phase with time, the consequence being that at the end of a bit time T seconds long,

$$\langle \phi^2(t) \rangle = 2\pi \Delta f_{\text{tot}} T \tag{8.81}$$

The implications are a bit startling. As the bitrate gets higher, the Wiener process has less time to act and the phase noise is less. So high bitrates are better. There is a second somewhat surprising thing about phase noise. One might expect that only phase-sensitive systems, such as those using PSK or DPSK, would be bothered by it, but not ASK systems. Yet coherent ASK systems, too, are affected by phase noise. The reason is that coherent detection of all forms requires that the IF filter (of bandwidth roughly $1/T$) see a pure single-frequency sinusoid *throughout* the integration interval T for full detectability.

The quantitative facts [36] are shown in Table 8-2. The "maximum permissible linewidth/bit rate ratio" is the minimum value of the product $\Delta f_{\text{tot}} T$ that can be tolerated in order to maintain the detectability loss at less than an equivalent 1-dB loss in light input.

Table 8-2. Effect of Laser Phase Noise (From [36]).

Coherent receiver type	Modulation format	Maximum permissible linewidth/bitrate ratio
Heterodyne	ASK	9.0×10^{-2}
	FSK	2.0–9.0×10^{-2}
	PSK	2.3×10^{-3}
	DPSK	1.7×10^{-3}
Homodyne	PSK	3.1×10^{-4}

One prediction that can be made from these numbers is that, if the coherent option is to be used widely, it will only be in those high bitrate systems where the economics allow great care to be taken with phase noise. Such systems are more likely to be intercity or intercontinental long haul links than networks, where low cost is a strong requirement and bitrates will probably not grow beyond Gb/s for some years.

8.14 Effect of Crosstalk in Semiconductor Amplifiers

Recall from Chapter 6 that the carrier lifetime in semiconductor laser-diode amplifiers is so short that, even at a bitrate of a gigabit per second, the gain experienced by a signal at one wavelength can be affected by the consumption of carriers that are amplifying some other signal at another wavelength.

We pick up this story [27] again where we left off at Section 6.6 and Figure 6-9, having meantime covered the material on distinguishing a "1" from a "0" required to understand the consequences of amplifier crosstalk.

Figure 8-35(A) shows, once again, the probability distribution of amplifier gain for a sample case of $N = 10$ wavelength channels. This was discussed in Section 6.6. Assume, as before, that at the amplifier, all N channels are bit synchronized; that is, the transitions between bits are simultaneous. While this situation would not occur in practice, it gives a worst case picture of the effect of crosstalk.

There is no closed-form solution for the optimum setting θ of the threshold. The error rate is given [27] by a sum of Q-functions, weighted by the relative frequency of occurrence of the spikes in Figure 8-33(B).

$$BER = \sum_{i=0}^{N-1} \left[\Pr\left[m_{i,0}\right] Q\left(\frac{\theta - m_{i,0}}{\sigma_{i,0}}\right) + \Pr\left[m_{i,1}\right] Q\left(\frac{m_{i,1} - \theta}{\sigma_{i,1}}\right) \right] \qquad (8.82)$$

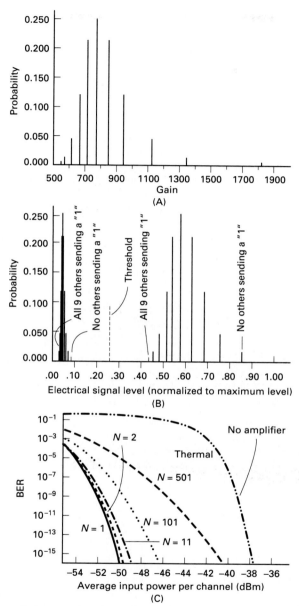

Figure 8-35. Crosstalk in a semiconductor laser diode amplifier can smear out the probability distributions. (A) Probability density of gain. (B) Signal-level probabilities of occurrence at the threshold decision point. (C) Bit error rate versus received power. The amplifier case with various numbers of channels N is compared with the no-amplifier thermal-noise case.

where i and j in the subscripts (i, j) on the means m and the standard deviations σ refer to the i th received signal level for either $j = 0$ or 1, depending on whether a "0" or a "1" was sent. The standard deviations are zero only if there are no random-noise sources.

Figure 8-35(C) shows a sample result for the case of various numbers N of 200-Mb/s. channels. The receiver bandwidth was 100 MHz, carrier relaxation time 1.0 ns, amplifier small-signal gain $G_o = 1800$, and saturation power $P_{sat} = -6$ dBm. The typical thermal noise-limited case of no amplifier (and therefore no crosstalk) is shown for comparison. This last curve is independent of number of channels N, since they do not interfere in that case.

It is clear that, even with crosstalk in the amplifier, considerable gain in performance is achieved by the use of such devices. For the erbium doped-fiber case, where crosstalk is absent for interesting data rates (above a few Kb/s), the case for amplifiers is even stronger.

8.15 What We Have Learned

The receiver is the place in the system "where it all comes together." Light comes in and electrical bits come out, and the reliability of this process depends on all the factors that have been introduced in the chapters preceding this one. The overall system consequences are the topics we shall soon turn to in Parts III and IV of this book, Chapters 11 through 16.

While the signal can be quite weak in terms of the number of photons actually arriving to signal a "0" or "1" bit, the wavelength is so short, compared for example to the perhaps more familiar radiowave case, that the energy hf is sufficient to cross the bandgap and produce almost one hole-electron pair for every photon arriving. The inevitable *Law of Edsel Murphy* [37] acts, however, to introduce noise. As if it isn't bad enough that the receiver possesses the inevitable thermal noise that contaminates any weak signal, plus other things such as internal amplifier noise or noise in the process of realizing avalanche gain, the signal itself arrives noisy. This is because of the quantized nature of the light, a property seen only at low light levels.

In attempting to use a threshold device to distinguish "0" and "1" reliably, the shot noise appearing at the decision point always places a theoretical lower bound on the received optical power required for a given bit error rate. While this quantum limit is less than 150 photons per "1" bit for idealized versions of almost all receiver types, the detectability limits are usually about two orders of magnitude higher using realizable values of (i) amplifier input capacitance, (ii) APD gain fluctuation, (iii) optical-amplifier stimulated-emission noise and (iv) laser linewidths. This list is a list of the principal limiting factors for receivers based, respectively, on (i) PIN diodes, (ii) APDs, (iii) optical preamplification and (iv) coherent detection, respectively.

At present, it appears that the most attractive option, particularly for low-cost systems, is to use ASK at the transmitter and optical preamplification at the receiver.

Since photonic amplifiers, particularly those using erbium-doped fiber, are quite simple and are decreasing rapidly in cost, one may expect that a common form of receiver front end in the future will be a packaged optical preamplifier/PIN/FET, and that this will take the place of today's PIN/FET and APD standard optical receiver components.

8.16 Problems

8.1 (*Comparison with microwave detectability*) Ideally, a well-designed microwave (1-10 GHz) receiver for 1 Gb/s service should be able to deliver 10^{-9} error rates for a received signal level approaching -74 dBm. This is if the front end runs at room temperature; even lower signal levels will suffice if, through the use of cryogenic cooling, the receiver front end sees lower temperatures. (These numbers come from setting the power signal-to-noise ratio of such receivers, E_b/N_o, to be 10 or more, where $E_b =$ signal energy in a bit, $N_o = KY$, $K = 1.38 \times 10^{-23}$ being Boltzmann's constant in joules per °K and Y being the absolute temperature). (A) How many photons per bit is this -74 dBm for a microwave system at 10 GHz? (B) How many photons per bit is -74 dBm at 1.5 μ? (C) Explain why there have to be so many photons per bit in the microwave case. (D) Explain why the optical receiver couldn't similarly be pushed to detectability levels of -74 dBm if there are so many photons per bit for the microwave case. (E) Why couldn't the principles that are used in the laser diodes of Chapter 5 and the laser-diode amplifiers of Chapter 6 be used to generate and amplify microwave signals, respectively?

8.2 You are a development manager, and two of your people come to you with proposals for improving the link budget of your optical network. Hatfield offers to get you a 3-dB improvement in the front-end transistor amplifier circuit for the same resources that McCoy will require to get you a 3-dB improvement in the performance of an optical coupler component. Which would you pick and why?

8.3 In designing a point-to-point intercity trunk of the TDM type being installed in the mid-1980s, which elements in the system diagram of Figure 8-4 do you think you could ignore, and why?

8.4 You are asked to design a TV distribution system in the form of the wavelength-division multipoint structure shown in the figure below.

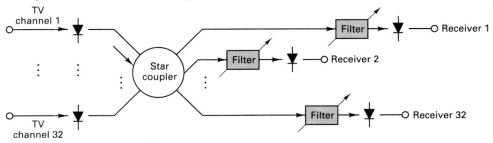

The components are as follows:

- 32 1.2 Gb/s digital HDTV television channels at the head end, each ASK-modulating a laser diode at a separate wavelength, with an extinction ratio of 0.2.

- Each laser diode has a peak power of 1 milliwatt.

- The 32×32 star coupler has an excess loss of 1.0 dB.

- The most distant receiving station is 20 km from the head end.

- Link attenuation is 0.25 dB/km.

- A tunable optical filter is used at each receiver. This filter is selective enough that there is negligible crosstalk, but has 2.0 dB of excess loss.

- Connector and splice losses are 2.0 dB.

- The receiver requires at least -30.0 dBm of signal power to produce the desired bit error rate.

How many decibels of margin are left between the actual arriving signal power and that required by the receiver?

8.5 (*Dynamic range in a network*) A network using dense wavelength division has fifty nodes connected by a 64×64 star coupler with 3 dB of excess loss. Two of the nodes are within 100 meters of the star, while the farthest node is 25 kilometers from the star. Each node has a transmit power level of 0 dBm. (A) What dynamic range requirement must be satisfied by each photodiode-amplifier combination if 1.3 μ is chosen as the center operating wavelength? How much better or worse would 1.5 μ be? (B) If this is a WDM network, with wavelengths from 1.520 to 1.565 μ, you had to use erbium-doped-fiber amplifiers in the network, and supposing these amplifiers had the $G(\lambda)$ spectrum shown in Figure 6-11, with 25 dB gain at 1532 nm, what would be the contribution to the dynamic range of the amplifier gain nonuniformity?

8.6 Why is one quadrant of Figure 8-17(A) unoccupied?

8.7 Given the reverse-biased photodiode whose characteristic curves are those shown in Figure 8-14. The responsivity is 0.75. If the supply voltage V_{bias} is 20 volts and the load resistor is 10^6 ohms, how large an optical input power can be tolerated before saturation of the preamplifier input sets in?

8.8 (A) Give the spectral density of the thermal noise power of a room temperature PIN photodetector having an input resistance of 5 kilohms. (B) The units "dBW/Hz" (dB below one watt per Hz of bandwidth) are sometimes used. Express the present answer in these units.

8.9 An InGaAs photodiode for use at 1.5 μ is followed by a transimpedance amplifier with a gain of 20 dB and a 20-kilohm feedback resistor. If the input capacitance is 5 pF, how thin must the total intrinsic region be kept in order to make the transit-time bandwidth constraint equal that of the electrical-circuit time constant? Assume that the hole mobility is the same as the electron mobility.

8.10 In Section 8.7, supposing there were a photonic amplifier in the link, why not just lump amplifier ASE in with the other noise components such as thermal noise, dark-current noise, and electrical amplifier noise?

8.11 (A) Rederive Equation 8.39, for bit error rate with a direct detector, when the effect of amplitude fluctuations at the transmitter are included, as specified by the *RIN* parameter (Section

5.5). Assume gaussian statistics. (B) If $RIN = -103$ dB/Hz, what is the BER at a bit rate of 1 Gb/s, assuming all other sources of noise to be negligible? (C) Would it help to increase the photodiode's responsivity?

8.12 (A) Rederive the BER expression, Equation 8.39, for the direct-detection FSK case, again assuming that the transmitting laser is peak-power limited. (B) Discuss the relative performance of ASK and FSK.

8.17 References

1. Y. Yamamoto, *Receiver performance evaluation of various digital optical modulation-demodulation systems in the 0.5 - 10 wavelength region*, pp. 1251-1259, 1980.

2. J. Salz, "Modulation and detection for coherent lightwave communications," *IEEE Commun. Magazine*, vol. 24, no. 6, pp. 38-49, 1986.

3. J. Barry and E. A. Lee, "Performance of coherent optical receivers," *Proc. IEEE*, vol. 78, no. 8, pp. 1369-1394, 1990.

4. S. E. Miller, I. P. Kaminow, ed., *Optical Fiber Communications—II*, Academic Press, 1988.

5. C. Lin, ed., *Optical Technology and Lightwave Communication Systems*, Van Nostrand Reinhold, 1989.

6. E. E. Basch, ed., *Optical Fiber Transmission*, Sams/McMillan, 1986.

7. G. Keiser, *Optical Fiber Communications*, McGraw-Hill, 1983.

8. T. L. Koch and U. Koren, "Photonic integrated circuits: Research curiosity or packaging common sense," *IEEE LCS Magazine*, vol. 1, no. 4, pp. 50-56, November, 1990.

9. P. E. Green and R. Ramaswami, "Direct detection lightwave systems: Why pay more?," *IEEE LCS Magazine*, vol. 1, no. 6, pp. 36-49, November, 1990.

10.

11. T. L. Koch and U. Koren, "Semiconductor lasers for coherent optical fiber communications," *IEEE/OSA Jour. Lightwave Tech.*, vol. 8, no. 3, pp. 274-293, March, 1990.

12. F. Tosco, ed., *CSELT Fiber Optic Communications Handbook - Second Edition*, TAB Professional and Reference Books, Blue Ridge Summit, PA, 1990.

13. J. Wozencraft and I. Jacobs, *Principle of Communication Engineering*, Wiley, 1965.

14. R. E. Slusher and B. Yurke, "Squeezed light," *Sci. Am.*, vol. 258, no. 5, pp. 50-56, 1988.

15. W. B. Jones Jr., *Introduction to Optical Fiber Communication Systems*, Holt, Reinhart and Winston, 1988.

16. R. J. Hoss, *Fiber Optic Communications Design Handbook*, Prentice Hall, 1990.

17. J. C. Palais, *Fiber Optic Communications*, Prentice Hall, 1988.

18. J. Gowar, *Optical Communication Systems*, Prentice Hall, 1984.

19. S.E. Miller and I.P. Kaminow, eds., *Optical Fiber Telecommunications—II*, Academic Press, chap. 18, pp. 689-717, 1988.

20. E. A. Lee and D. G. Messerschmitt, *Digital Communication*, Kluwer Academic Publishers, 1988.

21. R. K. Willardson and A. C. Beer, eds., *Semiconductors and Semimetals, Chap. 12*, Academic Press, chap. 12, 1977.

22. P. P. Webb, R. J. McIntyre, and J. Conradi, "Properties of avalanche photodetectors," *RCA Review*, vol. 35, no. 2, pp. 234-278, 1974.

23. T. Okoshi and K. Kikuchi, *Coherent Optical Fiber Communications*, Kluwer, 1988.

24. V. W. S. Chan, K. Nakagawa, and D.W. Smith, eds., "Special Issue on Coherent Communications," *IEEE/OSA Jour. Lightwave Tech.*, vol. 5, no. 4, pp. 413-637, 1987.

25. B. S. Glance, J. Stone, K. J. Pollock, P. J. Fitzgerald, C. A. Burrus Jr., B. L. Kaspar, and L. W. Stulz, "Densely spaced FDM coherent star network with optical signals confined to equally spaced frequencies," *IEEE/OSA Jour. Lightwave Tech.*, vol. 6, no. 11, pp. 1770-1781, 1988.

26. N. A. Olsson, "Lightwave systems with optical amplifiers," *IEEE/OSA Jour. Lightwave Tech.*, vol. 7, no. 7, pp. 1071-1982, 1989.

27. R. Ramaswami, "Issues in multi-wavelength optical network design," *IBM Research Report RC-15829*, May, 1990.

28. P. S. Henry, "Error rate performance of optical amplifiers," *Conf. Record, Optical Fiber Commun. Conf.*, February, 1989.

29. P. M. Gabla, E. LeClerc, J. F. Marcerou, and J. Hervo, "92 photons/bit sensitivity using an optically preamplified direct-detection receiver," *Conf. Record, Optical Fiber Commun. Conf.*, p. 245, 1992.

30. E. Hecht, *Optics - Second Edition*, Addison-Wesley, 1987.

31. S.E. Miller and I.P. Kaminow, eds., *Optical Fiber Telecommunications–II*, Academic Press, chap. 3, pp. 56-107, 1988.

32. N. G. Walker and G. R. Walker, "Polarisation control for coherent optical fiber systems," *Brit. Telecom Technol. Journal*, vol. 5, no. 2, pp. 63-76, 1987.

33. R. Calvani, R. Caponi, and F. Cisterno, "Polarization phase shift keying: A coherent transmission technique with different heterodyne detection," *Elect. Ltrs.*, vol. 24, no. 10, pp. 642-643, 1988.

34. G. L. Lee, V. W. Chan, and T. K. Lee, "A dual-detector optical heterodyne receiver for local oscillator noise suppression," *IEEE/OSA Jour. Lightwave Tech.*, vol. 3, no. 10, pp. 1110-1122, 1985.

35. R. Ramaswami, "Cancellation of LO fluctuations in a balanced receiver," *Private communication*, July, 1991.

36. L. Kazovsky, "Impact of laser phase noise on optical heterodyne communication systems," *Jour. Optical Commun.*, vol. 7, no. 2, pp. 67-78, 1986.

37. D. L. Klipstein, "The contributions of Edsel Murphy to the understanding of the behavior of inanimate objects," *Elec. Equipment. Design: Circuit Design and Engineering*, vol. 15, no. 8, 1967.

CHAPTER 9

Subcarrier Systems

9.1 Introduction

As we saw in Chapters 5 and 8, it is possible to make commercial-grade laser diodes and photodetectors that are capable of passing signal bandwidths of 10 or more GHz, in other words, frequencies up to K-band in the microwave region.

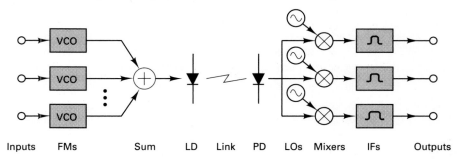

| Inputs | FMs | | Sum | LD | Link | PD | LOs | Mixers | IFs | Outputs |

Figure 9-1. Basic subcarrier multiplexing (SCM) function. Several signals individually modulate microwave subcarrier signals, the algebraic sum of which is applied to a single laser diode at wavelength λ.

This is obviously enough bandwidth to handle more than just one television signal or computer-network bitstream. Therefore, it is tempting to drive a transmitting laser diode with the spectral sum of many information streams, and at the receiver to recreate the entire spectrum at the photodetector output for further processing. This *subcarrier-multiplexed* (SCM) function, shown in Figure 9-1, is applicable to links, multipoints, and networks, and to both analog and digital signals [1-3]. If the transmitter and receiver local oscillators shown in the figure are fixed in frequency, a *multiplex* system results, but if either transmit or receive LO (or both) are made tunable, a dynamic *multiaccess* capability is provided.

If there are several wavelength channels, each carrying several SCM subchannels, we have an "SCM/WDM" function, providing additional flexibility. Calling the number of wavelengths C and the number of subcarriers per wavelength M, then the total is $N = MC$, and each bears the address (f_i, λ_j), with $1 \leq i \leq M$, and $1 \leq j \leq C$.

The advantages of SCM/WDM are severalfold. First, the cost of some microwave technologies has already reached low levels, particularly those associated with home TV. Such cost reductions will not occur for some time with many of the optical components discussed in Chapters 3 through 8. While some of the economic advantage can be lost, for example because lasers that have the modulation bandwidth and linearity to handle many frequency-multiplexed microwave carriers are expensive, low cost remains one of attractions of SCM/WDM.

Second, tighter channel spacing can be achieved because the stability of RF tuning is much greater than that of laser tuning, and because the RF conversions can be derived from quartz crystals.

Third, the tuning speed of RF local-oscillator circuits can be much faster than that of today's commercially available tunable optical filters. This fact can be exploited to provide fast packet switching, as will be discussed in Section 11.11.

Finally, it is easier at microwave frequencies than at optical frequencies to use modulation formats that achieve more than one bit per baud, such as QPSK, discussed in Chapter 7.

However, there are several disadvantages to the SCM/WDM approach. Both laser diodes and photodetectors have a finite dynamic range, as we have seen in Chapters 5 and 8. We shall see that the algebraic addition of M microwave signals at different frequencies requires that, if severe intermodulation distortion is to be avoided, the power radiated at each microwave subcarrier must be considerably less than $1/\sqrt{M}$ of the laser's peak power, a serious link budget penalty.

Another disadvantage shows up in applying the technique to networks, where in order to achieve any-to-any accessibility it is necessary to equip each receiver with both wavelength- and frequency-tunability. Also, at the transmitting end, the individual bitstreams (or their microwave counterparts) must be carried from individual nodes to the node where the shared laser diode resides. All these considerations place certain limits (described in Section 11.11) on how the network can be laid out topologically.

In this chapter, we shall discuss the carrier-to-noise requirements of the different television and data modulation formats, then the most important factors affecting the link budget, and conclude by comparing SCM/WDM with straight WDM. It will turn out that SCM/WDM is much more interesting for multipoints than for networks.

9.2 Carrier-to-Noise Ratio Required

The received optical power at a given wavelength can be written

$$P(t) = P_o \left[1 + \sum_{i=1}^{M} m_i \cos (\omega_i t + \phi_i) \right] \tag{9.1}$$

where m_i is the *modulation depth* of each subcarrier. We shall assume all the m_i to have the same value, m.

In dealing with RF subcarriers it has become traditional to refer to the power signal-to-noise ratio at the RF receiver input as *carrier-to-noise ratio CNR*, rather than SNR, the usage we employed in Chapter 8. Therefore we shall use the term CNR here to refer to the ratio of carrier to all forms of noise at the RF receiver *input*, and shall refer to the similar ratio at the *output* as SNR.

For a bit error rate of 10^{-9} and 10^{-15}, we saw in Chapter 8 that the Q-function argument had to be 6.0 and 8.0, respectively, which translates into CNR values of 36 (15.6 dB) and 64 (18.0 dB), respectively, for FSK.

For analog signalling the situation is somewhat more complex [2]. Reckoning the required CNR for television is particularly complicated, since it involves subjective questions of picture quality. For studio-quality 525-line NTSC television, the accepted number for video signal power to noise power is 56 dB, and therefore this is the number that must be achieved by the CNR for an AM system. FM, on the other hand, exchanges bandwidth expansion for an increase of the SNR at the FM detector output over the CNR seen at the input, the well-known phenomenon of *FM noise quieting*. Assuming frequency modulation by a single sinusoid, and defining β to be the ratio of maximum frequency deviation to modulating frequency, the relative performance of FM and AM against white noise is given by the following expression [4] for the ratio of their receiver output SNRs for sinusoidal modulation:

$$\frac{(S_o/N_o)_{FM}}{(S_o/N_o)_{AM}} = \frac{(S_o/N_o)_{FM}}{(S_i/N_i)_{FM}} = 3\beta^2 \tag{9.2}$$

so that one gains as the square of the ratio of bandwidths. (For AM, the input and output signal-to-noise ratios are equal). *White noise* is that noise type whose power spectral density in frequency is flat.

If one applies this principle to NTSC television transmission using the parameters of typical satellite links, one obtains a figure of 40 dB for this ratio of SNRs [2]. Therefore, a fiber optic FM subcarrier link using the same system parameters would require $56 - 40 = 16$ dB of CNR at the RF receiver input for studio-quality TV.

Thus it is seen that digital transmission and FM television transmission both require modest CNRs (15 to 18 dB), but AM television, in exchange for bandwidth efficiency, has a very severe CNR requirement (56 dB). Since consumer TV receivers use AM, this latter number is an important design parameter in considering future TV-to-the-home using fiber. In succeeding sections of this chapter, we shall include some effects that contribute noise to the denominator of the CNR equation that could easily be ignored for FM video and digital transmission but cannot be ignored for AM video.

Essentially three of these impairments must be discussed: (i) nonlinear distortion (intermod) within the transmitter laser, (ii) laser intensity noise fluctuations (as expressed by the *RIN* parameter), and (iii) signal-spontaneous noise products introduced in the receiver because of the spontaneous noise generated within any photonic amplifier that might be included in the system. There are some other sources of noise that we treated in Chapter 8, such as shot noise, APD gain noise, and thermal noise, but since these can be dealt with using the equations of Chapter 8, we shall not consider them further. The proper perspective on them is to say that, for a given number of microwave channels, the RF power per channel that can be used to drive the laser diode is upper-bounded by intermodulation and lower-bounded by the unavoidable sources of receiver noise.

9.3 Intermodulation

There are three ways that intermod products can arise at the transmitting laser diode: (i) *static intermod* due to the curvature of the laser *P–I* curve,   (ii) *clipping*, which occurs when the sum waveform of the subcarriers makes excursions below the threshold, and (iii) *intrinsic intermod*, due to the inherently nonlinear nature of the rate equations. The first two occur for any modulation frequency, and can be visualized clearly from considering Figure 7-2(A) for a curved $P - I$ characteristic. The third comes into play as the uppermost microwave subcarriers begin to approach the natural relaxation frequency of the laser diode.

We shall give a rough quantitative picture of each of these, sufficient to understand what is going on and to decode the jargon in the literature. Examples of complete system designs with consistent quantification of all these factors together can be found in [2, 3].

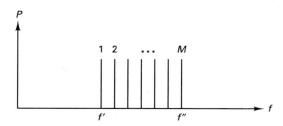

Figure   9-2. Frequency spectrum of M subcarriers, lying between f' and f''.

Static intermod

Imagine that a number M of simultaneously present sinusoids lying between f' and f'', as in Figure 9-2, are applied either to some nonlinear device such as a laser diode with a curved P–I curve, or to an external modulator with a sinusoidal curve of output versus applied voltage. There will be second-order intermod terms at frequencies $(f_i \pm f_j)$, and third-order terms at frequencies $(f_i \pm f_j \pm f_k)$, where i, j, and k range over M, the total number of frequencies present. Thinking of a series expansion of the device nonlinearity in the neighborhood of the zero-signal operating point, it is clear that the coefficient of each second-order intermod term will be larger than that of each third-order term, which tends to make the second-order intermod more harmful than third-order. However, if one can spare the RF bandwidth, it can be seen from Figure 9-2 that second-order terms at $(f_i \pm f_j)$ can be made to lie completely outside the band $(f'-f'')$ by making

$$f'' / f' \geq 2 \qquad\qquad (9.3)$$

that is, by causing all microwave frequencies to lie within a one-octave span.

The third-order terms that hurt are the ones between f' and f'', of which there are $M(M-1)$, of the form $(f_i + f_j - f_k)$, $(i = j)$. These are called *two-tone* third-order terms (because there are only two distinct frequencies), and the term *composite second-order (CSO)* is used to refer to the power ratio of carrier to these intermod products.

The other set that can lie within the band are those of the form $(f_i + f_j - f_k)$, $(i \neq j \neq k)$, of which there are $M(M-1)(M-2)/2$. These are called *triple-beat* third-order terms, and the CNR for them is called *composite triple-beat (CTB)*. (These CTB third-order terms are mathematically the same ones that were referred to in Section 3.14 as "three-wave mixing" or "four-photon mixing"). Of the third-order terms, CTB is the dominant factor, because there are roughly $M/2$ times as many such intermod products as with CSO.

If one can suppress second-order terms by confining the subcarriers to one octave, CTB is essentially the whole static intermod story. However, this is not always possible, in which case second-order terms can predominate. There are at least two reasons that one might allow some second-order intermod by violating the condition of Equation 9.3. First, enforcing the condition obviously wastes about half the available bandwidth. Second, the higher you push the top subcarrier frequency f'', the closer you come to the natural relaxation frequency of the laser, with consequences that we turn to next.

Intrinsic intermod

As long as the top microwave frequency f'' is considerably less than the relaxation frequency f_o of Equation 5.33, the rate equations can be viewed as static equations, as was done in Section 5.6 in deriving the threshold current (Equation 5.11) from 5.35 by inspection. But as f'' approaches f_o, the coupling between carrier density and photon density brings in terms in the rate equation that constitute a nonlinear coupling between the two, adding to the intermod-producing device nonlinearities as seen by the set of microwave signals.

The intrinsic intermod that results is sometimes called *resonance distortion*. A review of this effect and expressions for the effective receiver noise power introduced by it are given in an appendix of [3]. They show that the intermod power due to second-order interactions goes up as m^2 and that due to third-order as m^3. These factors are multiplied by another that goes up as the square of the ratio of subcarrier frequency to relaxation frequency, the latter typically being 1 to 4 GHz, even though special laser diodes are available with f_o up to 12 GHz. It is common practice to make the highest microwave carrier one or two GHz less than f_0.

Clipping

As laser diodes have improved, with more linear $P–I$ curves and higher relaxation frequencies f_o, the two forms of intermod just discussed have receded in importance for all but AM television applications. This has led to driving the laser with higher and higher microwave input power, i.e., increasing the modulation depth m to the point where clipping at the threshold current value begins to occur. Today, clipping is a prime consideration in most system designs. The situation is depicted in Figure 9-3, where the clipping is assumed to be ideally abrupt.

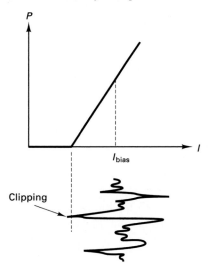

Figure   9-3. Clipping can occur when the mixture of M subcarrier signals makes an excursion below the laser $P − I$-curve threshold.

It is clear that under the condition

$$m \leq 1/M \tag{9.4}$$

there will be no clipping at all, but that this will be at the expense of the link budget. It is usual practice to make m large enough that some clipping occurs and to balance off that CNR loss against the CNR loss due to smaller modulation depth m.

For values of M less than, say 10, one can say that M, the total number of channels permitted, will not be much greater than m, because otherwise there will be frequent excursions past the threshold point. However, as M increases, under the realistic assumption of independent modulations, statistical averaging takes place, and the frequency of excursions past the threshold point decreases. One is therefore led to increase m to favor the link budget.

A quantitative analysis of clipping for the large-M case is given in [5]. The sum of microwave carriers is represented as a gaussian random process, as shown in Figure 9-3. From Equation 9.1, the photodiode output is

$$I(t) = I_o\left[1 + \sum_{i=1}^{M} m_i \cos(\omega_i t + \phi_i)\right] \tag{9.5}$$

where $I_o = \mathcal{R}P_o$ from Equation 8.18. The total mean-square current is

$$\langle I^2_t(t)\rangle = I^2_o M m^2/2 \tag{9.6}$$

The received mean-square current of the tail component of $I(t)$, i.e., the power in the clipped part of the signal, is approximately

$$\sqrt{2/\pi}\, I^2_o \mu^5 \exp(-\mu^2/2) \tag{9.7}$$

the RMS modulation index μ being defined as

$$\mu = m\sqrt{M/2} \tag{9.8}$$

From this, the carrier-to-noise ratio for large M, if only clipping is considered, is

$$(CNR)_{cl} = \sqrt{\pi/2}\, \mu^{-3} \exp(1/2\mu^2) \tag{9.9}$$

For example, if we want this CNR to be 20 dB, then $\mu = 0.48$, and one has a handy rule of thumb that the required modulation depth is

$$m = \sqrt{\frac{2\mu^2}{M}} = 0.68/\sqrt{M} \tag{9.10}$$

The exact calculation of the spectrum of the clipped tail is arduous; here we have simply assumed that it is flat and of the same bandwidth as the modulation.

To summarize, for small M the required modulation depth for each channel goes as M^{-1}, whereas for large M it decreases more slowly, namely as $M^{-1/2}$.

9.4 Laser Intensity Noise

As mentioned in Section 5.5, if reflections into the transmitting laser diode are not controlled to very low levels, the light output will fluctuate in amplitude due to an increase in spontaneous noise generated. Subcarrier systems are particularly sensitive to this. As we saw in Section 5.5, RIN runs typically between -110 dBW/Hz and -150 dBW/Hz, depending strongly on how well isolated the laser diode is. If laser amplitude fluctuations are the only noise component to be considered in the CNR equation, the calculation is easy, because RIN is nothing but a noise-to-signal power ratio defined as

$$RIN = \frac{\langle I^2 \rangle}{\langle I \rangle^2} = \frac{\langle I^2 \rangle}{I_0^2} \tag{9.11}$$

so that if the (amplitude) modulation depth is m, the CNR due to laser amplitude fluctuations only is

$$(CNR)_{RIN} = \frac{m^2}{2 \, RIN \, B} \tag{9.12}$$

where B is the microwave receiver filter bandwidth [6].

9.5 Photonic Amplifier Signal-Spontaneous Noise

For the case in which the only noise is signal-spontaneous noise from a photonic pre-amplifier, the CNR can readily be deduced from the results of Section 8.11, which treated purely optical ASK. The numerator (carrier power) is found by multiplying Equation 8.65 by $m^2/2$, and the denominator is given by Equation 8.66 as it stands. Using $B = 1/T$, as is appropriate for the bandpass case, the result is

$$(CNR)_{s-sp} = \frac{(mGPT)^2}{8hfG(G-1)\chi N_{sp}} \tag{9.13}$$

(analogously to the square root of the argument of the Q-function in Equation 8.67, which is for the optical ASK case).

9.6 Carrier-to-Noise Ratio Achieved

In Section 9.2 we discussed the CNR required for analog FM and AM television and for FSK data. In order to calculate the actual CNR that will be seen at the receiver input in a system in which intermod (static, intrinsic, clipping), laser intensity noise, and/or preamplifier signal-spontaneous noise can be present, it is merely necessary to combine all these factors into one equation in such a way that the denominators add:

$$(CNR)_{total}^{-1} = (CNR)_{si}^{-1} + (CNR)_{ii}^{-1}$$
$$+ (CNR)_{cl}^{-1} + (CNR)_{RIN}^{-1} + (CNR)_{s-sp}^{-1} \qquad (9.14)$$

This equation captures all the major effects that must be considered in an SCM or SCM/WDM link-budget design. Expressions for the last three terms have been given earlier in this chapter; the other two must be calculated numerically using techniques described in the references.

9.7 Comparison of SCM/WDM with WDM

If one combines the idea of subcarrier multiplexing and wavelength multiplexing, a larger total number of channels results, and, in the case of a network, an interesting approach to packet switching is opened up. The tunable filters and tunable lasers commercially available today require at least milliseconds to retune to a different wavelength, whereas RF tunability time can be in the tens or hundreds of nanoseconds. References [7-9] describe the theory and prototyping of an SCM/WDM network aimed at doing packet switching by exploiting fast RF tunability in a "multihop" logical topology. The topology implications are interesting and will be dealt with in Section 11.11 and the packet-switching protocol aspects in Section 13.10. The normal way of providing RF tuning is to use a voltage-controlled oscillator (VCO) whose tuning speeds are in the tens of microseconds, but submicrosecond tuning was provided in this study by switching between two oscillators.

Figure 9-4 was derived [7] using the results on clipping and doped-fiber preamplifier signal-spontaneous noise given in Section 9.3 and 9.5. It shows that, although one can achieve a given number of channels either with many wavelengths and a few subcarriers per wavelength, or with the converse, considerably more photonic amplification is required to support the latter, due to the need to reduce the modulation index to avoid excessive clipping. The figure shows that, if one has the options of achieving a given number of stations $N = \Lambda M$ either by large Λ or large M, the former is the better option. More detailed quantification of this advantage of WDM over SCM/WDM is given in [8].

The fundamental reason for the superiority of WDM is that the full optical power of a laser is available for each channel. For SCM, as we have seen, avoidance

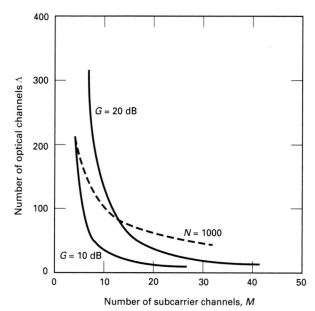

Figure 9-4. Tradeoff between maximum number of wavelengths Λ, and M, the maximum number of microwave frequencies per wavelength for a maximum bit error rate of 10^{-9}. Bitrate is 200 Mb/s and radiated power is 0 dBm. Two values of receiver optical preamplifier gain are given. The dashed line shows the total number of nodes $\Lambda \times M = N = 1000$. (Courtesy M. M. Choy).

of intermodulation does not allow this. The need to do this "power backoff" is a common situation, met in communication satellites for example. In order to avoid intermodulation, the power per channel has to be reduced far below the total power available from the transmitter.

Networks based purely on SCM have been considered [10]. They assume that every node uses the same wavelength. Unless extreme measures are taken to stabilize the laser frequencies to be identical, microwave-frequency beat products between them produce sufficient interference that usably low bit error rates are impossible to achieve.

9.8 What We Have Learned

In this chapter, we have reviewed the quantitative relationships that allow one to design a single SCM link or a single-wavelength set of M channels in an SDM/WDM system. For static and intrinsic (resonance) intermodulation terms, a more complex numerical analysis is required.

While SCM techniques are proving very interesting for links and multipoints, networks require SCM/WDM, which is attractive today only for packet switching. If the slower-tuned and therefore more easily available circuit-switching option is a suitable solution and can be made to provide the channel density desired, pure WDM networks appear to be a better alternative today.

9.9 Problems

9.1 Commercial FM broadcasting uses a maximum frequency deviation of ± 75 KHz, whereas in AM broadcasting the channels are 10 KHz apart, with negligible guard bands (zero-signal-power bands) between channels. How many dB better is the output SNR of FM than that of AM?

9.2 Equation 9.10 was derived assuming that the information being modulated was either digital FSK or FM television. Suppose it had been analog AM television that was being multiplexed. What would be the new version of Equation 9.10?

9.3 In deducing the CNR for only preamplifier signal-spontaneous noise present, why does P in the numerator of Equation 8.67 get multiplied by $m^2/2$ but not the denominator?

9.10 References

1. W. I. Way, R. Olshansky, and K. I. Sato, eds., "Special Issue on Application of RF and Microwave Subcarriers to Optical Fiber Transmission in Present and Future Broadband Networks," *IEEE Jour. Sel. Areas in Commun.*, vol. 8, no. 7, September, 1990.
2. W. I. Way, "Subcarrier multiplexed lightwave system design considerations for subscriber loop applications," *IEEE/OSA Jour. Lightwave Tech.*, vol. 7, no. 11, pp. 1806-1818, 1989.
3. R. Olshansky, V. A. Lanzisera, and P. M. Hill, "Subcarrier multiplexed lightwave systems for broadband distribution," *IEEE/OSA Jour. Lightwave Tech.*, vol. 7, no. 9, pp. 1329-1341, September, 1989.
4. F. G. Stremler, *Introduction to Communication Systems - Third Edition*, Addison-Wesley, 1990.
5. A. A. Saleh, "Fundamental limit on number of channels in a subcarrier-multiplexed lightwave CATV system," *Elect. Ltrs.*, vol. 25, no. 12, pp. 776-777, June, 1989.
6. P. M. Hill and R. Olshansky, "A 20-channel optical communication system using subcarrier multiplexing for the transmission of digital video signals," *IEEE/OSA Jour. Lightwave Tech.*, vol. 8, no. 4, pp. 554-560, 1990.
7. M. M. Choy, F. K. Tong, and T. Odubanjo, "An FSK subcarrier/wavelength network," *Conf. Record, Conf. on Lasers and Electrooptics*, 1991.
8. R. Ramaswami and K. N. Sivarajan, "A packet-switched multihop lightwave network using subcarrier and wavelength multiplexing," *Submitted to IEEE Trans. on Commun.*, vol. 39, 1991.
9. M. M. Choy, S. M. Altieri, and K. Sivarajan, "A 200 Mb/s packet-switched WDM-SCM network using fast RF tuning," *Submitted to IEEE Journal of Lightwave Technology*, March, 1992.
10. N. K. Shankaranarayanan, S. D. Elby, and K. Y. Lau, "WDMA/subcarrier-FDMA lightwave networks: Limitations due to optical beat interference," *IEEE/OSA Jour. Lightwave Tech.*, vol. 9, no. 7, pp. 931-943, 1991.

CHAPTER 10

Frequency Stability and Its Control

10.1 Overview

The optical frequency of a laser diode or a tunable filter is not only a parameter that can be controlled by design, but also one that is at the mercy of ambient variables such as temperature, vibration, and drive current or voltage. Neither the narrowband filters discussed in Chapter 4 nor the laser-diode narrowband sources of Chapter 5 always have sufficiently constant frequency to be usable in third-generation optical systems without some means of stabilization. Both coherent detection systems and incoherent systems using tunable-filter receivers face this problem.

Traditionally, the problem has been considered more severe in coherent than in incoherent systems, since in the former the frequency difference between transmitter and receiver local oscillator had to be accurate to less than the modulation bandwidth, whereas the optical filter of a direct-detection receiver usually need not have a bandwidth that narrow. However, in order for incoherent wavelength-division systems to offer the same number of densely packed channels as coherent systems, the optical-filter bandwidth must eventually be limited by the modulation bandwidth for such systems too. Therefore the tolerance on frequency is essentially no different for the two classes of system.

In this chapter we shall survey various aspects of the frequency-stabilization problem and describe various solutions that have actually been implemented.

10.2 Sources of Instability

Lasers

As discussed in Section 5.8, the main sources of frequency drift in a laser have to do with changes of injection current and of temperature. For a 1.5-μ device, the frequency dependence on temperature is roughly 13 GHz per °C, or one part in 15,000 per °C. The dependence on current varies from device to device, but a typical number is 130 GHz. per milliampere, that is, one part in 1500 per milliampere. The frequency should be stable to no more than around one-tenth of the modulation bandwidth. Clearly, in a gigabit fiber optic link, for which one-tenth the modulation bandwidth is 2.5×10^{-7} of its center frequency, the frequency cannot be left just to drift for itself.

Careful experiments on commercial-grade DFB lasers [1] have shown that once such devices have been placed in service, if the bias current is kept within 0.1 mA, conventional temperature control measures suffice to keep the wavelength stable to a few hundred MHz (hundredths of an angstrom). Many commercial laser-diode packages (e.g., that of Figure 5-3) have most of the required elements already packaged in the device for the purpose of maintaining the threshold current relatively constant, and these are usually able to stabilize the temperature to within less than 0.1°C. We shall discuss these elements shortly.

This all means that, to all intents and purposes, one can forget about transmitter laser drifts, compared to other sources of frequency instability in the the network. The exception could occur if one were trying to do coherent detection.

Table 10-1. Temperature coefficient of expansion in parts per million per degree Centigrade.

Material	Application	Temperature coeff.
Super invar	Various	+0.3
Silicon	MZI chains, gratings	+3
Glass	Etalons	+9
Metals	Various	+10 to 20
Lead molybdate	Acoustooptic	−161
Tellurium dioxide	Acoustooptic	175
Lead zirconate tantalate (PZT)	Piezo tuning	−4
Lithium niobate	Electrooptic	7 to 15

Filters

For filters, the dependence of frequency on ambient parameters is no better than it is for lasers. Temperature is the main culprit. The temperature coefficient of expansion is tabulated in Table 10-1 for most of the materials mentioned in Chapter 4 as candidates for tunable-filter structures. It is seen that a particularly good form of filter with respect to temperature stability ought to be one made of silicon.

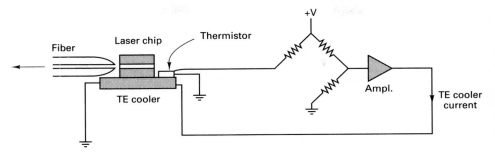

Figure 10-1. Temperature compensation circuitry for laser diodes.

Other effects can often enter the picture. For example, PZT material has a hysteresis effect in that the piezoelectric elongation of the material at a given voltage depends on past history – for example, whether that voltage operating point is approached from above or below.

10.3 Compensation

The simplest approach to mitigating the detuning effect of an ambient parameter is to try to cancel it out. The Fiber Fabry-Perot tunable filter [2] does this by playing off the negative temperature coefficient of PZT material against the positive coefficient of aluminum, with the relative thicknesses of the two materials set in inverse ratio of the magnitudes of their temperature coefficients.

Another common compensation scheme is that used in all distributed-feedback laser packages, and shown in Figure 10-1. Internal to the package and closely adjacent to the laser chip, which is mounted on a thermoelectric cooler element, is the thermistor, a resistor whose resistance is temperature-dependent. The TE cooler uses the Peltier effect to produce cooling that is a monotonic function of applied current. Reversing the current leads to regular ohmic heating. The arrangement of Figure 10-1 responds to an increase in temperature with increase of TE cooler current in the cooling direction.

10.4 Sources of Stable Frequency Reference

Some components to which one might reference a frequency-stabilization subsystem are so stable that they may be regarded as *primary reference standards*. Given the figures mentioned above for the needed degree of stabilization, certainly atomic lines and oven-controlled quartz crystals qualify for this role.

Secondary standards

These would include those that might drift, but less so than the laser or filter to be stabilized. Examples of secondary standards might be a temperature-compensated
capacitor or resistor or a fixed glass Fabry-Perot cavity,

Primary standards, by their very nature, are standards for one frequency only; the
other frequencies in a wavelength-division system have to be derived from this one
frequency. Secondary standards, on the other hand, include several that produce a
comb of regularly spaced frequencies, so that, once one tooth of the comb is stabilized,
so are all the others. Such reference standards are the Fabry-Perot cavity resonances,
the various orders of a diffraction grating, the spectral lines of a mode-locked laser,
and the comb of frequencies represented by the modulation sidebands produced by
modulating a laser with a sinusoidal source, .

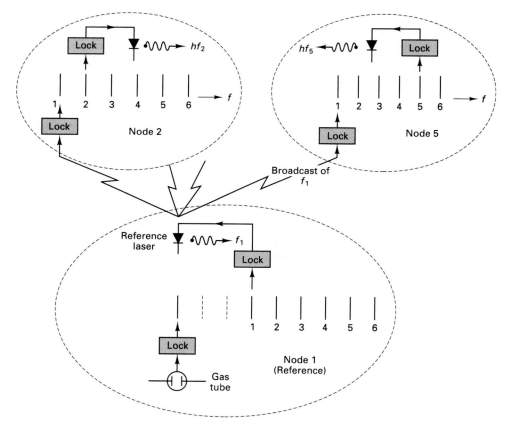

Figure 10-2. The AT&T Bell Labs scheme for network-wide locking of equally spaced laser
wavelengths to a primary atomic frequency source.

Of these, the one that has received the most attention is the Fabry-Perot cavity.
Figure 10-2 summarizes one approach [3, 4].

There is an etalon in each node, all nominally identical, and each local laser is locked to a different specific resonance tooth by means to be described in the next section. Then, to make sure the various etalons are together, the same tooth from each is locked to a CW reference signal that is broadcast from one of the nodes serving as the source of frequency reference for the entire network. This reference laser is locked to one of the teeth of an etalon identical to those in the nodes. If this last reference etalon is itself insufficiently stable, then one of its teeth can be locked to a primary standard in the form of an atomic line.

Use of an atomic line is much simpler and cheaper than one might expect. It has been shown [5] that when light at one of the atomic lines of a gas passes through a simple 25-cent indicator light bulb containing that gas, the current drawn by the bulb decreases slightly, since the ionization is being helped along by the incident light. Thus, the current through the bulb may be used as the variable that determines whether the laser is tuned to the atomic line. The many atomic lines of the widely used gases neon, argon, krypton, and xenon that lie in the 1.3- and 1.5-μ bands are tabulated in [5].

Whereas the approaches using the teeth of an etalon are using a comb of passive resonances, those involving mode-locked sources or modulation sidebands have the advantage that they actually generate the reference frequency comb themselves. If one wants lines, say 3 GHz. apart, amplitude-modulating a laser with a 3 GHz. waveform rich in harmonics, for example, a square wave, will produce them [6]. The other way of producing them is to feed the 3 GHz microwave signal to an element within the laser cavity that starts and stops the lasing at the 3 GHz. rate. This *mode-locking* technique produces extremely sharp pulses, which thus have a high harmonic content and considerable power in distant sidebands. This has been applied to optical networking in the time domain [7], but is equally applicable to the frequency domain.

10.5 Methods of Locking

The approach taken in locking the frequency of some target device to that of another more stable reference depends on whether each is active or passive.

If the device to be stabilized is a laser and there is a source of the standard frequency, the laser may be *injection-locked* to the standard simply by injecting a certain amount of power. Let us designate as B_p the *pull-in bandwidth*, the maximum frequency difference between the reference frequency f and the natural lasing frequency of the laser within which the laser will lock in to the reference frequency. B_p is given by [8]

$$B_p = \frac{\pi f}{Q} \sqrt{\frac{P_{inj}}{P}} \tag{10.1}$$

where Q is the ratio of the center frequency of the laser to its natural linewidth, P is the laser power, and P_{inj} is the injected power. The injection-locking method has been used successfully [6] to lock DFB lasers to various sidebands of a master laser by injecting power from the latter.

If the standard is not a source, but some observable sharp peak or minimum in some function of frequency, then the standard approach is the *dithering* technique shown in Figure 10-3.

A small dither signal is used to sinusoidally vary the frequency of either the reference or the target device over a small range. The figure illustrates the the latter case. When the device is exactly tuned, the small variation of the multiplier input on one side of the peak has a positive sign and that on the other a negative sign, so that the integrator (low-pass filter) produces an output with a zero average. If now, there is a frequency drift between the device and the standard, the smoothed multiplier output will be of one sign or the other, depending on the direction of the discrepancy. This smoothed signal is added to the dither signal and steers the target device back to coincidence with the reference.

The dither technique has been applied to lock a laser to an atomic line in a gas tube [5], to bring a tunable Fabry-Perot filter into coincidence with an incoming signal [2], and to lock one of the lasers in the scheme of Figure 10-3 to the appropriate Fabry-Perot resonance [3]. In this last context, a clever trick can be used if the information modulation format is FSK. Since the FSK waveform is a known waveform at the transmitter, it can substitute there for the dither oscillator of Figure 10-3. Incidentally, the reason that the dither technique can be used even for very weak standard signals, such as the arriving transmission, is that the integration time of the locking loop can be made many orders of magnitude longer than the bit time. Since one must develop from the arriving signal a certain signal-to-noise ratio for a usably low bit error rate, it follows that the longer integration time of the locking circuit will more than compensate for the fact that the locking circuit may need a higher output signal-to-noise ratio than the receiver.

10.6 Identifying the Channel

In a wavelength-division network, even after a number of equally spaced frequency values have been provided, there is often the problem of deciding which tooth of the comb to lock on. For some multiaccess protocols, such as the "circular search" protocol to be described in Chapter 13, it does not matter what the frequencies of the various channels are, so long as they do not crowd so close together that interchannel crosstalk is produced. The circular search protocol operates by scanning the entire available frequency band, looking for a signal bearing in its bitstream a certain tag.

However, such protocols are not very interesting from the performance standpoint, since they consume so much time examining each channel in turn. A much better approach is *tooth counting* [9], in which an initial scan is made once through the

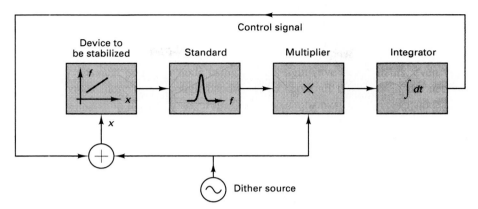

Figure 10-3. Dither scheme for stabilizing one device against a standard.

tuning range and the frequency of each channel noted and memorized in the memory of a microprocessor. (Actually, what is memorized is the control voltage necessary to set to that channel). When it is desired later to tune to that channel, the setting can be made without waiting, and then any necessary locking circuitry activated.

10.7 What We Have Learned

The requirements on frequency accuracy in lightwave systems are very exacting. Several orders of magnitude better frequency stability are demanded than are intrinsically available from tunable filters, or in some cases, even from fixed-tuned laser diodes. Fortunately, some fairly simple techniques suffice to stabilize either a laser or a filter to a very simple and inexpensive primary standard in the form of a gas discharge tube. From there, it can get fairly complicated to make sure that each node in the network occupies a different one of an equally spaced set of wavelengths or frequencies. While complicated, the available techniques are workable and should be capable of considerable cost improvement.

10.8 Problems

10.1 The manufacturer of a packaged DFB laser advertises that the thermistor temperature sensor and the TE cooler, operated with a suitable external-feedback circuit, can keep the temperature of the active region constant to ± 0.1 °C. If you must design a bias circuit for this laser diode, what is the tolerance on bias current that will allow your circuit to perform as well as the temperature control in maintaining constant radiated wavelength?

10.2 Assume that the network-wide frequency stabilization scheme of Figure 10-2 is used for a network of 1 Gb/s nodes, and that each node contains a reference Fabry-Perot etalon made of glass and having a free spectral range 10 GHz. There are 1000 channels. The first one is 10 GHz above the reference frequency that was broadcast from the reference node, and the last one being exactly 11/10 times the reference frequency. How accurately must the temperature of the etalon in any node be kept in order that that node's reference frequency drift no more than 10 percent of the modulation bandwidth (i.e., $0.1/2T$) ?

10.3 Suppose your car radio contains a local oscillator with two inputs. It is a free-running voltage-controlled oscillator (VCO), so it has an electrical input. In addition, it is injection-locked by tickling it with one of the harmonics of a quartz crystal-based digital synthesizer, these harmonics being spaced 1 MHz apart. This is the second input. The total synthesizer output power is 1 milliwatt and there are roughly 100 harmonics. The free-running oscillator at 10 MHz has an internal power level of 1 milliwatt, and in the absence of the injection signal it has an rms drift of 10 kHz. How near must the applied voltage tune the VCO to one of the harmonics of the digital synthesizer in order for injection locking to occur?

10.4 A certain 1-Gb/s optical receiver uses a tunable optical filter to reject unwanted wavelengths, and this filter has a very narrow bandwidth; actually, it is a matched filter, matched to the ASK modulation waveform. The bit error rate is 10^{-9}. If the time constant of the servo loop using the dithering technique on the tunable filter is 1 μsec, how much signal-to-noise ratio is developed in that servo loop? Express the answer in dB.

10.9 References

1. W. B. Sessa, R. E. Wagner, and P. C. Li, "Frequency stability of DFB lasers used in FDM multi-location networks," *Conf. Record, Optical Fiber Commun. Conf.*, p. 202, 1992.
2. C. M. Miller and F. J. Janniello, "Passively temperature-compensated fiber Fabry-Perot filter and its application in wavelength division multiple access computer network," *Elect. Ltrs.*, vol. 26, no. 25, pp. 2122-2123, December, 1990.
3. B. S. Glance, J. Stone, K. J. Pollock, P. J. Fitzgerald, C. A. Burrus Jr., B. L. Kaspar, and L. W. Stulz, "Densely spaced FDM coherent star network with optical signals confined to equally spaced frequencies," *IEEE/OSA Jour. Lightwave Tech.*, vol. 6, no. 11, pp. 1770-1781, 1988.
4. Y. C. Chung, K. J. Pollock, P. J. Fitzgerald, B. Glance, R. W. Tkach, and A. R. Chraplyvy, "WDM coherent star network with absolute frequency reference," *Elect. Ltrs.*, vol. 24, no. 21, pp. 1313-1314, 1988.
5. Y. C. Chung, "Frequency-locked 1.3- and 1.5- lasers for lightwave systems applications," *IEEE/OSA Jour. Lightwave Tech.*, vol. 8, no. 6, pp. 869-976, 1990.
6. K. Kikuchi, C. E. Zah, and T. P. Lee, "Amplitude-modulation sideband injection locking characteristics of semiconductor lasers and their application," *IEEE/OSA Jour. Lightwave Tech.*, vol. 6, no. 12, pp. 1821-1830, 1988.
7. P. R. Prucnal, M. A. Santoro, and S. K. Sehgal, "TDMA fibre-optic network with optical processing," *Elect. Ltrs.*, vol. 22, no. 23, pp. 1218-1219, November, 1986.
8. R. Adler, "A study of locking phenomena in oscillators," *Proc. Inst. Radio Engrs.*, vol. 33, no. 6, pp. 351-357, 1946.
9. B. Strebel and G. Heydt, "Multi-channel optical carrier frequency technique," *Conf. Record, IEEE Globecom*, pp. 18.3.1 - 18.3.5, 1987.

Part III

ARCHITECTURE

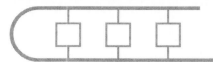

CHAPTER 11

Organizing the System Topologically

11.1 Overview

We turn now from the technologies out of which the total system is to be made to the total system itself.

At the outset, a distinction must be made between several topological aspects of the same network in order to keep track of what is going on physically and logically.

1. The placement of the actual links and nodes defines the *physical-network topology*, for example, a ring, a star, a bus, a mesh, a tree, or other choices.

2. On this physical-network topology we can impose a particular *physical-path topology* that defines where the traffic flows between different end users. The *routing* problem that has so occupied computer communication practitioners is basically that of imposing some sort of optimum physical-path topology on top of a given physical-network topology.

3. A representation that shows only which users can communicate with which other users, and suppresses the topological nature of the network layout, constitutes the *logical topology*.

The example of Figure 11-1(A), which is the same as that of Figure 1-4, serves to illustrate the differences between these three. The physical-network topology is composed of the four nodes and the physical links indicated by the solid lines. On this, a choice of physical-network topologies could be imposed, one of which is the set of dashed lines, a set of bidirectional routes or paths. The particular logical topology

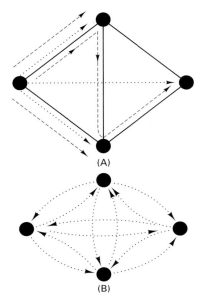

Figure 11-1. (A) Illustrating the difference between a network's physical-network topology (solid lines), its physical-path topology (dashed lines), and its logical topology (dotted lines). (B) Logical topology required in a network.

that is supported in turn by this physical-path topology happens to be a one-way broadcast from node 1 to all the other nodes, shown by the set of dotted lines.

Sometimes the desired logical topology will be a star, as in this illustration. However, for a network, as we found in Section 2.3, the requirement is usually for "any-to-any" logical connectivity; any node must be able to reach any other node, which is just another way of saying that the logical topology must be a set of bidirectional links connecting every node to every other node, in other words the *full mesh* of Figure 11-1(B).

The reader should note that in the lightwave networking literature, a careful distinction is not always made between the physical-path topology and the logical topology; sometimes the former is called a logical topology. As used this way, there is confusion between where the physical resources are deployed and where the messages are routed. By distinguishing between the two in a consistent way, and reserving the term "logical topology" to explicitly express who talks to whom, we shall avoid this ambiguity.

Perhaps a distinction with respect to time scale is also helpful in keeping the distinction in mind. The physical-network topology of a given network is semipermanent, being changed only on a time scale of months or years. The physical-path topology and logical topology, on the other hand, can be much more dynamic, changing in seconds or less.

The chapter will not deal with logical topologies, but only with the physical-network topologies and physical-path topologies relevant to fiber optic networks. We shall first discuss the topological characteristics that are desirable. We then point out how the most relevant lightwave network topologies today are those that compensate for some deficiency in the available technology. We then treat some physical-network topology questions: how to design stars and busses, where to place photonic amplifiers, should they be required, and how to use "wavelength routing" to steer messages through the network according to their wavelength. We then turn to the more dynamic physical-path topologies: how to make wavelength routing dynamic, so-called "linear" and "multihop" lightwave networks, and various hybrid schemes that address messages by varying some parameter in addition to wavelength.

Before proceeding with these matters, it is appropriate to put topology questions into the larger architectural context by giving a few words of preview of the next several chapters. In its most basic form, the logical topology of connections (*access paths*) between users is a topological tree in the case of a *multipoint*, and a full mesh for a *network*, as in Figure 11-1(B). A detailed dissection of the functions involved in using such access paths is given in the next chapter, Chapter 12. It will be seen that the functions that create and support an access path have a natural division into *protocol layers*, and into a sequence of *network-control stages*. Details of results to date on third-generation WDMA, TDMA, and CDMA protocols by which the connections are set up and taken down are given in Chapter 13, which also compares the performance of networks using these different protocols.

At the current stage of evolution of third-generation optical networks, far and away the biggest set of problems relate to devices, topological aspects rank second, and issues of layered architectures and protocols have subsidiary importance. So far, that is.

11.2 Requirements on the Physical-Network Topology

There are several requirements [1] on the desired physical-network topology, namely *scalability, modularity,* and *irregularity.* "Scalability" means the ability to expand the network to accommodate many more nodes than the number in the initial installation. We shall see that many designs have a modest upper limit on the number of nodes N, usually because of a limit on the available bandwidth. "Modularity" means the ability to add *just* one more node. Some of the topologies we shall encounter only work for integer values of N that are not very close together, and are therefore not very modular. "Irregularity" means that the topology should not be forced artificially into some unusual, highly stylized pattern that may not meet the user requirements.

There can be many forms of physical-network topology, the most important for optical LANs and MANs being the star, multilevel star, and the bus, as shown in Figure 11-2(A), (B), and (C), respectively. On any of these, a physical-path topology can be imposed, as shown by the dashed-line examples. These dashed lines are labelled according to wavelength λ, indicating *wavelength-division multiaccess* (WDMA) communication, that is, addressing the bits of a message on the basis of different wavelengths. The labels could just as easily have referred to different *time-division multiaccess* (TDMA) slots or different code-division multiacess (CDMA) "pseudonoises."

These three basic options for modulation and multiaccess will be discussed in more detail in Chapter 13. A fourth option would be to use one or more large photonic switches to form a *space-division network*, as discussed in Section 1.5. This would constitute *space-division multiaccess*, or SDMA.

Experience has shown that no one physical-network topology is universally favored for networks. However, the *multilevel star* of Figure 11-2(B) is a physical arrangement that seems to have proven widely practical. Consider first- and second-generation token-ring LANs. Each LAN is more often than not built as a series of spokes radiating from a central hub or "wiring closet." The messages and tokens travel in a logically circular way, but on a physical path that makes repeated traverses in and out along the N lobes radiating from the hub. This is done so that failed nodes may be centrally bypassed, so that maintenance and troubleshooting is facilitated, and so that nodes may be added to and deleted from the ring most easily. Similarly, an Ethernet bus installation often has the attachments of the nodes made to the bus in a single central physical location. Another reason for preferring star connections of at least subgroups of nodes out of the total population of nodes is that users are usually arranged in clusters to accommodate an organized group, a department, a floor of a building, and so forth. Other examples include the double-star arrangement of telephone central offices and private exchanges (PBXs) and the two-level topology typical of cable television installations.

If a star, or something else, is a convenient choice of physical-network topology in some particular instance, then why not build the LAN or MAN in that form? The reason is that one does not have a free choice of topology, because one does not have a free choice of the device technology to use. If technology were no limitation, one could build the system to have any topology that was dictated by the application and such other things as route availability, clustering of users, and tariffs. Unfortunately, as is clear from the last eight chapters, those technologies likely to be available set certain constraints on what can be built. When one imposes the further condition that the technology must be economically available today, the constraints are even tighter.

11.3 Trading Off Topology Against Technology

We shall find that there are clever tricks one can play with both physical-network and physical-path topologies that relieve some of the technology constraints. In fact, at this date, a large portion of the recent system literature on lightwave networking focusses on just such topological solutions to problems caused by component inadequacies.

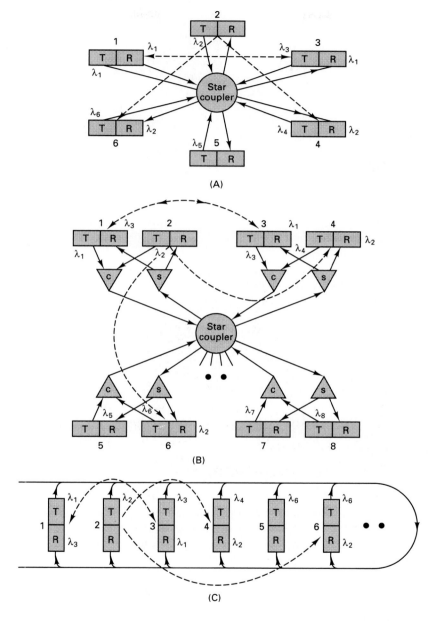

Figure 11-2. Some basic physical-network topologies favored for third generation lightwave networks. (A) Star, (B) Multilevel star, (C) Reentrant bus.

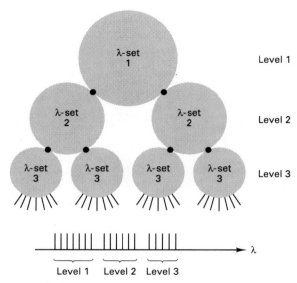

Figure 11-3. Example of wavelength reuse. Set 3 can be in use in four places simultaneously, Set 2 two places, and Set 3 is not reusable. The dots represent wavelength-selective physical links between hierarchical levels.

Most prominent among the technical constraints are:

- *Link-budget problems.* Unlike long-haul telephone interoffice links, the problem with providing enough photons per bit at the receiver in a third-generation multipoint or network is not usually the attenuation accumulated with distance. This is because the area covered by the system is rarely that large. For WDMA, TDMA, and CDMA (but not for SDMA), the problem is the *splitting loss* in couplers or taps. The splitting loss may be controlled up to a point by controlling the splitting ratio between passive coupler outputs, by introducing photonic amplification, and by using star rather than bus topologies.

- *Insufficient number of wavelengths.* The 200-nm window of silica fiber could prove to be insufficient to provide enough wavelengths if the number of nodes N is too large, if the channel spacing required is too large (due to insufficient selectivity, laser chirp, or component instabilities). The topological gambit called *wavelength reuse*, illustrated by the example of Figure 11-3 [2], allows a particular wavelength used in one place in the network also to be in use simultaneously in one or more other places. The term is perhaps a misnomer, since it implies sequentiality in time. It isn't that one wavelength is being used time after time in sequence, but rather at many places at the same time. Perhaps a better term than "reuse" would be "co-use," but we shall stick with the standard term "reuse" as implying simultaneity.

- *Tunability limitations* on WDMA. Tunability is available today at receivers only and is slow at that. Eventually, one would like to have extremely fast tunability at one's choice of transmitting end or receiving end (or both), but this is possible today only for research-grade components. Tunability requirements can be relieved by multihopping.

- *Photonic switching limitations.* Only small-dimensionality crosspoint switches are available in purely photonic form. One can combine wavelength selectivity with small-dimension space-division switches and the ease of bundling fibers together in single cables to build large networks using today's photonic switches.

11.4 Basic Comparison of Star and Bus Topologies

The wavelength labels on the transmitters and receivers in Figure 11-2 show how three physical-network topologies that have been widely studied and implemented (star, multilevel star, and bus) can be exploited to form a wavelength-division network. In principle, these same labels could have signified different time-division slots or different code-division pseudonoises. The figure shows a particular physical-path topology by means of dashed lines, each of which could be implemented using fixed-tuned transmitters and tunable receivers, or vice versa. One bidirectional (full-duplex) path connects node pair 1-3. Also shown as part of the physical-path topology, is a one-way multipoint from node 2 to nodes 4 and 6. The logical topology can be still another arrangement.

In the example, it is seen that with both physical-network topologies, almost all the transmitted energy is wasted because it is broadcast to all nodes, only one of which will use it (unless a total broadcast physical-path topology is implemented). This is the largest drawback of most all-optical networks; as was said earlier, there is essentially no shortage of bandwidth, but there is a shortage of photons. (With first- and second-generation networks, the situation was reversed). This being the case, an ideal third-generation network architecture would be one in which each receiver received only the signal energy it was intended to receive for that particular connection. This idealized situation can be approached in principle by bus architectures that use *tunable taps*, of the type discussed in Section 4.11.

Barring the invention of an ideal low-loss tunable tap, one of the reasons the star physical-network topology should continue to be favored over the bus is that the star accommodates more nodes than can either of the two bus topologies, given a fixed link budget. This can be seen from the following development.

Viewed in the simplest terms, when a lightwave signal passes through the central star coupler, it suffers a loss (in dB) of

$$L = 10 \log_{10} N - 10 \log_{10} \beta \log_2 N \tag{11.1}$$

where the first term is the splitting loss and the second is due to the excess loss, β, ($0 \leq \beta \leq 1$), of each 2×2 coupler stage, arranged as in Figure 3-23. At each 2×2 stage, the fraction of the total input power that appears as total output power is β. For a star, the first term of Equation 11.1 usually predominates, because β is typically 0.70 to 0.90, i.e., -0.1 to -0.5 dB.

For any linear arrangement of N taps, there is an end-to-end loss of

$$L = -10N \log_{10} (1 - \alpha) - 10N \log_{10} \beta \quad \text{dB} \tag{11.2}$$

where again the first term is the splitting loss and the second accounts for the excess losses. It is seen that as β decreases from unity, the effects are much more serious than with the star.

The fact that the loss L in dB builds up only logarithmically with N for the star but linearly for the bus has meant that bus networks with achievable values of β but no amplifiers have been limited to 20-30 nodes, while those using star couplers but no amplifiers can support hundreds of nodes.

The introduction of optical amplifiers changes this picture, and at first glance one would think argues in favor of bus topologies. One might recover the accumulated loss by placing an amplifier every M nodes along the bus, thus requiring fewer amplifiers (N/M of them) than with a star network, which requires N amplifiers. Moreover, such amplifiers on the bus would need to be of only modest gain, of the order of $-10M \log_{10} \beta$ dB, whereas for the star each amplifier gain would have to be of the order of $10 \log_{10} N$ dB, from Equation 11.1. However, as we shall see in Section 11.6, this all turns out to be much too optimistic in favor of the bus, because it ignores the gain-saturation effect of having to amplify many simultaneously active channels in the same amplifier.

11.5 Optimizing the Splitting Ratio for Bus Networks

Figure 11-4 shows two promising bus topologies, the *single reentrant bus*, in which all nodes transmit into the top half of the bus and receive from the bottom half, and the *double bus*, in which each node sends to a node farther to the right on the upper bus but to a node farther to the left on the lower one. The reentrant bus has the very large advantage that it requires only one transmitter and receiver per node, but suffers from the fact that the average connection passes through twice as many taps as with the double bus, and therefore cannot support as many nodes per network as can the double bus.

We analyze these two situations first in the absence of photonic amplification and then introduce amplification in the next section. Adding amplifiers seems a natural thing to do with busses, one reason being that, since busses are often implemented as a series of reentrant spokes from a central "wiring closet," the amplifiers can share a common power supply and physical housing.

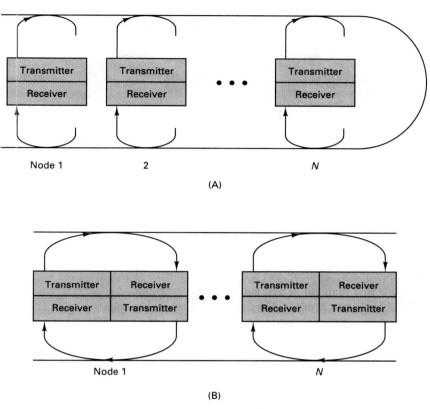

Figure 11-4. Two bus structures: (A) Reentrant single bus, (B) Double bus.

Again, at each tap, the ratio of total output power to total input power is β. Assume that all taps have the same power splitting coefficients $|s_{11}| = |s_{22}| = (1 - \alpha)$, which means that $|s_{12}| = |s_{21}| = \alpha$. These parameters were discussed in Section 3.11. Given the number of nodes N on the bus, Reference [3] shows how the overall link budget is affected by these splitting and excess losses. It even shows how the overall link budget can be optimized by allowing the values of α to differ from tap to tap. We shall assume that this option is not of practical interest, in spite of the fact that the overall loss is reduced, because to use this option would mean stocking either adjustable taps or a variety of different "part numbers" corresponding to the different tap coefficients. Even more seriously, the tap values would need to be changed whenever nodes were added or removed.

For the reentrant bus of Figure 11-4(A), the worst-case loss will occur on the path from the leftmost node to the next-to-leftmost node. We seek to adjust α so that the ratio of output to input power over this path is maximized. If α is made too large in an attempt to couple into and out of the bus efficiently at each tap, this will degrade $(1 - \alpha)$, the in-line coupling coefficient at each tap. If α is too small, not enough power is delivered to each local receiver. The worst-case ratio of output to input power is

$$R = \alpha(1 - \alpha)^{(2N - 2)}\beta^{(2N - 1)} \tag{11.3}$$

because a fraction α of the transmitted energy from the leftmost transmitter enters the bus, and at each tap thereafter, $(1 - \alpha)$ of it is passed on. Meanwhile, every one of the $2(N - 1)$ taps accumulates a ratio of output to input power β, less than unity.

   The optimum value of α, gotten by differentiating and setting to zero, is $\alpha_{opt} = 1/(2N - 1)$, from which the corresponding R is

$$R_{max} = \frac{1}{2N - 1}\left(1 - \frac{1}{2N - 1}\right)^{(2N - 2)}\beta^{(2N - 1)} \tag{11.4}$$

which for large N, is

$$R_{max} = \frac{1}{2eN}\beta^{2N - 1} \tag{11.5}$$

   For the double bus of Figure 11-4(B), the worst-case loss will be that between the two end nodes, left and right,

$$R = \alpha^2(1 - \alpha)^{N - 2}\beta^N \tag{11.6}$$

which is maximized when $\alpha_{opt} = 1/N$, whereupon

$$R_{max} = \frac{1}{N^2}\left(1 - \frac{1}{N}\right)\beta^N \tag{11.7}$$

which, for large N, is

$$R_{max} = \frac{1}{eN^2}\beta^N \tag{11.8}$$

   It is seen that for the reentrant bus the number of dB of total loss due to β is twice that of the double bus. This also means that the dynamic range that each receiver at the extremities of the bus will have to handle is twice as bad for the reentrant bus. This dynamic-range requirement arises because sometimes the receiver is listening to the nearest node and sometimes to the farthest. However, the double bus requires twice as many transmitters and receivers.

11.6 Amplifier Placement

Four trade-off factors

Chapter 8 described how the bit error rate in a system decreases as the ratio of received signal to received noise increases. In the absence of photonic amplification, the dominant form of receiver noise was found to be thermal noise appearing at the input of the electrical amplifier. Earlier, in Chapter 6, we saw how photonic amplifiers can increase signal power to combat such forms of noise as receiver thermal noise, but unfortunately they also add their own form of noise, amplified spontaneous emission (ASE). We also saw that a third factor is at work, too: The amount of gain produced by the amplifier was seen to be decrease with the total applied input power, the phenomenon of *gain saturation.*

The value of input power P_{in} that caused the gain G to drop to half its small-signal value was defined as the saturation input power P_{sat}. The amplifier gain is given by Equation 6.23 as

$$G(f) \;=\; 1 + \frac{P_{sat}}{P_{in}} \; \ln\left(\frac{G_o(f)}{G(f)} \right) \tag{11.9}$$

and the spectral density of the ASE noise by Equation 6.9

$$N(f) \;=\; hf\chi\, n_{sp}\,[G(f) - 1] \tag{11.10}$$

In order to include optical amplifiers in a link, multipoint, or network in the correct way topologically, we need to examine the tradeoffs between four factors: gain, gain saturation, ASE, and thermal noise.

Transmitter power amplifier and receiver preamplifier

In Chapter 6, we distinguished three roles for optical amplifiers, namely as transmitter power amplifiers, receiver front-end preamplifiers, and line amplifiers.

The receiver-preamplifier case was examined at length in Section 8.11, where we saw that it is desired to make the gain $G(f)$ high enough to overcome thermal noise. The arriving signal power P_{in} will usually be low enough that the preamplifier need not be a very high-power amplifier; that is, P_{sat} need not be made very high. Putting these two considerations together, we can say that, for receiver-preamplifier purposes, high gain is a requirement but high P_{sat} is not. For a transmitter amplifier, however, the converse is true. We want high saturation power, and not necessarily large gain.

The effect of ASE on a receiver can be greatly reduced by bandpass filtering following the amplifier, and this filter can be made narrower and narrower until the modulation bandwidth is reached (matched-filter condition), whereupon further narrowing actually loses detectability.

Single line amplifier on a point-to-point link

For service as a line amplifier, we must decide where to place it. There will be a compromise between placing it near the transmitter (so that the ASE noise generated in the amplifier suffers the maximum attenuation before reaching the receiver) and placing it nearer the receiver (so that thermal noise is overcome and so that the signal has been maximally attenuated before reaching the amplifier, thus minimizing gain saturation).

When the noise enters the receiver, the nonlinear nature of the photodiode leads to beat products of the noise with itself and with the arriving signal. As we saw in Section 8.11, the latter cross-products are the significant ones, under the usual condition that there is some sort of optical filter preceding the photodiode. Since the other important source of signal contamination is thermal noise, we expect to find that if the amplifier is placed near the transmitter so that ASE is significantly attenuated, thermal noise will be the dominant receiver noise. On the other hand, we expect that if the amplifier is too close to the receiver, ASE will dominate.

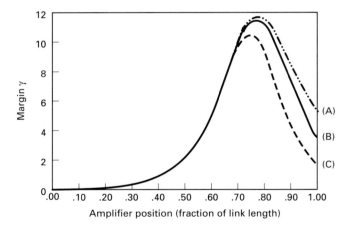

Figure 11-5. Amplitude signal-to-noise ratio y (margin) as a function of relative position of a single optical amplifier. 0.0 = next to transmitter. 1.0 = next to receiver. Transmit power = 1.0 mw, bitrate = 200 Mb/s, link length = 250 km, $\lambda = 1.5\ \mu$, thermal noise = 3.8 pA/$\sqrt{Hz}$. Results for three different optical filter bandwidths are shown, (A) 200 MHz, (B) 10 GHz, and (C) 100 GHz. (From [4]).

Given a single point-to-point link, it is possible to put all these factors together and determine the optimum point along the link at which to place the amplifier. Unfortunately, a closed-form solution is not possible. Figure 11-5 shows the results of calculations [4] for a particular choice of parameters. The ordinate is y, the argument of the Q-function of Equation 8.9, i.e., the amplitude signal-to-noise ratio. It is seen that there is a rather narrow optimum for the placement of the amplifier.

Several amplifiers on a point-to-point link

The question of where to place a number of amplifiers, say K of them, along a single link of length L is even more analytically intractable than the single-amplifier case. One difficulty is that it is not easy to decide what the constraint of the problem is. Is it the total cost, and if so, how does cost of an amplifier increase with amplifier gain? Is it total overall gain, to be partitioned among the K amplifiers in some optimized way [5]? Or does one fix the amplifier gain and the power level into the first amplifier and then see how many amplifiers can be cascaded before the accumulated ASE gets too large [6]? Or should there be some other optimization?

Perhaps the most realistic approach is the following. Consider Figure 11-6. Suppose there is available a standard amplifier component, and one wants to know the total link length $Z = KL$ that can be achieved by using one of these amplifiers as a receiver preamplifier, and some number $(K-1)$ of others as line amplifiers, all being spaced L kilometers apart. What should the gain G of these amplifiers be, and what is the desired K? We shall choose the spacing L such that the gain G of each amplifier exactly compensates the link loss [$\exp(\alpha L)$] between them. We shall also assume given values of link attenuation α, bitrate $1/T$, transmitter power P_T, amplifier excess noise factor χ, population-inversion factor n_{sp}, operating optical frequency f, and required bit error rate.

Figure 11-6 shows along the top of the chain of K amplifiers the signal power level, with P_{in} and $P_R = GP_{in}$ being, respectively, the input signal power to the first amplifier and to the receiver. Along the bottom of the diagram are the corresponding values of ASE noise spectral power density. Note that the signal power is the same at the output of each amplifier, but the noise power increases along the chain. This would seem to argue that the number of amplifiers K should be small, but we shall see now that quite the opposite is true.

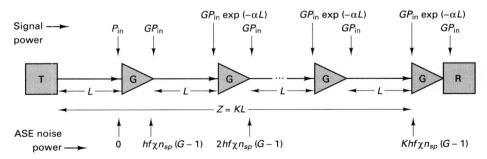

Figure 11-6. Several identical amplifiers spaced equidistantly by L. Signal powers are shown along the top and amplified spontaneous emission (ASE) noise power densities are shown along the bottom.

We make the arbitrary assumption that the length of the first span from transmitter to amplifier is L, as with the other spans. Because

$$G \exp(-\alpha L) = 1 \tag{11.11}$$

then

$$Z = \frac{K}{\alpha} \ln G \tag{11.12}$$

Also,

$$P_R = P_T \tag{11.13}$$

Now, given the required bit error rate, we can use Figure 8-7 to find y, the argument of the Q-function, i.e., the electrical amplitude signal-to-noise ratio. Then, assuming that the receiver performance is limited by signal-spontaneous rather than by thermal noise, we know from Equation 8.67 for the ASK direct-detection case, that the receiver must receive a signal power level of

$$P_R = GP_{in} = \frac{2\,y^2\,hf(G-1)\,K\chi\,n_{sp}}{T} \tag{11.14}$$

in order to achieve this desired bit error rate. From Equation 11.12,

$$Z = \frac{P_R T}{2\,hf\chi\,n_{sp}\,y^2\alpha}\,\frac{\ln G}{(G-1)} \tag{11.15}$$

Noting that in the last term the denominator changes much faster than the numerator, this equation tells us that in order to maximize the total link length Z, we should use a low value of gain G, which then implies a large number of amplifiers K.

Figure 11-7 shows the total link length Z of a 1-Gb/s link, assuming some typical values for the different parameters. The curves show clearly that if circumstances permit, it is more desirable to have a number of low-gain amplifiers spaced closely than a few of them spaced farther apart.

Remote pumping on a single link

One of the important advantages of erbium doped-fiber amplifiers that use 1480 nm as the pump wavelength is that the amplifier can be remotely pumped. This offers the attractive possibility that a repeater station can be built that requires no local electrical power. This is of considerable potential importance for undersea links, such as the one shown in Figure 11-8. A few meters of doped fiber are spliced into the main fiber at

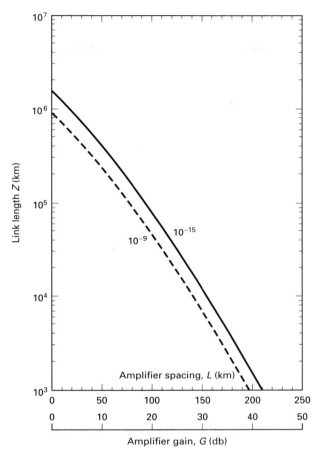

Figure 11-7. Total point-to-point link length versus amplifier gain for two values of bit error rate at 1 Gb/s. The abscissa is also labelled to show the spacing between amplifiers. Operation at 1.5 μ is assumed. Other parameter values are: transmitted power $P_T = GP_{in} = 1$ mw, excess noise factor $\chi = 1$, spontaneous emission factor $n_{sp} = 1.4$, and link attenuation $\alpha = 0.2$ dB/km.

a suitable point along the underwater path, and the pump source is on shore, at either the transmitter or the receiver. The determination of the optimum amplifier location in the case of pumping from the receiver end is similar to that of the locally pumped single amplifier assumed for the data of Figure 11-5, except that the attenuation of the pump power has to be taken into account, which moves the optimum point to the right.

Bus network

The single point-to-point link that we have just discussed is one in which the loss along the path is evenly distributed, being caused by the attenuation processes

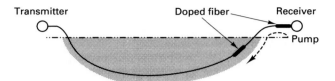

Figure 11-8. Remote pumping of a doped-fiber amplifier.

described in Chapter 3. When the path between transmitter and receiver travels across a network, the loss comes in big pieces, as shown in Figure 11-9 [5]. At (A) is shown the situation corresponding to a single 16-way star coupler made as shown in Figure 3-23. (B) shows the situation corresponding to Figure 3-24 where the 16-node network is divided into four four-node clusters, each served by a 4×1 combiner and a 1×4 splitter. In the bus network of Figure 11-9(C), each tap has a dB loss of $-10 \log_{10} [(1 - \alpha)\beta]$. These losses usually dominate any losses due to attenuation.

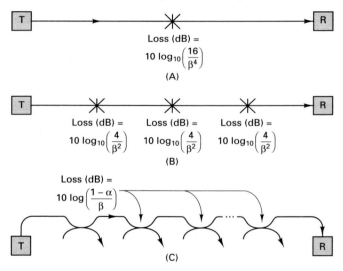

Figure 11-9. Showing how, in a network, the loss can occur in large increments. (A) Single 16-node star using one star coupler, (B) Two-layer star using a 4×4 coupler preceded by 4×1 combiners and followed by 1×4 splitters, respectively. (C) Bus.

Let us consider the optimum-amplifier-placement problem, first for the bus and then for the star. Both situations have been analyzed and compared in [7]. Typical results from this study are summarized in Table 11-1. For the star, a preamplifier at every receiver is assumed. For the bus, a distinction is made between two cases. In the "single-channel" case, only one transmission is present anywhere along the bus at any instant of time, and in the "multichannel" case, all N transmissions are active everywhere on the bus at all instants of time. For busses, amplifiers are assumed to be some optimized value of M taps apart.

In the single-channel case, the addition of amplification at intervals along the bus, one amplifier every M taps, leads to two or three orders of magnitude increase in the number of nodes that can be supported on a bus. It also provides better performance than a star topology. However, the single-channel case is not practical for gigabit networks, as we shall see in some detail in Chapter 13. The reason is that when the propagation time across the network is many bit times, then forcing a situation where any amplifier sees only one transmission at any time implies an extremely inefficient use of the network. The multi-channel case, in which many time-overlapped transmissions are allowed, is the more realistic situation, and it is seen from the table that the necessity to accommodate many simultaneously active transmissions tips the balance back in favor of the star topology with respect to both the single folded bus of Figure 11-4(A) and the double bus of Figure 11-4(B).

Table 11-1. Comparison of number of nodes supported with and without optical amplifiers, for three topologies.

Topology	Single folded bus	Double bus	Star
Without amplifiers, optimized α	23	29	335
Without amplifiers, $\alpha = 1$ dB	5	11	NA
With amplifiers, single-channel, $G_0 = 15$ dB, $\alpha = 1$	4350	8700	2575
With amplifiers, multichannel, $G_0 = 15$ dB, $\alpha = 1$	482	795	2375
No. amplifiers required	482/M	795	2375
No. transmitters required	482	1590	2375
No. receivers required	482	1590	2375

Star network

Consider Figure 11-9(A) and (B), and assume for the moment that we are using direct-detection WDMA receivers with tunable filters. If we assume that the links are short enough that link attenuation is not a major consideration compared to splitting loss, there are essentially six places that we can think of placing a single amplifier:

1. Immediately after the transmitter

2. Immediately preceding the star coupler

3. Within the star coupler (thus forming an *active star coupler*)

4. Immediately following the star coupler

5. Immediately preceding the receiver optical filter, and

6. Between the optical filter and the photodiode.

It can be shown [4] that Option 5 is usually the best one.

Option 1 would be unsatisfactory because of gain saturation, unless the amplifier is specifically designed to be a power amplifier, that is, to have a higher P_{sat} than the power out of the laser diode that drives it.

As for Options 2 and 4, note that the total powers into and out of an ideal lossless star coupler, splitter, or single-mode combiner are equal, as can be seen by following the paths in Figures 3-23 and 3-24. Therefore, if the amplifier is of the doped-fiber type, so that one can ignore the crosstalk due to gain fluctuations discussed in Section 8.14, then neither of these options has an advantage over the other. The active star, Option 3, will be discussed in the next section.

Options 5 and 6 are more attractive than 2 and 4, because all the N signals have had a chance to undergo link attenuation, thus minimizing gain saturation. With Option 5, all the N signals are present, but only the ASE noise spectrum in the neighborhood of one of the wavelengths reaches the photodiode, whereas with Option 6, only one signal is present, but all the ASE noise reaches the photodiode. Usually, the latter effect must be avoided, so Option 5 proves to be the best one. Of course, if one were willing to pay for two optical filters, one before and one after the amplifier, that would be even better. This is an interesting possibility for receivers based on such filters as the multiple Fabry-Perot filters of Section 4.5 or the Mach-Zehnder chain of Section 4.6.

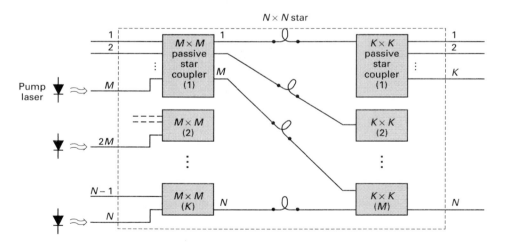

Figure 11-10. Active $N \times N$ star coupler made from K ($M \times M$) stars, K pump laser diodes, N lengths of doped fiber, and M ($K \times K$) stars. $N = KM$. (From [8]).

Active star coupler

Figure 11-10 shows how an active star coupler can be made from a number of smaller star couplers in such a way that each $M \times M$ star coupler in the first column serves not only to evenly distribute M of the input signals to its M outputs, but also the signal from a single pump laser, added to one of the signals. We can then interpose between the first and second column $N = MK$ lengths of doped fiber and be sure that all of them will be pumped. This scheme saves greatly on the number of pump diodes that would otherwise be required for Options 2 and 4, discussed earlier, since only $N/M = K$ of them are required. However, these pump lasers must develop enough power that $1/M$-th of the pump laser output power will still be sufficient to provide adequate inversion of the the ions in the doped-fiber segments. Also, if the various star couplers are made up of fused tapered biconical elements, one must pump at 1480 nm, rather than at the more efficient 980 nm, because otherwise the wavelength-selective properties of such couplers, discussed in Section 3.10, will severely attenuate either pump or signal.

This limitation is avoided, (and also presumably the cost is decreased) by using a planar coupler of the type shown in Figure 3-26. The reason 980-nm pumps can now be used is that the insertion loss is essentially wavelength-flat, with only a slight increase at longer wavelengths due to directivity pattern broadening and another at shorter wavelengths due to radiation losses from bends.

Both types of active star couplers have been realized experimentally [9, 10]. The planar-coupler approach should prove to be a very successful and economical system option in the future, as planar waveguide fabrication costs drop, as pump diodes of ever-higher power become available, and as doping of larger concentrations of atoms of rare-earth elements into silica material is achieved.

11.7 Wavelength Routing

A number of physical-network topologies have been suggested in which the nodes contain passive wavelength-selective structures that function in such a way that signals arriving at different input ports will be routed onward from different output ports according to wavelength. In order to vary the path, tunable transmitters and receivers are required at originating and termination nodes, respectively. We shall discuss such networks in this section, and then in the next section treat the converse situation in which the wavelength-routing structure can be varied while the transmitters and receivers do not tune.

Most *wavelength-routing* structures are based on fixed-tuned grating multiplex/demultiplex components, which we discussed in Section 4.12. An M-way grating, used as a demultiplexor, has the property that it converts a multiplexed set of M optical signals, each at a specified wavelength on the common input fiber, into M spatially distinct outputs. Used as a multiplexor, it collapses M spatially separate optical signals, each at the proper wavelength, onto one common output fiber. Such components can be viewed as SDM/WDM transducers.

One advantage of gratings over the conventional wavelength-insensitive couplers, splitters, or star couplers is that the grating avoids the splitting loss. In an ideal grating, all the energy entering the device at one of the fixed-tuned wavelengths passes to a single output port rather than being split among all the output ports. (In real devices, of course, the excess loss remains). The price paid involves cost, the fact that the grating is a fixed-tuned device, and fact that the wavelengths of the lasers in the system must thus correspond to those of the gratings.

The node illustrated in Figure 11-11 allows one to provide point-to-point paths from any one of N incoming transmissions to any one or more of N output ports, but without the splitting loss associated with star couplers, and with fewer wavelengths [11, 12]. For example, input 1 connects to port 3 simply by tuning the transmission to λ_3, and the receiving node served by output port 3 also tunes to λ_3. Meanwhile, some other pair of ports could also be using λ_3.

Thus, wavelength routing networks offer considerable wavelength reuse. Note from Figure 11-11, that each of M wavelengths is used M times. The node shown in the figure is a purely intermediate routing node in that there is no attached local user shown. However, one input port and one output port could easily be used for this function.

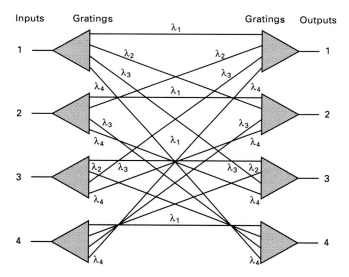

Figure 11-11. A wavelength routing node that is a pure intermediate node (no local user attached). Any transmission incoming on any port can be directed to any output port by changing the wavelength.

In practice, one might not interconnect all the nodes in the network in a mesh, but might take advantage of the ease and economies of bundling multiple fibers in the same envelope to lead the M input and M output fibers of the node of Figure 11-11 to one or more centralized *patch panels* where the necessary cross-connections would be made.

An interesting possibility for two-port wavelength-routing component [1] is to use the surface-wave acoustooptic tunable filter described in Section 4.10 and Figure

4-25 in an unusual way. By driving the transducer with multiple RF signals, instead of just one, multiple wavelengths can be directed to output 1, all others appearing at output 2. Although at present the number of wavelengths that can be separated in this way is only 7 (compared to many tens for gratings), the approach does have the advantage of allowing a switching capability without resorting to the use of photonic switches combined with gratings, a subject to which we now turn.

The wavelength-routing approach has been of considerable interest to common carriers, not for multiaccess networking, but in a static application where existing central offices are retrofitted with nodes such as that in Figure 11-11. This allows increased capacity and considerable routing flexibility without requiring the installation of new fiber. We shall discuss an example in Section 16.7.

11.8 Hybrid Wavelength-Space Topologies

The principle drawback of the wavelength-routing example of the previous section using static routing structures is that it requires both tunable transmitters and receivers in order to build a multiaccess network. Much more interesting is the option of making the internal connections of the wavelength-routing element variable, thus avoiding tunable lasers and receivers altogether.

The node structure of Figure 11-12 does this by employing photonic switching inside the routing element [13]. It thus forms a switchable version of the wavelength-routing node of Figure 11-11.

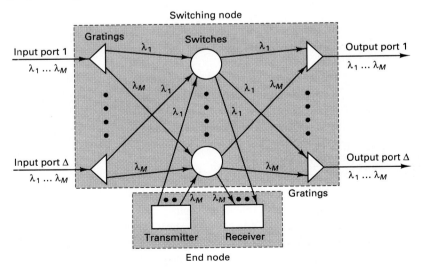

Figure 11-12. One node of a wavelength-space division hybrid network.

As the figure shows, we can think of each of the N nodes in the network as consisting of a *switching node* and an *end node*, the latter containing M fixed-tuned transmitters and an equal number of fixed-tuned receivers. As with the static wavelength-routing network, the same set of M wavelengths is used throughout the network. Each adjacent switching node contains Δ receive ports and the same number of transmit ports. Each receive port feeds a $1 \times M$ grating demultiplexor, and each transmit port is fed by a $M \times 1$ grating multiplexor. In between, there is a set of M switches, each of size $(\Delta + 1) \times (\Delta + 1)$.

There are many ways in which the input and output ports of N such nodes can be connected. For each there is a different accumulated end-to-end loss for every path in going through several photonic switches, and there is a different number of wavelengths employed. For example, by using one particularly regular topology [13], four transmit and four receive ports per node, and by limiting the number of hops between end nodes of any connection to $H_{max} = 5$, one may connect 1024 nodes with a 10^{-5} blocking probability employing only five wavelengths. Each node would have eight 5-way gratings and five 5×5 photonic switches. The maximum accumulated loss would be $78L$ dB, where L is the dB loss per switch.

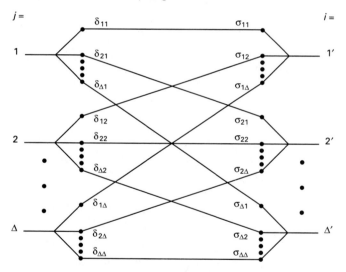

Figure 11-13. The LCD (linear combiner/divider) element of a linear lightwave network. (From [2]).

Combining the multi-wavelength dimension with fast-acting photonic switching avoids three technology bottlenecks: that only small-dimension photonic switches are available, that only small-dimension gratings are available, and that rapid tunability is not widely available today. Also, by controlling H_{max}, one can keep the loss across the network within modest limits. Furthermore, the approach requires a number of wavelengths M that is much less than N, since each wavelength can be reused at several different physical locations in the network simultaneously. The principal impediments to constructing such a network today are the component costs and the fact that the significant loss that today's photonic switch modules present depletes the overall link budget suffers excessively.

11.9 Linear Lightwave Networks

This is the name given to a class of circuit-switched networks in which couplers with adjustable coupling coefficients are used in place of switches. The term is somewhat confusing. Actually any all-optical network will be linear in the sense that linear superposition holds on the all-optical path between nodes. On such paths there are no intermediate operations such as frequency conversion, modulation-demodulation, coding-decoding, and so forth, associated with conversions from photonic to electronic form. As used in the present context, the term "linear" can be associated instead with the idea that coupling and splitting coefficients within each coupler are allowed to be continuous variables instead of either only 0 or 1, as with "nonlinear" switches. Linear lightwave networks were invented by Stern, and the discussion in this section is based on the work of him and his students [2, 14, 15].

Single-waveband case

Figure 11-13 shows the adjustable $\Delta \times \Delta$ coupler, or *linear combiner/divider (LCD)* that is used instead of a more conventional $\Delta \times \Delta$ switch. By conservation of energy, the splitting coefficients δ_{ij} at the input side are related by

$$\sum_{i=1}^{\Delta} \delta_{ij} \leq 1 \quad \text{for every } j \tag{11.16}$$

and the combining coefficients σ_{ij} at the output side by

$$\sum_{j=1}^{\Delta} \sigma_{ij} \leq 1 \quad \text{for every } i \tag{11.17}$$

The equalities hold only when the excess losses are zero.

We discuss first the case in which the LCD is wavelength-flat, so that no attempt is made to keep different transmissions at different wavelengths separated within the LCD. We then proceed to the more interesting case in which the LCD is changed to keep a number of *wavebands* separated from one another, each waveband containing a number of wavelengths.

Figure 11-14 shows a typical linear lightwave network for the case where each node has degree $\Delta = 2$. The LCD now becomes an adjustable form of 2×2 coupler, with splitting coefficient α, as discussed in Sections 3.10 and 11.4. In terms of Figure 11-13,

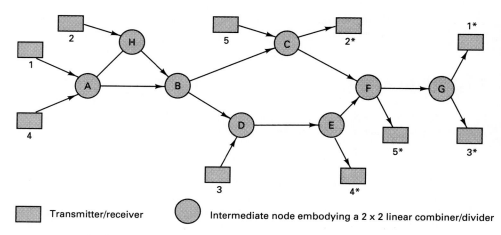

Figure 11-14. A linear lightwave network.

$$\delta_{11}\sigma_{11} = \delta_{22}\sigma_{22} = (1 - \alpha) \tag{11.18}$$

and

$$\delta_{21}\sigma_{21} = \delta_{12}\sigma_{12} = \alpha \tag{11.19}$$

The coefficients can be managed by a central controller, not shown, that responds to call requests by commanding the various couplers to assume new values of α. An alternative, used in peer decentralized computer networks [16], is to maintain in each node a *topology database* that contains complete up-to-date knowledge of all the LCD settings, thus avoiding a central controller in favor of completely distributed control. Either way, the necessity for networkwide coordination of the tap coefficients makes linear lightwave networks appropriate for a circuit-switched but not a packet-switched mode of operation.

We can use the example of Figure 11-14 to illustrate the key points about these networks. Suppose, first, that each node, instead of using LCDs, was obliged to use a 2×2 photonic switch, i.e., a 2×2 optical coupler with $\alpha = 0$ or 1. The network would still be "linear" in the sense that linear superposition applies to any all-optical network, but it would not be a linear lightwave network in the sense meant here. The connectivity provided would be very restricted. For example, if, while 1 is talking to 1☆ via A-B-C-F-G, 2 wants to talk to 2☆, 2 must do so via H-B-C, and is therefore out of luck, since the switch at node B has already been set to send to D all the power from H ($\alpha = 1$), and therefore cannot send any to C.

If, in order to connect 2 to 2★, we now let $\alpha < 1$ at node B, connectivity becomes much better. By choosing suitable values of α at B and C, we can send from 2 to 2★ while at the same time sending from 1 to 1★. This requires that the 1-1★ and 2-2★ connections use different wavelengths, otherwise they will interfere at nodes B and C and on the BC segment. By adding a third wavelength, we can also make it possible for 3 to talk to $3^{☆}$ at the same time, using suitable values of α at F and G.

One price paid for substituting LCDs for switches is that the essentially lossless connection that originally existed from 1 to 1★ is now augmented by several *fortuitous* connections that bleed some of the signal power intended solely for 1★ to other nodes. This happens at nodes B, C, F, and G. For example, at node F some of the power intended for 1★ goes instead to 5★, and at G some goes to $3^{☆}$. These fortuitous connections are not considered a serious problem; think of the energy that is wasted in a purely star-coupler-based topology.

What is worse than fortuitous connections is topological "circuits." A circuit is a violation of the condition that the topology of all paths from the origin node must be entirely a *tree*. If any path from an origin node (root of the tree) reconverges to meet another path that started at the same origin, this constitutes a circuit, and we no longer have a tree. Such a situation exists at node F for the connection from 1 to 1★. Once the split at B allows messages from 1 to go to D and then to E and then to F (to support the connection of 2 to 2★, for example), then the link from F to G contains two time-delayed copies of the bitstream from 1. This multipath condition produces interference nulls and intersymbol interference. It is conceivable that the link attenuations and coupling coefficients could be played off against each other to attenuate the interfering signal to the desired extent and thus live with topological circuits, but this is an extremely complex combinatorial problem. Therefore the rule adopted in setting up routes is usually to prohibit circuits altogether. Even with these restrictions, when the network gets sufficiently large, it is possible to gain considerable reuse of the same wavelengths in different parts of the total network.

A solution exists for the splitting and combining coefficients at each node, assuming that the routing has been done in such a way as to avoid circuits, and assuming the wavelength-flat $\Delta \times \Delta$ LCD of Figure 11-13. Given the desired network topology (e.g., Figure 11-14), and the conditions of Equations 11.16 and 11.17, then

$$\delta_{ij} = \frac{D_i}{D_j} \tag{11.20}$$

and

$$\sigma_{ij} = \frac{S_i}{S_j} \tag{11.21}$$

where S_i is the total number of superimposed signals sharing the fiber at output i, S_j is the same for input j, D_i is the total number of destinations (including fortuitous ones) to which the power on output fiber i goes, and D_j is the same for input j.

Given a fixed physical-network topology, a new physical-path topology is set up by computing available routing and wavelength assignments using these equations, and then commanding the LCDs to take up the proper settings. Thus, for large networks one must be willing either to disrupt and reroute existing connections in order to satisfy a new connection request, to block the new connection request, or to provide a large pool of free routes or free wavelengths. The question of "how large" is an unsolved analytical problem. Several numerical examples of a single-waveband network of the type shown in Figure 11-14 have been computed.

Use of wavelength bands

Up to now, we have considered the LCDs to be wavelength-flat. A great increase in the number of nodes that can be supported can be gotten by either adding more independent parallel fiber paths between nodes, or equivalently by using many wavelength bands. The "λ-sets" of Figure 11-3 are just such adjacent sets of individual wavelengths. The multiple waveband approach requires the use of wavelength-sensitive LCDs, such as the one shown in Figure 11-15. Such a device allows different σs and δs to be set up for different wavebands.

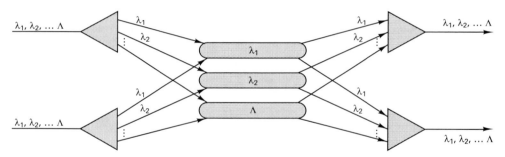

Figure 11-15. Realization of a wavelength-sensitive version of the LCD of Figure 11-13 for the case of $\Delta = 2$ ports and Λ wavelengths.

A hierarchical network of the sort illustrated in Figure 11-3 can be built by using λ *set 3* for all communication within each third-level subnetwork, λ *set 2* to go to a node within another nearby third-level subnetwork or a node at level 2, and so forth. In Section 11.13 we shall discuss the use of hierarchies and wavelength bands as an all-optical means of interconnecting individual all-optical networks.

Relationship to WDM/SDM hybrid topology

The linear lightwave network can be considered, in some sense, as a compromise between networks that broadcast all the energy (Figure 11-2), and the hybrid wavelength- space-switching architecture of Section 11-8 and Figure 11-12 where there are no fortuitous paths. The broadcast networks waste energy and have no wavelength reuse, but are simple and involve simple wavelength control functions. The WDM/SDM hybrid, on the other hand, provides great savings in energy and provides considerable wavelength reuse, but at the expense of control complexity.

The linear lightwave network preserves some features of both. The presence of fortuitous paths constitutes a partial broadcast which adds considerable flexibility to the assignment of wavelengths, wavelength bands, and routes. This is achieved at the expense of some power loss, complexity of control, and possible temporary disruption of existing connections when new ones are to be put in place.

So far, quantitative understanding of linear lightwave networks is incomplete as is their comparative performance relative to pure broadcast networks or hybrids.

11.10 Hybrid Wavelength-Time Topologies

The first third-generation lightwave network to be implemented at the prototype level was the Lambdanet [17], depicted in Figure 11-16. There were no wavelength-tunable components. Each of $N = 18$ transmitters transmitted TDM frames on its own wavelength channel, the first slot to receiver 1, the second to receiver 2, and so forth. At each receiver, there was an 18-wavelength grating demultiplexor feeding 18 TDM receiver chains. Thus the ith receiver received the bitstream intended for it by first looking at the TDM receiver fed by that grating output whose wavelength corresponded to the transmitter sending the desired bitstream. It then looked in the ith slot of that received TDM frame.

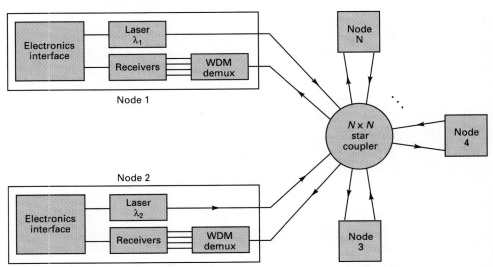

Figure 11-16. The topology of Lambdanet, a hybrid wavelength- time-division network.

Although tunability was avoided, this was done at the expense of providing electronics that was required to run at N times the bitrate of the connection. Lambdanet did provide a capability, should it be needed, for every node to talk simultaneously to every other node.

11.11 Hybrid Subcarrier-Wavelength Division Topologies

As was discussed in Section 9.7, there are several attractions to the idea of using the subcarrier-division multiplex/wavelength-division multiplex (SCM/WDM) alternative to pure WDM. A typical SCM/WDM network [18] is organized into clusters of workstations, each subordinate to a *cluster controller*, as in Figure 11-17. There are Λ clusters, with $N/\Lambda = M$ workstations per cluster. Each workstation owns a different microwave transmit frequency f_i and each cluster owns a different optical wavelength λ_j. Therefore, the address by which a given workstation is identified is the pair (λ_i, f_j). The transmit side of each workstation is extremely simple, involving the modulation of a single RF subcarrier, but the receive side requires both a tunable optical receiver and a succeeding tunable microwave receiver. Each cluster controller consists, on the receive side, of only a passive $1 \times M$ splitter, and on the transmit side of a wideband electrical network that sums all the microwave inputs, followed by a single wideband highly linear laser diode operating at the prescribed λ.

This SCM/WDM option for packet switching has been explored at the theoretical level in [19], and has been prototyped in its essential components [20].

11.12 Multihop Physical-Path Topologies

The *multihop* form of lightwave network was invented by Acampora [21] as a means of avoiding the need for either transmitter or receiver tunability in WDM networks. The multihop physical-path topology can be imposed on any physical-network topology having several ports per node. The concept can perhaps be most easily explained by starting with Figure 11-18(A). That diagram shows a classical first-generation packet-switched network of seven nodes, with unidirectional links connecting various nodes. This is an example of the same mesh topology that was discussed in Chapter 1 as being typical of earlier wide-area networks (WANs) such as the ARPAnet, and its scores of descendants. There is a buffer at each receive port in case the node cannot handle all arriving messages as fast as necessary.

A multihop lightwave network is one that results from taking any packet-switch physical-network topology and imposing it as the physical-path topology on a real physical-network topology that might be something quite different, say a star or a bus. In the example of Figure 11-18, the seven nodes could actually be arranged physically in a star topology with a star coupler in the middle or in a bus arrangement with taps leading to and from each node, or in some other arrangement. The paths, however, would be arranged as in the figure. We know that, since couplers and taps are linear devices, if each physical path between nodes corresponds to a different wavelength, then each receiver will receive traffic from only one particular transmitter. For example, the left receive port of node 1 can receive only from the transmit port of node 7, so that if 5 wants to send a message to 1, 5 can send it first to 7 who will then forward it to 1. Or 5 could send it to 3 who could send it to 6 who could send it to 1.

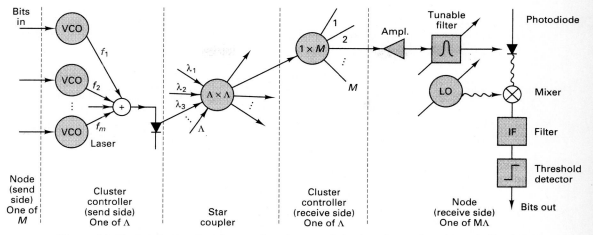

Figure 11-17. A hybrid subcarrier-wavelength network having Λ wavelengths and M subcarriers per wavelength.

Note that, in this example of an underlying star or bus, the real physical-network topology is *all-optical*, and it is *single-hop* in the sense that every node can, in principle, receive any transmission from any other node directly without any intervening optical/electronic conversion.   However, the physical-path topology is not always single-hop, it is often multihop (e.g. Figure 11-18(A)), and therefore involves intermediate optical-to-electronic conversions.

The basic idea, then, is that each node i is provided with not necessarily one port per node but Δ_i transmitting ports per node (the "out-degree").   We shall arbitrarily assume that there is an equal number Δ_i of receiving ports at node i (the "in-degree"). Each transmitter port goes to a different other node, and the same is true of each receiver port.   A distinct wavelength (of which up to $\Delta_1 + \Delta_2 + \cdots + \Delta_N$ are required) is dedicated to every transmitter-receiver pair, and each receiver in a node receives from a different node's transmitter.   Thus, a multihop network has the advantage that, even though no tunability has been provided, a connection exists from every node to every other node, in some cases in one hop, more often by relaying indirectly through one or more other nodes.   At each intermediate node there is a conversion between optical and electrical format.   The packet is electronically buffered while the address is recognized, and then, based on the address information, forwarded to the correct output port.

Among the disadvantages of the multihop strategy is the fact that the delay across the network is exacerbated by the series of "deliberate misroutings," that many wavelengths are required, and also the fact that the electronics in each node must handle multiples of the per-node bitrate.

There are several reasons that one might be willing to go to the trouble of providing the extra ports and having to accept the extra delay:

- A tunable port may be so expensive that Δ fixed-tuned ports are cheaper.

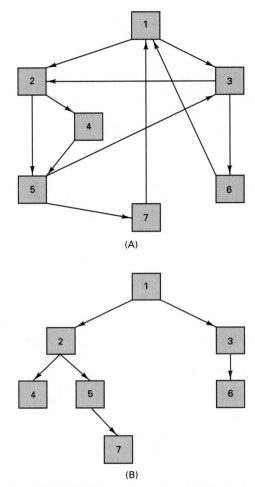

Figure 11-18. An irregular multihop lightwave network. (A) Physical-path topology, and (B) Spanning tree of (A) with respect to node 1.

- Since, on a given wavelength only one node can send to a given other node, packets can be sent around the network, and it is guaranteed that there will be no *receiver collisions*, the simultaneous arrival at a receiver port of packets from more than one transmitter.

- Only some of the nodes may need to have tunability.

- Tunability may be too slow for packet switching, but fast enough to reconfigure the physical-path topology frequently, thus gaining useful increases in capacity.

Some of these variants will be touched on after discussing regular multihop topologies quantitatively. These regular topologies are interesting from the analytical point of view, but have the practical disadvantages that they lack modularity and the capability of irregularity.

Given Δ, the number of send and receive ports per node, and some requirement on H, the number of hops that must be traversed on a connection, the problem is to determine how many nodes N can be supported. The constraint on number of hops arises from a desire to keep the delay across the network as small as possible. This constraint might be a constraint on the mean over all paths $(\overline{H})$ or on the length of the longest path (H_{max}). This last quantity is what a topologist would call the *diameter* of the graph, (the maximum over all node pairs of the shortest path between the nodes of the pair). Good multihop topologies are those that will support many nodes with small H_{max}. We shall consider cases with very constrained conditions, namely (1) *uniform traffic*, that is, the traffic level is identical between all possible node pairs, (2) the graph of the physical-path topology is *regular* (that is, every node has exactly Δ paths emanating from transmitters, Δ paths leading to receivers, and all these paths are disjoint), and (3) the number of nodes is highly constrained to just certain integers. The network of Figure 11-18(A) is not regular, and therefore not as easy to analyze.

The *spanning tree* of a node is a graphical representation of the set of paths from that node to each of the other nodes, some by one hop, others by more than one. For the irregular graph example of Figure 11-18, diagram (B) shows the corresponding spanning tree for node 1 transmitting. There are two one-hop paths emanating from node 1, then one from node 2 and one from node 5, and so forth until all nodes are reached once, the farthest one (3) being reached in three hops.

Of the class of graphs having a given diameter H_{max}, and for which all nodes have an identical degree Δ, the *Moore bound*, not usually achievable, is an absolute upper limit N_{max} on number of nodes that could be achieved for any graph. It can be deduced easily from the spanning tree of Figure 11-18(B), in the version (Figure 11-19) that applies to any node in a regular graph. Δ nodes will be reached in one hop, Δ^2 in two hops, and so forth, so that

$$N_{max} = 1 + \Delta + \Delta^2 + \cdots + \Delta^{H_{max}} = \frac{\Delta^{H_{max}+1} - 1}{\Delta - 1} \qquad (11.22)$$

the Moore bound.

Two highly regular topologies have been suggested for multihop lightwave networks, the *shufflenet graph* [22], and the *deBruijn graph* [13]. Figure 11-20 shows an example of each for the number of ports per node (degree) $\Delta = 2$ and longest path (diameter) $H_{max} = 3$. In this simple case, both the shufflenet and the deBruijn topologies have eight nodes, but for higher values of Δ, the deBruijn physical-path topology supports many more nodes, as we shall see.

Each edge (link) of each of these graphs represents a wavelength. Thus, to get from node 0 to node 7 for the shufflenet of Figure 11-20(A), one way would be to route over one hop via λ_1 to node 5, thence by λ_{11} to node 3, and finally by λ_7 to node

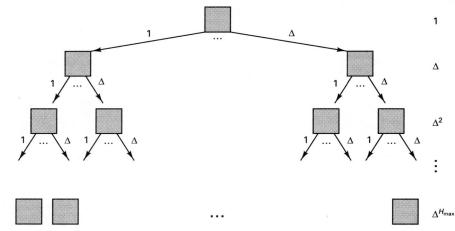

Figure 11-19. Spanning tree for any node of a regular graph of degree (number of ports) Δ and longest path H_{max}.

7. For the deBruijn graph of (B), a connection might be by λ_0 to node 1, then by λ_7 to node 3, and finally by λ_{12} to node 7.

Table 11-2 compares both topologies against the Moore bound for some typical values of Δ and H_{max}. The table shows N, the number of nodes supported, and the mean number of hops $\overline{H}$, under the uniform-traffic assumption, and assuming that one always routes by means of the shortest path. For shufflenets,

$$N = k\Delta^k \qquad (11.23)$$

where k is an integer given by

$$k = \frac{1}{2}(H_{max} + 1) \qquad (11.24)$$

so that shufflenets exist only for odd H_{max}. For the deBruijn graph

$$N = \Delta^{H_{max}} \qquad (11.25)$$

The shufflenet topology is easy to synthesize, given Δ and H_{max}. The nodes are to be thought of as laid out in a cylindrical arrangement of k columns of Δ^k nodes each. Figure 11-20(A) (for $k = 2$) displays this cylindrical arrangement by providing a fictitious $(k + 1)$th column added to represent a round trip around the cylinder and back to the first column. Then, going counterclockwise around the cylinder, each of the Δ^k nodes in a column has Δ links that go to Δ nodes in the next column according to the following rule: Numbering the nodes in a column from 0 to $\Delta^k - 1$, node i has edges directed to nodes $j, j + 1, \ldots, (j + \Delta - 1)$ in the next column, where

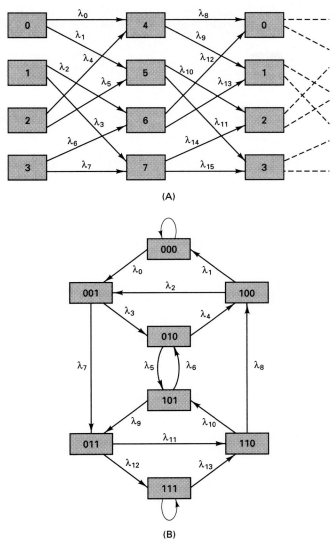

Figure 11-20. Two multihop physical-path topologies for number of ports Δ and maximum number of hops H_{max} of 2 and 3, respectively.   (A) Shufflenet graph, (B) deBruijn graph.

$j = (i \bmod \Delta^{k-1}) \Delta$.   The resulting topology of Figure 11-20(A) is called a *perfect shuffle* [23], hence the name "shufflenet." The perfect shuffle is widely used in patterns of processor interconnection to form multiprocessors.

Table 11-2.   Comparison of shufflenet and dBruijn physical-path topologies.

Δ	H_{max}	N_{max}	Shufflenet				deBruijn	
			k	N	$\overline{H}$		N	$\overline{H}$
2	3	15	2	8	2.0		8	2.1
2	5	63	3	24	3.3		32	3.6
3	5	364	3	81	3.6		243	4.3
6	5	9131	3	684	3.8		7776	4.8
10	5	111,111	3	3000	3.9		10^5	4.9

The average number of hops $\overline{H}$ of a shufflenet is [21]

$$\overline{H} = \frac{k\Delta^k(\Delta - 1)(3k - 1) - 2k(\Delta^k - 1)}{2(\Delta - 1)(k\Delta^k - 1)} \tag{11.26}$$

To synthesize the deBruijn graph, given Δ and H_{max}, one proceeds as follows. Imagine a Δ-ary shift register of length H_{max} stages, this being a binary shift register of length three in the example of Figure 11-20(B). Assigning one state of such shift register to each node, there will obviously be $\Delta^{H_{max}}$ nodes. There is a link joining node i to node j if node j can be reached from node i with one shift that brings in a new input digit (either a 0 or a 1 in the example) from the right. In other words, the deBruijn graph is identical with the state-transition diagram of the shift register, which goes from state to state by shifting in a new digit from the right. (An equally valid deBruijn graph could have been gotten by always shifting in from the left).

It is seen that several of the nodes (Δ of them), there is a shift that leads to no change at all in the shift-register contents. In the example, this would be the case with node 000 shifting in a 0 and 111 shifting in a 1. This implies a connection from the node to itself, obviously not meant to be implemented. One consequence of this is that Δ of the N nodes will have only $\Delta - 1$ transmit ports and an equal number of receive ports.

Whereas the calculation of $\overline{H}$ was only a little tedious with shufflenets, it is considerably more difficult for the deBruijn graphs. An exact recursive procedure has been given in [13], along with the following bounds:

$$\frac{H_{max}N}{N - 1} - \frac{\Delta}{(\Delta - 1)^2} + \frac{H_{max}}{(\Delta^{H_{max}} - 1)(\Delta - 1)} \leq \overline{H}$$

$$\leq \frac{H_{max}N}{N - 1} - \frac{1}{\Delta - 1} \tag{11.27}$$

In summary, both the shufflenet and deBruijn graphs are quite interesting mathematically as neat and efficient examples of multihop physical-path topologies. The deBruijn graph is particularly efficient, even though it is slightly unbalanced because Δ of the nodes are slightly deficient in connectivity. However, both these topologies are clearly too regular to be of much use for practical network topologies. Nevertheless, by analyzing them and presenting the quantitative results of Table 11-2 we can get some sort of bound on the way in which degree, number of wavelengths, number of nodes, and mean number of hops might trade off in more realistic topologies.

Several interesting variants of the multihop physical-path topologies have recently been studied. First of all, it is obvious that one could use any of the multihop topologies as an actual physical-network topology by running Δ transmit fibers and Δ receive fibers from each node to a suitably connected central "patch panel" at which all the cross-connections are made. As a second application of the multihop idea, the rapid tunability of RF components, compared to many wavelength tunable devices, can be exploited to achieve packet switching by giving the entire system the tunability speed of the RF nodes at the expense of adding some extra hops to many connections, as was discussed in Section 11.11.

The multihop networks we have described, whether highly regular or not, are not particularly well-suited to randomly varying traffic. To most efficiently keep up with changing traffic demands, the physical-path topology would have to be modified from time to time, and in a way that does not disrupt existing active connections. This question has been studied and a specific set of algorithms developed for making the changes in a systematic way [24]. The procedure, *delegated tuning and forwarding*, is practical when one is merely reconfiguring the physical-path topology infrequently on a time scale slow with respect to message durations. This technique provides significant decreases in the delay across the network by reducing $\overline{H}$.

11.13 Interconnecting Optical LANs and MANs

To be of more than limited local use, fiber optic LANs and MANs must somehow be provided with some means of interconnection to one another. There are two directions in which this can proceed, the "business-as-usual" approach of using electronic common carrier resources, and the more aggressive "all-optical" approach.

Business as usual

If long-haul *dark fiber* is not available, one might use more conventional wide area point to point connections of the types to be described in Section 14.2, for example the Synchronous Optical Network (SONET) or Synchronous Digital Hierarchy (SDH) framing conventions. Using this book's arbitrary division into "generations" of systems, one would describe this approach as connecting third-generation networks with second-generation means. There are several suitable gigabit WAN backbones today, notably the U.S. National Research and Education Network (NREN) [25], which uses the 0.6 Gb/s and later the 2.4 Gb/s rates of SONET. Clearly, such bitrates will be quite insufficient if each of the many users demand gigabit rates.

Completely optical interconnection

It was mentioned earlier that dark fiber has been installed first over LAN distances, then MAN, and only later over WAN distances. Where dark fiber is available for wide area interconnection, there are still several problems to be solved.

First, the total number of distinguishable wavelengths available in a fiber is large, but it is not infinite. If there is only infrequent traffic outside each subnet, then perhaps there are enough wavelengths available that the simple hierarchical filtering system depicted in Figure 11-3 can be used. Such a situation is not expected to be the general rule, however. Generally, some or all of the wavelengths available in one network (its *address space*) will have to be reusable in others. This is the standard "address translation" problem that all large networks of subnetworks possess. We must somehow translate from one wavelength to the other within the same address space when traversing the gateway. In the dark fiber case, the translation at the optical gateway takes place by remodulating the bitstream on a wavelength arriving at the gateway from within the first network onto another wavelength used for the desired destination in the second network. Technologies for doing this without converting to electronic form were discussed in Section 7.5.

Second, it may be necessary to boost the power level so that the N_2-way split of energy in the second network of N_2 nodes does not have to be compounded with the N_1-way split that the signal will already have suffered inside the first network of N_1 nodes.

11.14 What We Have Learned

In this chapter we covered the variety of physical-path and physical-network topologies that underlie today's third-generation lightwave networks. While, for a network, the logical topology is usually any-to-any point-to-point connections, the physical-network topologies are seen to be considerably different in many cases from the more familiar rings, busses, and meshes of first- and second-generation networks. These differences are driven more by technology insufficiencies than by a natural user requirement that the topology be laid out in some particular way.

With several of the topologies discussed, there are sub-cases of considerable analytical interest, but these sub-cases are unlikely to be the ones used in practice, since they provide limited modularity and capability for serving irregular groupings of users.

To build networks supporting a number of nodes much greater than Λ, the total number of wavelengths available, requires wavelength reuse, a promising approach for realizing very large networks, but only when the resetting of connections is relatively infrequent. In order to avoid conflicts when any new connection is to reach distant parts of the network, network-wide coordination of wavelength use is required. Quantitative studies are needed to determine just how much wavelength reuse can be gotten in such situations.

In spite of the differences in physical-network topology, the structures we have discussed are still, after all, just the foundation upon which the significant system aspects of the network are built. These system aspects have to do with providing function, much of it implemented in software, that exploits the physical media for moving bits around by means of pulses of light, in order to serve the requirements of the end users of the network.

The next two chapters describe what these functions are. Chapter 12 presents the layered structure of these functions and Chapter 13 the protocols with which these layered functions are invoked to provide the connection that the communicating end users need between them.

11.15 Problems

11.1 Compare the number of nodes that can be supported on a single reentrant bus and on a double bus (Figures 11-2 (A) and (B), respectively), if $\beta = -1$ dB and all the tap splitting coefficients are uniform and optimized. Assume that direct-detection receivers and ASK are used, that the transmitted peak power is 1 milliwatt, and that the desired bit error rate is 10^{-9}

11.2 A network at 1.5-μ wavelength is to consist of N 1-Gb/s nodes arranged in a circle of diameter 10 kilometers, and it is desired to maximize N, keeping the bit error rate below 10^{-9}. No photonic amplification is available. Each node transmits at 0 dBm (1 milliwatt), and APD receivers are to be used. Compare the maximum number of nodes for the following two cases:

(A) The topology is a star, using a single N-fold star coupler, where N need not be a power of two.

(B) The topology is a double bus running in an arc between node 1 and node N, using a uniform optimized value of tap coefficient α throughout the network.

In both cases, assume that the incidental loss per 2×2 coupler or per tap is 1.0 dB.

11.3 A multihop network is to be built in which each node has three input ports and three output ports. If the maximum allowed number of hops that the misrouting must produce is 3, compare a shufflenet graph and a deBruijn graph with respect to (A) Number of nodes supported, and (B) Average number of hops traversed, assuming uniform traffic between node pairs. (C) Draw a picture of each topology.

11.4 In a wavelength-routing network, how would you support broadcasting of the same data simultaneously to all receivers?

11.5 Suppose an 8-node packet-switched multihop network is to be built that is all-optical in that all the switches are photonic rather than electronic. If each switch has 5 dB of attenuation, and each link attenuation is 2 dB, how large a network can be built, keeping the total transmitter-receiver path loss below 45 dB?

11.6 Consider the linear lightwave network of Figure 11-14, operating in a wavelength-flat mode. What are the values of the splitting coefficient α at nodes A and B? Assume that the link from E to F has been deleted.

11.16 References

1.   C. A. Brackett, "A perspective on scalability and modularity in multi-wavelength optical networks," *Workshop on WDM Technologies, Systems and Network Applications*, Optical Fiber Communications Conference, 1992.

2.   T. E. Stern, "Linear lightwave networks," *Research Report, Columbia U. Ctr. for Telecom. Res. CU/CTR/TR 184-90-14*, 1990.

3.   J. O. Limb, "Fiber optic taps for local area networks," *Conf. Record, IEEE Intern. Commun. Conf.*, vol. 3, pp. 1130-1136, 1984.

4.   R. Ramaswami, "Issues in multi-wavelength optical network design," *IBM Research Report RC-15829*, May, 1990.

5.   H. D. Lin, "Gain splitting and placement of distributed amplifiers," *IBM Research Report RC-16216*, October, 1990.

6.   N. A. Olsson, "Lightwave systems with optical amplifiers," *IEEE/OSA Jour. Lightwave Tech.*, vol. 7, no. 7, pp. 1071-1982, 1989.

7.   R. Ramaswami and K. Liu, "Analysis of optical bus networks using doped fiber amplifiers," *Submitted to IEEE/OSA Jour. Lightwave Tech.*, vol. 9, 1991.

8.   A. E. Willner, A. A. Saleh, H. M. Presby, D. J. DiGiovanni, and C. A. Edwards, "Star couplers with gain using multiple erbium-doped fibers pumped with a single laser," *IEEE Photonics Tech. Ltrs.*, vol. 3, no. 3, pp. 250-252, 1991.

9.   A. E. Willner, I. P. Kaminow, M. Kuznetsov, J. Stone, and L. W. Stulz, "1.2 Gb/s closely-spaced FDMA-FSK direct-detection star network," *IEEE Photonics Tech. Ltrs.*, vol. 2, no. 3, pp. 223-226, 1990.

10.   H. M. Presby and C. R. Giles, "Amplified integrated star couplers with zero loss," *IEEE/OSA Jour. Lightwave Tech.*, vol. 3, no. 8, pp. 170-173, 1991.

11.   G. R. Hill, "A wavelength routing approach to optical communications networks," *Conf. Record, IEEE Infocom*, pp. 354-362, 1988.

12.   G. R. Hill, "A wavelength routeing approach to optical communication networks," *British Telecom Technol. Jour.*, vol. 6, no. 3, pp. 24-31, July, 1988.

13.   K. Sivarajan and R. Ramaswami, "Multihop lightwave networks based on deBruijn graphs," *Submitted to IEEE Trans. on Comm.*, vol. 39, 1991.

14.   T. E. Stern, "Linear lightwave networks: How far can they go?," *Conf. Record, IEEE Globecom*, pp. 1866-1872, 1990.

15.   K. Bala and T. E. Stern, "A minimum interference routing algorithm for a linear lightwave network," *Conf. Record, IEEE Globecom*, pp. 35-4.1 - 35-4.6, 1991.

16.   A. E. Baratz, J. P. Gray, P. E. Green Jr., J. M. Jaffe, and D. P. Pozefsky, "SNA networks of small systems," *IEEE Jour. Sel. Areas in Commun.*, vol. 3, no. 3, pp. 416-426, May, 1985.

17.   M. S. Goodman, H. Kobrinski, M. P. Vecchi, R. M. Bulley, and J. L. Gimlett, "The LAMBDANET multiwavelength network: Architecture, applications and demonstrations," *IEEE Jour. Sel. Areas in Commun.*, vol. 8, no. 6, pp. 995-1003, 1990.

18.   M. M. Choy, F. K. Tong, and T. Odubanjo, "An FSK subcarrier/wavelength network," *Conf. Record, Conf. on Lasers and Electrooptics*, 1991.

19. R. Ramaswami and K. N. Sivarajan, "A packet-switched multihop lightwave network using subcarrier and wavelength multiplexing," *Submitted to IEEE Trans. on Commun.*, vol. 39, 1991.

20. M. M. Choy, S. M. Altieri, and K. Sivarajan, "A 200 Mb/s packet-switched WDM-SCM network using fast RF tuning," *Submitted to IEEE Journal of Lightwave Technology*, March, 1992.

21. A. S. Acampora, "A multi-channel multihop local lightwave network," *Conf. Record, IEEE Globecom*, pp. 37.5.1 - 37.5.9, November, 1987.

22. M. G. Hluchyj and M. J. Karol, "ShuffleNet: An application of generalized perfect shuffles to multihop lightwave networks," *Conf. Record, IEEE INFOCOM*, pp. 4B4.1 - 4B4.12, 1988.

23. H. Stone, "Parallel processing with the perfect shuffle," *IEEE Trans. on Computers*, vol. 20, no. 2, pp. 153-161, 1971.

24. J. S. Auerbach and J. Pankaj, "Delegated tuning and forwarding in wavelength division multiple access networks," *Submitted to IEEE Trans. on Commun.*, August, 1991.

25. "Gigabit network testbeds," *IEEE Computer Magazine*, vol. 23, no. 9, pp. 77-80, 1990.

CHAPTER 12

Layered Architectures and Network Control

12.1 Overview

In the first chapter, it was made clear that there are three principal factors at work determining the character of a network. The *communication technology* is the first. The second is the *demands of the user*. The third is the *logic and memory technology* out of which the nodes of the network are made, whether they be telephone switches, large mainframes, workstations, terminal controllers, or whatever. This book is concerned with the first two of these three areas.

The purpose of this chapter is to try to put some of the large mass of detail in the three categories into an orderly framework so that one may reason clearly about lightwave networks as total structures.

When we get to the communication pieces, we shall see that there are today two basic ways of organizing them. The more familiar one is in terms of *protocol layers*. After working through a tutorial on layered communication architectures, we proceed to another way of looking at things, the *network-control* viewpoint, which deemphasizes the division of communication functions into layers and focusses more on the time sequence of activations, deactivations, and parameter settings involved in getting the layers to work. These can be considered to be carried out by a *control point* in each node.

The organization into layers, controlled by the control point, is appropriate for any network, past present or future, but the particular content and control of the layers in use today evolved in an era dominated by voice-grade telephone and satellite connections. Therefore, after discussing the picture today, the various layers and the control point are reexamined in the light of the new developments in fiber optic technology.

At the grossest level, one can think of a single *node* in a network as being a box (Figure 12-1) whose implementation details involve a complex mix of hardware and software of a nature that is strongly influenced by the three factors: communication technology, user requirements, and computer technology.

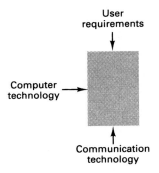

Figure 12-1. Three interacting factors in a typical node in a network.

A *network* consists of a number of such nodes connected together as in Figure 12-2, for example, to serve a collection of users and resources at different locations. The example shows a terminal user at node 1 accessing a database at node 4 over an *access path*, whose physical course is traced out by the dashed line, but which is logically the direct connection shown by the dotted line. So the physical-path topology of that particular connection is given by the dashed line, but the logical topology by the dotted line. All these boxes involve communication technology and computer technology, but those whose only function is to forward or switch traffic (*intermediate nodes*) do not have the third component, the user-driven component. An example is node 3 in the figure.

12.2 The Three Sets of Layers in Every Node

It has proved very useful to pick apart and somehow organize all the things going on in a typical node, and to divide the complex set of functions into layers. The notion of *layering* has become widely useful in understanding all manner of complex concepts. When we talk on the telephone, the audio voltages or bit patterns on the line are only the most worm's-eye sort of view. At a higher level there is the world of speech syllables, with the details of the waveforms or bit patterns hidden. Higher still are the words of the language being spoken, which suppress the phonemic detail of the syllables. Eventually one comes to the information being imparted, and again all the underlying messy details are no longer evident.

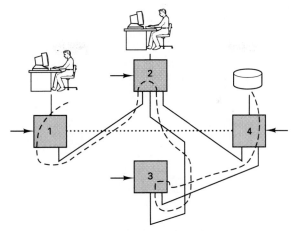

Figure 12-2. Typical physical interconnection of nodes. The dashed line shows an *access path* between a human user at a terminal and some data on a storage device. This is part of the physical path topology. The corresponding part of the logical topology is shown dotted.

This is essentially the process of *abstraction*, or *hiding*. The details of some lower level are collected into a viewpoint that captures all the relevant functions of the lower layer and expresses them (either as individual functions or some particularly appropriate groupings) in a way that is especially useful, while suppressing the irrelevant underlying details. Another often-used word for this is *virtualization*. A virtual memory system hides the fact that some of the memory may be real (semiconductor main memory) and some may not be (rotating magnetic storage); the user sees what looks like a very large main memory.

We already saw this process at work in Chapter 11. The physical-network topology supported a physical-path topology, which in turn supported a logical topology. At each level, the details of the underlying levels were suppressed.

Let us apply this process of abstraction, virtualization, or layering to the three technologies in a communicating node. The process is shown in Figure 12-3, an "exploded view" of Figure 12-1 that gives a more detailed view of the gross three-way subdivision of a network node.

Computer technology

The layers of computer technology [1] are shown at the left side. At the most basic level, that of individual logic and memory gates, computer technology appears as silicon or gallium arsenide logic elements. But in order to be useful, the irrelevant details are suppressed and the relevant details expressed by the construction of an assembly language, which in turn, by means of a compiler or interpreter, hides even these details and presents a service in the form of some programming language, like C or FORTRAN. Computer people and communication people alike call the set of func-

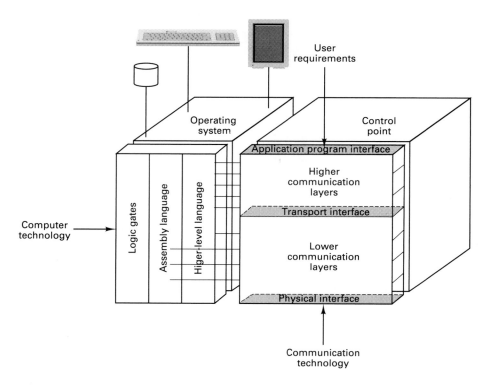

Figure 12-3. A more detailed look at a typical node.

tions abstracted and presented to the higher level an *interface*, or *service interface*, which in the case of a computer language would be the set of verbs (commands, responses) and nouns (data structures, such as parameters or tables) that comprise its syntax and which would also include the semantics (what these entities do).

The *operating system* of the node is the piece of software that manages all this I/O (input/output). As the diagram shows, the operating system, considered in its relationship to the hardware, controls the functions that are realized in computer technology, optical and magnetic storage, CRT displays, etc. It is implemented in assembly language or some higher-level language. Those communication functions that underlie the *transport interface* are also managed by the operating system, which treats them just like any other part of its responsibilities. All the I/O, including the lower communication layers, is usually implemented in the very fastest and tightly coded part of the operating system, the *kernel*, that part of it that the computer user is not permitted to get into. Examples familiar to the reader might include the UNIX kernel, the BIOS (basic input/output system) of PC/DOS as used in the IBM PC and relatives, and so forth.

The application program can touch both the services provided by the operating system and those provided by the computer and communication hardware, but the operating system stands guard over the latter set of services to see to it that one application does not interfere with another or bring the entire system down.

Communication technology

In our specimen communication network node, the base transmission technology, whether the fiber optic technologies described in Part II of this book or anything else, has a number of layers of communication abstraction overlaying it. As with those that overlay the basic computer technology, the further one progresses through the layers, the more is hidden about the raw physics. Passing through a series of communication layers, one eventually comes to the *transport interface*, whose details consist of a set of verbs (connect, disconnect, send, receive, etc.) and data structures (in the form of permitted message formats and parameters) that can be presented to it in order to pass information across the network to a party at the other end.

User requirements

Such a party is the *user*, shown in Figure 12-3 as connected into an application-program via an *application-program interface*. This nomenclature reflects the fact that in a real network, each of the parties being connected is either a piece of code supporting a human user (e.g., a display driver, as in node 1 of the example in Figure 12-2), or a real application program, for instance a matrix-inversion program or a database program such as the one shown in node 4 in the example. The application program must provide the services required by the user. Going downward, there will be a series of several layers, the bottommost of which requires from the transport interface a certain set of services. Proceeding all the way to the bottom, the properties of the physical medium must constitute a set of services at the *physical interface*. Chapter 1 discussed what these are: a physical connection to another node having a certain bandwidth, error rate, and propagation delay.

There is a significant split in the communication functions that lie above the transport interface and those that lie below. Above this interface, the layer content is driven by the idiosyncrasies of the user; below, the layer content is driven by the idiosyncrasies of the physical communication network. A fiber optic communication medium is so fundamentally different from a more traditional one that everything up to the transport interface can change. As applications evolve that exploit the capabilities of third-generation lightwave networks the upper layers will change, too. We described the state of some of these applications in Chapter 2.

Caveats

In the next section we shall proceed to pick apart all the functions that lie along the access path between applications in Figure 12-2, and will find that between the physical interface and the application interface there is a natural arrangement into layers, each of which spans multiple nodes. First, however, it is necessary to point out several things about Figure 12-3. The first is that it is easy to be misled by the result of this exercise into thinking that real communication nodes are built from software and hardware that dutifully pass each message physically from a piece of code representing one layer to another piece of code representing another. What really happens is that the central processor of the node is time-sliced to perform all the functions of the node in small pieces, visiting one after the other in some priority scheme in which higher-priority functions can interrupt lower-priority ones. And the communication functions are just a small part of the list of resources in a node that must be managed in this way. Figure 12-3 shows some of the others – the keyboard, displays, rotating storage, and printers.

There is another point about Figure 12-3. Although *communication architectures* (documented sets of rules of discourse for guiding the *implementation* in hardware or software) often present specified interfaces between all the many layers, in real-life implementations there are usually only three that are exposed and truly visible in the implementation, the specific three shown in Figure 12-3: the physical, transport, and application interfaces.

There is one transport interface at one node at one end of the access path and another one in a second node at the other end. A useful sorting of the types of connection commonly provided by transport interfaces into several classes is given in [2]. The least complicated is is the *datagram* class. Datagrams are isolated single *packets*, a packet being defined as some message of length up to several thousand bytes. In the datagram class, the packets need not arrive in serial order. Such a class is called *connectionless*, the remaining classes being referred to as *connection-oriented*. These enforce serial ordering of packets that are associated with one another into a *virtual circuit*.

At the other extreme from datagrams is the *stream* class of virtual circuit, in which each full-duplex message can be of arbitrary and presumably unlimited length, with no record boundaries, but with enforced serial ordering and losslessness. The TCP (Transport Control Protocol) portion of TCP/IP (IP for Internet Protocol) is an example of this class. TCP/IP is a set of protocols developed in the United States under government sponsorship.

In between the two extremes is the *sequenced packet* class of virtual circuit, which differs from the stream class in providing visible record boundaries. Examples are the *virtual circuits* of SNA and the transport interface of OSI (Open System Interconnection). SNA is System Network Architecture, the communication architecture of IBM products. OSI (Open System Interconnection) is the outcome of an international standardization activity aimed at providing an alternative to such proprietary or national architectures as SNA, DECnet, TCP/IP, and others [3].

12.3 The Communication Layers

It is traditional to break up all the layers leading from the physical copper or fiber to the application interface at the top into a series of seven layers [3-5], the bottom four dealing with the network and the top three with the application. Since the early 1980s, this relatively clean picture has evolved to the point where many layers have split into several sublayers. In addition, many options have sprung up for each of the layers of the original standard model. While it might be claimed that these developments have made the original subdivision [6] obsolete, this seven-layer model still makes a great deal of pedagogical sense.

Table 12-1. Typical functions carried out in connecting two end users.

Function carried out	Using
Make sure that a physical path exists to the next node	A local or common carrier link made from copper, fiber or radio technology.
See that the two ends of the link talk in bits at the correct speed	A modem or other form of driver
Move individual messages error free between adjacent nodes	A data link protocol that asks for retransmissions if a message is received in error
Send message to correct node and to correct subaddress within a node. Bypass a failed link or station.	Addressing, routing
Accommodate speed of communication network resources in presence of network congestion	Buffering and network flow control (with messages such as "start the flow" or "stop the flow")
Accommodate buffer length dictated by error rate, propagation delay	Packetizing and depacketizing (breaking longer messages into packets and reassembling them, respectively)
See that flow as seen by the two ends of the connection is error-free and adheres to one of the prescribed classes	End-to-end error checking, if required; organizing the message flow
Accommodate intermittency patterns expected by the application	Dialogue management; application session flow control
Accommodate context (format, code, language and security) requirements of the application	Code conversions, compression, encryption

Consider the list of functions given in Table 12-1 that must be provided in order for the access path (dashed line of Figure 12-2) to be completed.

Suppose one looks at Figure 12-2 and traces the progress of a given message from the application in node 1 to the application in node 4, and makes a sequential list of the occurrence of each step in Table 12-1. Unwinding the access path in such a way produces the sequence shown in Figure 12-4.

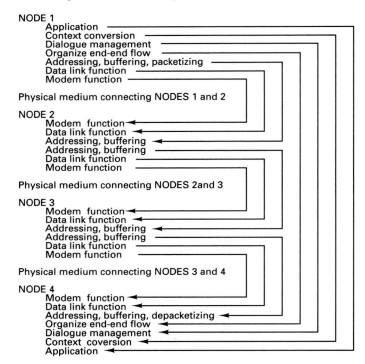

Figure 12-4. Sequential list of functions (Table 12-1) along the dotted-line access path from node 1 to node 4 of Figure 12-2.

Notice that the functions in the sequence always come in pairs, as indicated by the arrows. Also notice in Figure 12-4 that, in the case of nodes 2 and 3, none of the functions on the list in Table 12-1 after the third are used , and also that only the two ends of the access path, in nodes 1 and 4, use functions higher than the third on the list. Note also that only the third item on the list (routing, buffering, packetizing) appears in every node, and even then, the packetizing/depacketizing function is usually effected only at the two ends.

All of this suggests that the network of Figure 12-2 be redrawn as in Figure 12-5, where each node consists of a sequence of layers (*protocol stack*), network-wide. All nodes have layers 1-3 (the *intermediate nodes* 2 and 3 having only these), whereas only the *end nodes* have 4 - 7. The names of these layers, in OSI parlance, are given in Figure 12-5. All four nodes along the access path are shown in more detail in Figure 12-6.

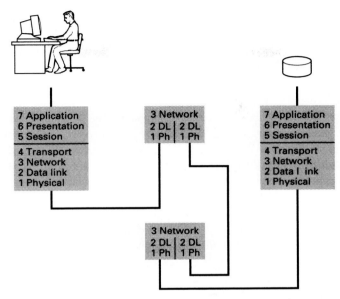

Figure 12-5. Network of Figure 12-2 drawn in such a way as to emphasize the three important interfaces and the seven OSI layer designations.

Since the two members of an interacting pair of layer instances in different nodes lie at the same layer level, they are called *peers*. The internal logic of the part of a node at each layer level is often described formally as a *finite-state machine* (FSM), a specification of a set of possible states, what it takes to make the machine switch from one state to another, and what happens when it makes that transition. Usually what makes one FSM change to another state is the arrival of a *command* from the other machine within the other member of the peer pair, and one of the things that can happen when it does switch states is the issuance of a *response*. So, the way the two FSMs do their work to carry out the function of that protocol layer is by sending and receiving verbs (commands and responses) and nouns (parameter values or tables) to and from each other over the very same access path that they are supporting. The stylized set of rules that define these exchanges is called a *peer protocol*, or commonly a *protocol*.

All the Level 3 instances in the *N* nodes in the network function together as a coordinated set of *N* finite-state machines, rather than a pair, as is the case for all other layers, both those below layer 3 (node-to-adjacent node) and those above (end-to-end).

The peer protocols are indicated by the horizontal dashed lines in Figure 12-6. These dashed lines are simply the arrows of Figure 12-4 drawn in a different way. Usually the way the two FSMs at the same level in different nodes communicate with each other is by means of a *header*, a reserved string of bit positions in each packet. The header is owned by just that layer and the commands, responses, and parameters are encoded as bit patterns in that header. This is all information for the use of the FSM at the same level in the other node. Since layers higher than the one using the header have no use for that information, matters proceed as shown in the figure, where the zigzag line represents the sequence of bits that would be observed on the connecting link between nodes. For an outbound message propagating downward in the stack, the layer will add its header ahead of what is already there. When the message reaches the peer in the other node and propagates upward in the stack, the header information is stripped off and used before the rest of the packet is forwarded upward.

In any one node at a given layer level there can be than one *instance* of that layer. For example, in Figure 12-6 there is one instance of the physical level for each port and there is more than one port. Each instance of the physical level serves an instance of the data link level. If there is more than one application running at the same time, then there are that number of instances of all the levels above the third. Usually there is only one instance of the third *network* layer in each node.

Thus, the number of instances can split and merge as one goes up the protocol stack. The single instance of the network level can be supported by as many instances of data link and physical function as there are physical ports. Above, the network level can support as many instances of higher layers as there are applications. In a large node such as a mainframe or large server the number of active application instances can be in the thousands; for a desktop personal computer the number is usually less than ten.

The treatment of layers we have just given sets the stage for consideration of the many options for the content within each layer, an extensive topic that is well covered elsewhere [3-5] We shall use only a small part of this extensive body of information, that having to do with the bottom three layers, and will wait to do so until the next chapter.

12.4 Network Control

It is quite possible to adopt a completely different viewpoint. We can focus more on the network-control functions of activation, deactivation, and parameter setting [7], while still keeping in mind a conceptual subdivision into layers.

Returning to Figure 12-3, it will be noted that something called the *control point* is shown supporting the function of all the communication layers. What the control point does can best be described by going to the time domain and distinguishing between *transient* and *steady-state* exchanges of commands, responses, parameters and data. We can usually associate the transient communication with network-control messages and the steady-state with transmission of user data. The steady-state functions are those that take place after the access path has been set up and is being used for the useful exchange of data between users. They were enumerated grossly in the layer labels of Figure 12-6.

If the access path has yet to be set up or is to be taken down, then the control point must activate or deactivate the FSMs that execute the protocols in the various layers, supplying them with the proper parameters or setting the proper tables. To pick

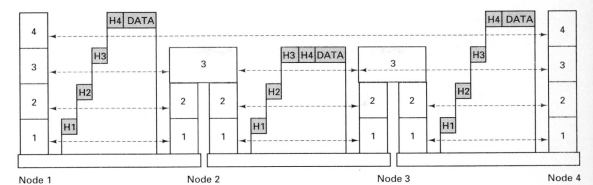

Figure 12-6. Layers in the four nodes of the previous example, using the OSI terminology. Horizontal arrows indicate the peer protocols. The zigzag line shows the bitstream (including the headers) that would actually be observed on the link while node 1 is sending to node 2 and vice versa. Note the presence of more than one instance of some of the layer functions.

just one illustrative example of these parameters for each layer, there are phone numbers for the physical level, link addresses for the data-link layer, routing tables for the network layer, class of connection information and flow control limits for the transport layer, session parameters such as passwords for the session layer, and crypto keys for the presentation layer.

In setting up and taking down an access path, all the transient activation functions presumably need to be done only once and the deactivation done once, so they are not as time-critical as the actual steady-state delivery of bits over the access path. In some architectural descriptions, such as OSI, the activations and deactivations are shown rippling sequentially through the protocol stack. Specifically, in OSI, the activation of a session is specified as emanating in the application layer, being passed down to the transport layer, which in turn gets the network layer to look into its routing table and set up a series of network, data-link, and physical connections, and so forth down the protocol stack, across the network, and back up again. For each of these layer-to-layer interactions across the interface in question there is an exchange of commands and responses. In a real network, this is not an effective way to handle network control, and instead all the network-control function in a node can be considered to be gathered up into one control point, as shown in Figure 12-3. This control point has hooks into the different layers by which it activates and deactivates them itself instead of letting them do so to each other. The control points in different nodes communicate directly with each other, usually over some extremely terse and efficient flow at the transport interface, such as that provided by a special datagram transport interface dedicated to this purpose.

Since the transient network-control functions of activation, deactivation, and parameter passing are less time-critical than the steady-state functions, most have usually implemented today in some higher-level language and executed as an application using the execution cycles of the node's central processing unit (CPU). The steady-state transmission functions, however, are especially time-critical, and will become even more so as faster and faster fiber optic technology is introduced at the physical level. As physical link speed has increased over the years, most of the physical-layer and data-link layer steady-state functions have been implemented outside the operating system kernel on special boards or chips, because of this time-criticality. As faster steady-state function is achieved, there is corresponding pressure to speed up the transient network-control function by such measures as outboarding.

After delineating what network control must accomplish, it will be interesting to speculate later in this chapter about how the network-control function might evolve so as to best take advantage of fiber optic technology.

The operation of the control point can be broken up into phases. The ordering of these phases varies today from one implementation to another. For example, in one architecture, routing tables are computed and loaded as almost the first step in using the network, while in some other architectures, this step is left until just before a connection between two end users is to be put in place. Therefore, while the steps in the ordered list to be presented next happen to occur in that order in one architecture, IBM's Advanced Peer-to-Peer Networking (APPN) [7], it's new version of SNA, they occur in a completely different sequence in others.

Consider the physical-network topology example of Figure 12-2. Imagine that the person at node 1 (running an application named CLIENT) wants to access the database program in node 4, named SERVER. Suppose further that node 1 isn't even connected to the preexisting network that happens to consist of nodes 2, 3 and 4, and even if it were, is in total ignorance of what the physical-network topology looks like, where SERVER is, or even whether there is such an application. The phases of network control that eventually set up a session (access path) between CLIENT and SERVER are, at least in APPN:

- **Connectivity services**. One or more links are activated, e.g., to nodes 2 and 3, and the revised topology of the network is discovered by all nodes by exchanging probe messages around the network.

- **Directory services**. Again by probing, it is learned that there is only one copy of SERVER and it is in node 4.

- **Route selection services**. Using the knowledge of topology, a good route to 4 is calculated at 1, in this case by way of 2 and then 3.

- **Session services**. A request for an access path (session), as shown in the dashed line in Figure 12-2, is issued on behalf of the relevant application CLIENT at 1 and responded to on behalf of the relevant application SERVER at 4.

If all this is successful, the transient network-control-activation phase ends and the steady-state data-transmission phase begins. The control point gets out of the way of these more time-critical processes and only gets reinvolved if a deactivation of the access path is requested by CLIENT or SERVER or if there is some error condition

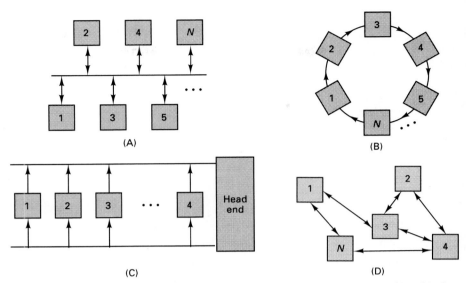

Figure 12-7. Four widely used LAN, MAN and WAN topologies: (A) Bus, (B) Double bus, (C) Ring, and (D) Mesh. (A), (B) and (C) are LAN and MAN topologies, while (D) is a WAN topology.

from which the steady-state link layer or transport-layer error-handling protocols cannot recover.

All this may sound straightforward enough, but can become quite complex in practice. In APPN, the four phases are handled by exchange of network-control messages using fail-safe protocols designed to require a minimum of such message overhead. In other architectures, much of the information can be prestored, thus saving on message overhead at the expense of flexibility. Also, the order of the phases can be significantly different.

Thus we see that, for the steady-state functions, it is appropriate to think in terms of layers, while for network control, it is more revealing to forget about the individual layers and think in terms of the overall functions that must be set up.

12.5 The Communication Layers of Classical Networks

How the lower layers evolved

The layer structure undergoes an interesting break point at geographical distances smaller than a few kilometers, i.e., the break point between LANs and MANs on the one hand and WANs on the other. One of the things that changes at this breakpoint is the physical network topology. For LAN and MAN distances the nodes are arranged along a bus (as with Ethernet), double bus (as with DQDB), or ring (as with token

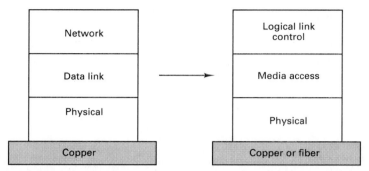

Figure 12-8. Evolution of (A) the lowest three layers of WANs into (B) the three lowest layers of LANs and MANs.

rings), as shown in Figures 12-7(A), (B), and (C), respectively. For WANs, the arrangement is more likely to be a mesh, as shown at (D).

The layered architectures of WANs evolved several years before LANs and MANs were introduced. With the WAN of Figure 12-7(D), a routing decision must be made on every hop, and a fairly rich routing (network) layer therefore exists, as shown in Figure 12-8(A). On the other hand, the connection function of the physical layer is very simple, that of making a leased or dial-up connection (voice-grade phone line, ISDN connection, and so forth). ISDN is the Integrated Services Data Network of the world's telephone companies.

The converse is true with the later LAN and MAN architectures, as shown in Figure 12-8(B). For such networks, there is no routing decision to be made, because either every node hears every other node, since they are only one hop from one another, as with (A) and (B) in Figure 12-7, or there is only one exit port, as in (C). In both these cases, when one node has a message for another, it seizes some of the capacity of the network and addresses a message to another directly using some *multi-access protocol*. Therefore the third (routing) layer is essentially absent for LANs and MANs, but, since the medium is shared, the setting up of a physical connection, which is the function of the multiaccess protocol, becomes quite complex. This is reflected in the separated MAC (multiple-access-control) layer shown in Figure 12-8(B). The Logical Link Control layer (LLC) takes the place of DLC.

The original OSI seven-layer model of Figure 12-5 is based on WAN architectures. For the more modern LANs and MANs, Figure 12-8(B) is a better model for the bottom three layers. Today's LANs and MANs [3, 5, 8] are similar to those all-optical networks that have been investigated to date, since the latter are all based on star couplers or busses.

As interconnection of separate all-optical LANs and MANs proceeds, the routing layer reappears.

Preview of the MAC layer

The MAC layer is of particular importance in lightwave networks, since it is so directly influenced by the technologies employed. We shall devote the entire next chapter to this topic. There we shall roughly quantify the performance deficiencies of two of the most widely used classical multiaccess protocols, and then discuss specific protocols for third-generation optical networks. Here, we give some background on multiaccess techniques as preparation. We introduce the first- and second-generation MAC-layer protocols as they appear in the context of the total set of layers and the control point.

The *shared-medium* network topology of Figure 12-8(A) and (B) has a long history, going back to telephone party lines. In computer communication it has been common for years to attach up to several dozen terminals or controllers in a *multidrop* configuration along leased voice-grade telephone lines, for example between banking terminals in different branches in the same city. In the multidrop architecture, one node is the primary, and each secondary can send to or receive from only that primary. The multiaccess protocol is usually one involving *polling*, in which the primary solicits messages from the secondary in some serial order. This was all handled in the DLC layer. The international standard is called High-Level Data Link Control (HDLC).

Later, *bus-oriented LANs*, such as those that use one of the large family of *contention* multiaccess protocols or the *controlled-access* token bus, have become widespread.   In LANs and MANs using contention or tokens, all nodes are peers, as distinguished from the master-slave pattern of the earlier multidrop systems.

All the many contention multiaccess protocols, of which the Ethernet LAN is the most famous, are basically derived from the original *Aloha* protocol. In Aloha, a node wanting to transmit to another does so, and if there is a *collision* (the message arrives garbled because it overlapped in time with someone else's attempt to use the shared medium) each node colliding eventually learns that fact and tries again later. Ethernet uses the *carrier-sense multiaccess with collision detection (CSMA/CD)* protocol, which we shall discuss in the next chapter.

In token busses, the right of access is controlled by preventing any node from transmitting until it has been passed a special bit pattern ("token") by some other node after the latter has finished using the shared medium. The token is passed in some fixed round-robin order (for example, in increasing order of the value of each node's address), so that access is "controlled" rather than being contended for. One can think of the node's passing the token to one another in some circular order so that the nodes are arranged in a logical ring while being physically connected in a bus topology.

These bus protocols were found to have severe performance deficiencies as users began to insist on longer distances, higher bitrates, and more nodes. The reason is that the number of bit durations taken for a signal to travel the length of the network began to consume so large an overhead, compared to the length of a packet, that the fraction of time that the medium was transferring useful bits became intolerably low. For multidrop systems, it took too long for the primary to receive an acknowledgement from the secondary and for a secondary with a message to send to be visited in the polling sequence.   For contention protocols such as Ethernet, it took a long time to sense whether someone else was using the medium, or to find out if one's own packet was going to suffer a collision, even when one could do the sensing while transmitting at the same time.   For token buses, it took a long time to pass the token up and down the bus.

This led to the development of the *token ring* LAN and MAN.   In these networks the nodes are ordered physically around a ring, rather than in some logical order on a bus.   Not only is access controlled rather than chaotic as with contention (thus gaining security, fault isolation, and maintainability advantages over contention protocols), but also when the load builds up there is more efficient usage of the medium.

In principle, even token rings are much less efficient than optical networks using wavelength-division multiple access or space-division switching, because in principle all nodes in the latter can be kept busy most of the time receiving messages over different wavelengths or different physical paths, respectively.   For token rings, there is only one node sending at any one time; all others with messages to send are waiting.

MAC layer standards

As we have seen, the introduction of LANs and MANs caused the older definition of the bottom layers to change in the way shown in Figure 12-8.   The physical layer split into two pieces, the one that makes the connection (media-access control – MAC) and the one that is specific to the technology used (physical).   In earlier protocols they were merged into one physical layer.   For example, RS-232, the familiar multiwire physical protocol used with every phone-line modem, specified the service interface (a 25-pin plug) between the physical-layer modem and the higher level.   It also defined the peer protocol as the usual sequences of verbs and nouns exchanged with the other end, in this case the other modem.   But it also went on to specify voltages and impedances for the service interface between the modem and the copper wires of the telephone system.   The practice today is to split all this into the MAC layer (which says nothing about the technology used for transmission of signals) and the physical level (which does).

To date there are a number of standards [9] for the MAC layer that the reader is quite likely to encounter:

- IEEE 802.3 for Ethernet-like contention protocols, termed CSMA/CD (carrier sense with collision detection)

- IEEE 802.4 for token bus protocols such as used in manufacturing-plant-floor applications

- IEEE 802.5 for IBM token ring-like protocols

- IEEE 802.6 for a MAN protocol using the DQDB (Distributed Queue Dual Bus) protocol

- ANSI (American National Standards Institute) X3T9.5 FDDI (Fiber Digital Data Interface) LAN/MAN protocol

- ANSI X3T9.3 Fiber Channel Standard, for serial interconnection of computers to storage peripherals using space-division switching. This is to be the serial replacement for the all-copper point-to-point HIPPI (High Performance Parallel Interface), widely used in the supercomputer world. This standardization effort is quite new and is particularly relevant in the present context because it is the first MAC-layer standard to propose providing gigabit-per-node service and multigigabit total network throughput using fiber.

There is also

- IEEE 802.1 for the control point, and

- IEEE 802.2 for logical link control.

In the process of LAN development and standardization, the data-link control function was renamed the logical-link control function Since the LAN node now has only one physical port per node (the two of each token ring node being lumped into one), there is only one instance of LLC per node. LLC does the following functions of earlier DLC protocols: (1) recovering from errors by error detection and retransmission, and (2) flow control on the link by returning negative acknowledgements from the receiving end whenever the transmitter should withhold the next packet. In addition, several new functions are added: (1) providing an addressing function so that several instances of higher levels can sit on top of the single instance of LLC, and (2) for each of these instances, providing one of two classes of service, a connectionless "datagram" class, and a connection-oriented class of sequenced error-free packets.

This list of four functions that LLC performs may be interpreted to mean that LLC provides a primitive transport service interface, as discussed in the concluding paragraphs of Section 12-2. The datagram class is represented, and the sequenced packet class is represented. This does not mean that the fourth layer is absent in one-hop LANs and MANs, since there are many verbs that must be supported in such service interfaces as that of TCP or OSI Level 4 that are not present in LLC. However, it does serve to show, once again, how much simpler things get when all nodes are only one hop away from each other.

12.6 Lightweight Transport Protocols

As fiber optic links became more widespread in the early 1980s, and the trends illustrated in Figure 1-2 came more clearly into view, it became evident that all the old expectations about computer performance evolving faster than communication performance were no longer valid. In response, a growing number of efforts have taken place to shorten the *software pathlength* of the communication layers by devising *lightweight protocols*, that is, lower-layer functions that can be completed quickly with just a few machine instructions.

Communication pathlength is defined as the number of instructions that must be executed in forwarding a single incoming packet from the arrival of its leading edge at the bottom of the physical layer to the transmission of its trailing edge upward across the transport interface. Figure 12-9 shows the progress that will be necessary in keeping the pathlength short in order for performance of future applications not to degrade with respect to present-day performance. A representative communication pathlength for large computers in 1990 was 3000 instructions. Even with such a seemingly large pathlength, a typical such machine spent only one percent of its time executing communication code, even when each attached link was 100 percent utilized. Figure 1-2 shows that communication-link bitrates are now increasing by a factor of 100 every eight years, while computer execution rates continue to increase by a factor of 10 every ten years. From these two rates, plus the 3000-instruction figure, we can draw the curve of Figure 12-9 that shows very dramatically what is needed from the lightweight protocols of the future in order that less that one percent of a single processor's cycles will be spent executing communication code.

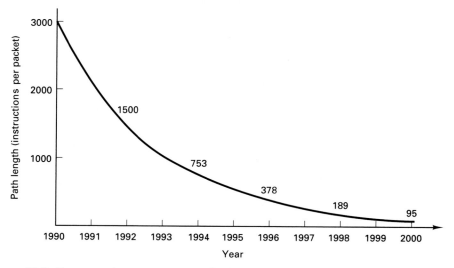

Figure 12-9. Future requirement on communication software path length.

A number of these lightweight protocols have emerged as second-generation lightwave networks have proliferated [10, 11], and work continues in improving them. Some of these efforts involve completely changing the content of the communication protocol layers themselves; others involve taking well-established protocols, especially TCP/IP (Transmission Control Protocol/Internet Protocol) [3] and improving the execution environment. We shall touch on some of these improvements in the next section, since these measures will be even more necessary in order to realize the performance benefits of third-generation networks.

12.7 Lower Layers of Third-Generation LANs and MANs

Given that third-generation LANs or MANs possess at least the architectural simplifications of the one-hop classical networks just discussed, what additional simplifications does the technology allow? This question is just beginning to be answered and there is no existing body of knowledge on the subject. The ideas that follow are therefore speculative.

As emphasized earlier, the principal physical-level factors are bandwidth, error rate, and propagation latency. A ten-order-of magnitude increase in both bandwidth and infrequency of errors is offered by lightwave technology, but (barring a repeal of the laws of physics) there is not much to be done about the latency except to hope to be clever enough that the delay seen by the end users is pared down almost to this irreducible minimum.

If wavelength division is used, the physical layer of a lightwave network depends on whether tunability is at the transmitter, the receiver, or both. If tunability is at the receiver only, then there is an electrical service interface at the transmitting member of the peer pair such that when a sequence of bits is presented to this interface, the other receiving member of the peer pair sees light at some wavelength owned by the transmitting station. The peer-to-peer interaction is simply light transmission. At the other end, the service interface of the physical layer will be some electrical port that accepts the bits or voltage that controls the tuning, and another that delivers the bitstream. Analogous statements could be made for other tunability choices, and for cases for which the mode of transmission is not wavelength division but composite wavelength-subcarrier transmission, code division, or space division.

The media access layer changes completely, a topic meriting a separate chapter, Chapter 13.

The LLC layer is significantly affected. The principal difference is the potential for almost complete absence of a requirement for error recovery. Today's coax and copper-pair transmission produces such high error rates that the errors have to be handled by using some redundancy at the transmitter (in the form of a *cyclic redundancy check* field, usually 16 or 32 bits long) and error detection at the receiver, so that the receiver can solicit from the transmitter a retransmission of any packet found to be in error. Even with this protection, the probability of the error-checking procedure failing to detect an error is not zero.

It seems reasonable to expect that third-generation systems will exhibit a "bimodal" distribution of bit error rates: when the system is being set up or is in a problem condition, error rates will be high, but when the system is operating normally, error rates will be as low as 10^{-15}, or perhaps even lower.

The very low bit error rate has some important consequences in the third and fourth protocol layers. They have to do with the old argument between packet switching and circuit switching and whether packet switching always had to mean short packets, e.g., less than a thousand bytes. There seem to be three reasons for using packet switching rather than circuit switching:

- **Economic savings**, because the statistical interleaving of many traffic streams utilizes the link capacity more completely. Today, this argument is only of historic interest and obviously goes away for fiber optic media.

- **The data streams needed to be broken up into packets anyhow**, because the bitrate was low enough, the bit error rate high enough, and the propagation latency high enough that one could not make the packets much longer than, say, one thousand bytes. If the packets were much larger, the probability of packet error would become excessive, as would the overhead in time to retransmit and in buffer resources required at the transmitter to save a copy of the packet until successful receipt had been confirmed.

- Packet switching provides **many logical ports per physical port**. One physical stream of packets can support many concurrent *virtual circuits*, logical connections to many other addresses in the network.

Of these motivations for packet switching, only the third is relevant in a lightwave network, but it is an important one. The time-slicing of a physical connection to handle many logical connections in a way that appears to be simultaneous to each using application is exactly analogous to the way the processing power of a computer is time-sliced to handle all the different applications, systems tasks, and input-output, including the various layers of communication processing.

However, the difference between choice of packet length in the past and that with fiber optic links is that the duration of packets need not be constrained by the communication variables, in particular the error rate and bandwidth. It is no longer necessary to have the short packet sizes that were dictated formerly by 10^{-5} to 10^{-8} bit error probabilities of voice-grade phone-line connections. The packet size can now be matched entirely to the application, and different application instances in a given node can potentially use different packet sizes. If the application is record-by-record file transfer, the packet length provided by the transport interface to that application instance can be that of the record, perhaps only a few hundred bytes. At the other end of the spectrum, if the application is to support an uninterrupted file transfer or a full-screen high-resolution workstation with continuous video refresh, the streaming type of service class can be provided (always subject to the need to make each physical port available to more than one application). For integrity purposes the application must be willing to take responsibility for error control and flow control or to dispense with them.

Today, when one physical port must be shared across a number of applications, a given buffer must be shared by several applications. This means that, even if the flow control carried out by the higher levels completely avoids per-application buffer overflow, this is not enough to guarantee that per-port buffer overflow will not occur. Therefore, the ideal situation in which flow control is relegated entirely to each application is not really feasible, short of the utopian situation of one physical port per logical port (one physical port per application instance).

Estimates for the number of logical ports per node needed in future nodes range from less than ten for small desktop workstations to thousands for large mainframes. As seen from Part II, the current cost levels are not yet so low that it will be easy to provide many physical ports per node using today's lightwave components, even where the fiber provides more than enough bandwidth to serve all the ports in the network. However, it is possible that as costs drop, the number of physical ports per node could grow until it actually becomes as large as the number of logical ports (communication-based application instances). The distinction between packet and circuit switching would then largely vanish.

In the past it was bandwidth as well as cost that prevented the realization of this ideal, but with lightwave, bandwidth is not an impediment. The rate at which any one computer can handle bits is certainly unlikely to soon approach the rate at which one fiber can transfer bits (10,000 Gb/s). More relevantly, when one considers that individual LANs or MANs consisting of more than several hundred nodes have proved difficult to administer and troubleshoot, if one reinterprets the capacity projections of Figure 1-7 in terms of number of ports per network rather than number of nodes per network, there is still ample fiber optic network bandwidth to support several physical ports per node.

From the protocol layering point of view, this means that, if future economics ever permit it, there can be one protocol stack per application. This is to be compared with today's situation in which the number of instances of the physical and data-link level (i.e., the number of physical ports) is one or two orders of magnitude smaller than the number of instances of the higher layers. In the future, all error recovery can be done in a way that matches the requirements of an individual application type; if little is required, as with display support, then little is implemented. Flow control to prevent buffer overflow is now much simplified, since it may now take place entirely on behalf of the needs of individual applications. In today's packet-switching networks, buffers in the lower layers are not dedicated to individual applications but are instead dedicated to individual links that are shared across all applications.

12.8 Network Control for Lightwave Networks

By the same token that the communication layers (Layers 1 through 4) are greatly simplified in one-hop networks, particularly lightwave one-hop networks, network control, too, can be simplified.

There is no intrinsic reason that the four network-control stages discussed in Section 12.4 (connection, directory, route selection, and session binding) cannot be collapsed into a single exchange. Instead of the application issuing one verb requesting a session and then having many control points successively manage a series of time-consuming exchanges that accomplish the four phases separately, the following simplification should be possible for the example treated in Section 12.4:

- CLIENT issues a request across the transport interface (of the appropriate class) for a session with the resource named SERVER somewhere in a network of which CLIENT's node may not yet even be a member.

- The control point of CLIENT's node causes the node to join the network. In the special case in which the MAC protocol does not require *a priori* knowledge of wavelengths (as with the code-division or "circular-search" protocols to be discussed in Chapter 13), no prior coordination is required; the control point simply picks some unused wavelength or code-division waveform, respectively, and connects.

- The request from CLIENT is sent into the network in the form of a session-bind request.

- Every control point in the network hears this request, and if any node contains SERVER, that control point commands completion of the MAC protocol connecting CLIENT's node to SERVER's node. (If there is more than one instance of SERVER, a suitable priority scheme picks one).

- SERVER's control point also completes the LLC connection to provide SERVER with the correct class of transport service interface. The access path from CLIENT to SERVER is now complete.

- A session bind-acknowledgement is returned across the same access path from SERVER to CLIENT.

- Steady-state function now commences.

As single-hop lightwave networks evolve, it might be expected that network control will be managed in this way.

12.9 What We Have Learned

This chapter presented a tutorial introduction to two schemes of organizing all the complex functions that are executed in the software and hardware of a computer network: the traditional layer approach and the network-control approach. Both ways of looking at things are at work in any real network.

The software and hardware functions are built on a technology base composed of computer technology (memory and logic) and communication technology (transmission and switching). Different pieces of these technologies are invoked in rapid serial order by the operating system of each node so that the many things in the network appear to be going on simultaneously, whereas inside the node they are rapidly time-sliced.

With further evolution of networks that are based on the true capabilities of photonic technology, network architects will be able to exploit the ten-orders-of-magnitude improvement in error rate and bitrate to achieve many new capabilities. One improvement that was emphasized in this chapter was the idea of making the communication technology as invisible as possible. Traditionally, the layers below the transport level allowed a number of communication idiosyncracies show through to the higher levels: error rate, bitrate, and propagation latency. In many cases, the application had to adapt to these idiosyncracies. In the architecture of future third-generation networks, it should be possible to provide completely user-driven transport interfaces, with the communication technology showing through only to the extent that the latency is only slightly larger than the natural propagation time.

12.10 Problems

12.1 Which of these ideas, if any, is an example of the abstraction (or virtualization or layering) that occurs with communication architectures: (A) The musicologist Heinrich Schenker's method of analyzing long-range effects in music looked for "pivotal points" in a given piece. First he reduced the movement into very short sections between pivotal points, each of which might be thought of as "sentences," and then did another reduction to form "paragraphs." He continued this process until the entire piece was represented by what looks like a sequence of a dozen or less very long notes. (B) The Interstate Highway System in the United States serves and is served by other Federal highways, which are in turn served by state highways, county highways, city streets and ultimately by individual driveways. (C) The linguist Noam Chomsky analyzed language into layers, the superficial one we use, its syntax (the rules by which we say it), a layer of semantics in turn that captures what we mean, and so forth.

12.2 Which of the following functions would you assign to (1) the operating system, (2) the control point, (3) the upper communication layers, or (4) the lower communication layers: (A) Log-on to a remote database server application. (B) Making sure, using flow-control messages that the spreadsheet program in your small node did not get overdriven by the large high-speed database server. (C) Processing keyboard input to your spreadsheet program. (D) Making sure, using flow-control messages, that the fiber-optic link does not overdrive packet buffers that are part of your communication driver.

12.3 Briefly discuss what you consider the relative advantages of datagrams and virtual circuits with respect to: (A) Time delay before data starts flowing, (B) Reliability and integrity, and (C) Processing load.

12.4 If the fifth layer that lives on top of the four shown in Figure 12-6 uses a header, which nodes would touch the bits in this header, and where within a typical packet would the header normally be inserted?

12.5 DEC's Digital Network Architecture (DNA) sets up a flow of packets in the following sequence: Knowing an address in a directory, a node sends a datagram-style message to the desired partner node, and this message sets up a virtual circuit between the two. Each succeeding packet flowing on the virtual circuit is routed "on the fly" at intermediate nodes using traffic-dependent routing tables. Compare the ordering of network control phases with the ordering given for APPN in Section 12.4.

12.6 Discuss the relative advantages and disadvantages of the topologies of Figure 5-7 from the message security point of view.

12.7 Suppose one were able to halve the communication code that is assumed to execute on the node's processor in Figure 12-9 by executing it elsewhere, say, in a RISC (reduced instruction set computer) chip on the communication line adaptor. How short would the pathlength of the communication software have to be in 1996 in order to keep the processor busy less than one percent of the time executing communication code?

12.11 References

1.   A. S. Tanenbaum, *Structured Computer Organization*, Prentice Hall, 1976.
2.   J. S. Auerbach, "TACT, a protocol conversion toolkit," *IEEE Jour. Selected Areas in Commun.*, vol. 8, no. 1, pp. 143-159, January, 1990.
3.   A. S. Tanenbaum, *Computer Networks - Second Edition*, Prentice Hall, 1988.
4.   P. E. Green, Jr., ed., *Computer Network Architectures and Protocols*, Plenum, 1981.
5.   J. D. Spragins, J. L. Hammond, and K. Pawlikowski, *Telecommunications: Protocols and Design*, Addison-Wesley, 1991.
6.   CCITT, OSI Reference Model, Commite Consultatif International de Telefonie et Telegraphie, 1981.
7.   A. E. Baratz, J. P. Gray, P. E. Green Jr., J. M. Jaffe, and D. P. Pozefsky, "SNA networks of small systems," *IEEE Jour. Sel. Areas in Commun.*, vol. 3, no. 3, pp. 416-426, May, 1985.
8.   D. Bertsekas and R. G. Gallager, *Data Networks - Second Edition*, Prentice Hall, 1987.
9.   D. C. Hanson, "Progress in fiber optic LAN and MAN standards," *IEEE LCS Magazine*, vol. 1, no. 2, pp. 17-25, May, 1990.
10. W. A. Doeringer, D. Dykeman, M. Kaiserswerth, B. W. Meister, H. Rudin, and R. Williamson, "A survey of light-weight transport protocols for high-speed networks," *IEEE Trans. on Commun.*, vol. 38, no. 11, pp. 2025-2039, November, 1990.
11. H. Rudin and R. Williamson, eds., "Special Issue on High Speed Network Protocols," *IEEE Commun. Magazine*, vol. 27, no. 6, pp. 10-53, June, 1989.

CHAPTER 13

Multiaccess, Switching and Performance

13.1 Overview

Earlier chapters have introduced two main themes about lightwave communication systems: the technical evolution from first generation (pre-fiber optics) to second generation (business as usual using fiber to replace copper) to third generation (systems based directly on fiber properties), and a progression in complexity from simple links to multipoints to networks. It is time now to focus on the processes that effect the any-to-any connectivity that makes a network a network, and to discuss how these processes differ in the third generation from their predecessors.

Chapter 11 talked about physical resources and introduced various physical-network topologies that offered various degrees of *scalability* to very large numbers of nodes, the *modularity* required to add just one node easily, and the support of *irregularity* in the physical location of users. We shall now turn from the less time-sensitive configuration or reconfiguration of link and node physical resources into some pattern to those fast-acting protocols required to set up the access path which was the subject of Chapter 12. At any given instant, a snapshot of the physical routing of all access paths constitutes the physical-route topology and a snapshot of the set of associations between end users constitutes the logical topology.

As we saw in Section 12.5, a network that requires routing decisions at each node has a functioning third (routing) layer in each node. A wide-area network (WAN), such as that of Figure 12-7(D), would be an example. Another example would be the interconnection of individual networks in which routing decisions are required at each gateway node. As Figure 12-8 indicates, such networks do not have a

417

multiaccess (MAC) layer, but the third layer is occupied instead. For LANs and MANs, there is usually no third layer, but instead a significant MAC layer.

Three of the third-generation architectures that we discussed in Chapter 11 include intermediate-routing nodes, and so there is an implied third layer in each. These are wavelength routing (Section 11-7), hybrid wavelength-space division (Section 11-8) and linear lightwave networks (Section 11-9). In the case of wavelength routing the routing decisions are made implicitly at origin and destination nodes only, since the wavelength-dependent routing is frozen in for all the intermediate nodes.

To date, little thought has been given to the exact content of the third layer for the intermediate nodes of wavelength-space division hybrids and wavelength routing or for the end nodes of linear lightwave networks. The same is true of the gateways in network interconnection discussed in Section 11-13.

Therefore, our focus in this chapter will not be on networks containing the third layer, but instead on the directly-connected physical topologies (either busses, stars or trees) appropriate to those third-generation optical LANs and MANs that are under active consideration and prototyping today.

We shall first explore what the requirements are on MAC protocols, classify them (for example, according to circuit- or packet-switched service), then discuss circuit-switched, then packet-switched multiaccess protocols, and finally optical CDMA and fixed-assignment TDMA.

13.2 Requirements on Multiaccess Protocols

In deciding which protocol to use, we must decide which requirements to try to meet, but these requirements are often mutually conflicting. For example, in designing the network to support some particular set of applications, it might be desired to accommodate the widest possible range of bit rates and framing conventions, but the choice of circuit switching to do this might come into conflict with the desire to support many concurrent connections from the same physical port, which packet switching is designed to do.

The desired logical topology could be the set of single point-to-point access paths (Figure 11-1(B)) that is most often the case for networks, or it could be a one-to-many multicast logical topology. We shall not deal specifically with this latter case, assuming that a number of individual access paths could be set up for the purpose, using the multiaccess protocols to be described.

We start by listing the principal requirements that these protocols must meet so that later in the chapter we may understand the relative advantages and disadvantages of the candidate access protocols for third-generation lightwave networks.

- **Delay** of two kinds must be considered, that required to set up the access path, and that which occurs in sending traffic over the path after it is set up. The latter delay will be defined as the time elapsing between the instant that the beginning of a transmission (for example, the leading edge of a packet) enters the transmitting end physical protocol level from the higher layer user issuing it and the instant that it leaves the receiving end physical level on its way to the user at that end.

Propagation delay usually plays the dominant role. Whether a given bit is a network control bit or a traffic bit, we would rather have it arrive usefully at the receiver instead of having the propagation medium serve as a place to buffer large amounts of information in transit. Yet, this is exactly what happens at gigabit rates where one bit time represents about eight inches of single-mode fiber propagation distance.

- **Throughput** is the rate at which payload information bits can be delivered over the access path, once it is is set up. We shall be interested in the maximum throughput that a given protocol can provide. In third-generation lightwave networks, the bandwidth is so large that throughput is often less of an issue than delay.

  Note that being able to send at gigabit rates helps speed up the rate at which bits enter the communication "pipe," but does not get those bits to the other end any faster. This is analogous to moving dirt on a construction job from one place to another by doing it in a large truck rather than a wheelbarrow at a time; the truck might not travel any faster than the wheelbarrow, but the user sees a much greater throughput with the truck.

- **Fairness** is the extent to which the various users have equal-priority access to the communication resources. In the absence of some deliberate *priority-class* distinction between users, the protocol should work in such a way that all users at all nodes are equally favored.

- **Reachability** means flexibility to access a dynamically changeable subset of a very large population of other users. The telephone network is the ultimate example. Although each subscriber seldom makes calls outside a group of "frequently-called numbers," this list is freely chosen from millions of available other subscribers worldwide and constantly evolves for each user.

- **Tell-and-go**. Ideally, the protocol should require essentially no preparatory information and no preparatory coordination in order to start communicating with a partner. The sending node should be able to inform the receiver that a transmission to it follows immediately and then send that transmission. This tell-and-go capability [1] is an important ease-of-use parameter and can become particularly important for packet-switch situations when the round-trip propagation latency is much longer than the packet transmission time.

- **Support of heterogeneity**. While the protocols of single links and multipoints have been fairly well standardized worldwide, because they originated mostly in the common-carrier community, most network architectures are mutually incompatible, having originated separately and competitively within the rapidly evolving computer industry. The heterogeneity of computer architectures touches every level of the protocol stack, even down to such simple physical level questions as bit rate and framing convention.

There are three approaches [2, 3] for dealing with this "protocol zoo" problem: (1) standardization on one protocol stack (for example, OSI), (2) construction of protocol-converting gateways [4] between incompatible subnetworks, and (3) making the access path *protocol transparent* in the first place, i.e., insensitive to the users' choices of bitrates, framing conventions, or other details of the protocol stack. This last feature is one of the things that have made copper telephone lines and WDM lightwave networks so useful.

13.3 Classification of Multiaccess Protocols

WDMA (FDMA), TDMA and CDMA

A distinction can be made between space division (one physical link for each access path) on the one hand and time division, frequency division, and code division on the other. In the latter three cases, each physical link can carry a number of access paths concurrently, and therefore these access paths must be distinguished from one another. We shall ignore space division and concentrate on the other three.

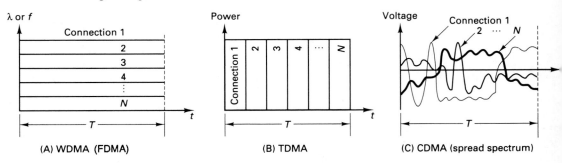

Figure 13-1. Three general classes of multiacces protocols: (A) Wavelength (frequency) division (B) Time division (C) Code division (spread spectrum).

Figure 13-1 shows at the top some ways the capacity of the medium can be subdivided according to different frequencies, different time slots, or different waveforms, respectively. Multiaccess protocols based on these three approaches are called wavelength- (or frequency-) division multiaccess (*WDMA*, often called *FDMA*), time-division multiaccess (*TDMA*), and code-division multiaccess (*CDMA*), respectively. The idea is that the signals agreed upon by the two communicating partners at the physical level are distinguishable from those used by other connections by virtue of occupying different parts of the wavelength (frequency) domain, different parts of the time domain, or by having suitably different waveshapes.

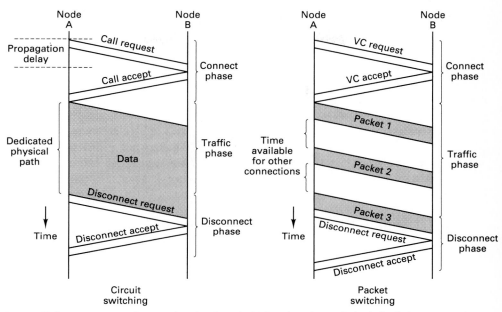

Figure 13-2. Time-space diagram for circuit switched and packet switched single-hop networks, such as LANs.

Although all three multiaccess classes are widely used in first- and second-generation optical networks, TDMA is the most prevalent. Variants within all three classes are being studied for third-generation optical networks. For TDMA and CDMA, one thing that must be taken into account is the fact that the speed of the electronic and photonic components must be much higher than the bit rate of one connection, whereas this is not true for WDMA. Referring to Figure 13-1(B), we see that with TDMA one bit time must be broken up into small subintervals, and then one connection can use this subinterval for its own information bit, allowing the rest of the bit time for interleaving of bits of the other connections. A CDMA connection (Figure 13-1(C)) requires that its two binary modulation waveforms be distinguishable not only from each other, but from those used by other connections   These two things can be done only by creating complex waveforms that execute many changes of value during one bit duration, as in Figure 13-1(C). These values could be two in number or a continuum of waveform values.

Circuit switching, packet switching and protocol transparency

We can also classify access protocols into *circuit-switched* and *packet-switched* protocols. From the WDM technology point of view, they are distinguished by the requirements on retuning time.

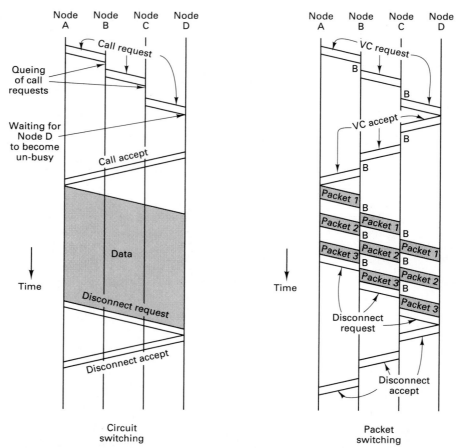

Figure 13-3. Time-space diagram for multiple-hop circuit switched and packet switched networks, such as WANS. (B = buffering delay).

Figures 13-2 and 13-3 illustrate the difference, both for a one-hop LAN or MAN network and a WAN involving a cascade of hops, respectively. In circuit switching the end-to-end physical path is established in the *connect* phase, user payload data is sent during the the *traffic* phase, and the circuit is disestablished during the *disconnect* phase. In circuit switching, as used traditionally with telephony, the access path is a real physical path that is protocol transparent during the traffic phase. Thus, a phone line can support analog telephones (*POTS*–"plain old telephone service"), modems, FAX, security alarms, etc.

With packet switching, there are usually the same connection, traffic, and disconnect phases as with circuit switching, but what is set up is not a real physical circuit but a *virtual circuit* consisting of a stream of associated packets that are all addressed to the same destination. Whereas in circuit switching the physical circuit was dedicated during the traffic interval to the designated pair of users only, in packet switching each link may be shared across many concurrent connections going to different places. Some of these "connections" can be only one packet long, a *datagram*. Such timesharing, whether using datagrams or virtual circuits, requires that packets be carefully identified (by suitable fields in the header of each packet), and for this reason packet switching is not as protocol transparent as is circuit switching during its traffic interval.

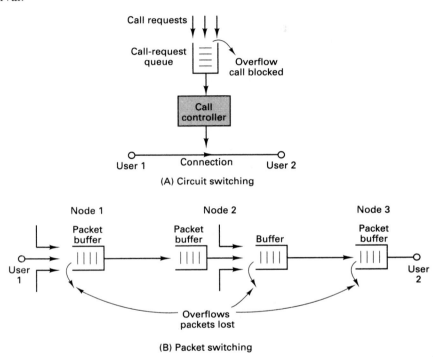

Figure 13-4. Randomness and loss in multiaccess occurs in circuit switching (A) because of queuing of call requests and in packet switching (B) because of queuing of packets.

Both packet switching and circuit switching are statistical. That is, there are random characteristics to the traffic they support. The way the randomness is manifested in both cases is indicated in Figure 13-4. For circuit switching (A), the call requests are usually queued at a single *switch controller*. As the queue gets full the calls are blocked, i.e., lost. Circuit switching therefore always has a nonzero *call-blocking probability* due to either the controller or the requested physical path or both being unavailable. For packet switching (B), packets are queued up in buffers at each node in the order in which they arrive, as shown in the figure, and since practical buffers are of finite size, there is always a nonzero *probability of packet loss* at any buffer due to overflow. Thus there are queueing and loss situations in both circuit switching and packet switching.

13.4 Circuit Switching

As we saw in Chapter 2, there many applications in which purely circuit-switched LANs or MANs are entirely sufficient so that the packet switching option is not required.   Partly for this reason, the first stage of Fiber Channel Standard, the first gigabit-per-node interconnection standard, is so-called Class I, a pure circuit-switched mode.   Classes II and III are packet-switching modes, and the standardization and implementation of these comes later.

For circuit switching, switching or tuning times of the order of many milliseconds can usually be tolerated.   Since, as we saw in Chapter 4, tunable components having this order of tuning times have become commercially available, the earliest WDMA networks have been circuit switched, as we shall see in Chapter 16.

We address, first, the question of whether to put WDM tunability at transmitter, receiver, or both.   This is a circuit-switching question when propagation latency is not included in the model.   We then go on to discuss a simple circuit-switch multiaccess protocol.

Tunability at transmitter, receiver, or both?

The question of where to place the tunability, given a choice, is always an interesting one in the architecture of WDMA systems.   There are component-availability reasons, as we have seen, for favoring tunable receivers today, and, as we shall see later in this chapter, there may be packet-switching protocols that require tunable transmitters, while others require tunable receivers.   Yet it is interesting to ask whether there is any deeper reason for preferring any one option.

The question of whether fundamental throughput advantages are to be had with one of the three options has been investigated in [5].   The analysis omits propagation latency as a factor favoring any one option, and is therefore appropriate for circuit switching.

If the number of available wavelengths Λ is greater than or equal to the number of nodes N, it is clear that no limitation on throughput is imposed due to lack of enough wavelength channels.   However, suppose there are not an unlimited number of wavelengths available.   What then?   This could occur because of limited tuning range, because of limited wavelength selectivity of filters or lasers, because of limited amplifier bandwidth, or for various other technological reasons.

We shall see later on that, given N, CDMA and TDMA require one wavelength, in-band WDMA protocols require N, and various WDMA packet switching protocols may require up to $2N$.   In Chapter 11, we saw that multihop packet switching requires $N\Delta$ wavelengths, $\Delta \geq 1$ being the number of bidirectional physical ports per node.

The following argument leads to the conclusion that having tunability at both transmitters and receivers provides a greater throughput than does tunability of either alone.   Figure 13-5 schematizes the situation in which there are N nodes but only Λ ($\Lambda \leq N$) wavelengths.   Imagine that time is continuous and that circuit-switched con-

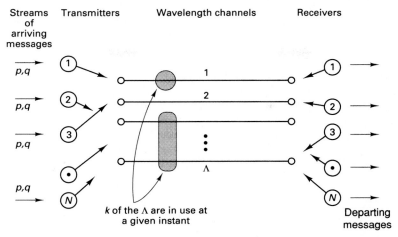

Figure 13-5. Situation when there are fewer wavelengths (Λ) than there are nodes (N).

nection requests are generated at each transmitter at a rate such that the probability that a new one starts in Δt is $p \Delta t$ and the probability that one in progress will end Δt after it starts is $q \Delta t$. The quantity of interest is the number k of the Λ wavelengths that are active, larger values of k representing higher throughput.

The process by which k grows and diminishes as connections enter and leave the system per unit time Δt can be represented by the diagram in Figure 13-6, a *Markov chain*. A Markov chain is a form of finite state machine (FSM). Each node represents a *state* and each edge represents a *state transition* from one state to another. A node in this particular FSM labelled with a value of k represents the state that exactly k wavelengths are in use. We can imagine a pointer that points to a node in the state diagram labelled with the number of wavelengths in use and moves right or left one node when a new wavelength is occupied or an old one relinquished, respectively. We would like the mean position of the pointer to be as far to the right as possible because that means higher throughput.

For every Δt, if k of the wavelengths are in use, any one of them could become free (a *state transition* to the left by one position) with probability $q \Delta t$ by virtue of the connection at that wavelength terminating. So the left-directed state-transition probabilities at state k are in general equal to $qk\Delta t$.

The probability of a right-directed state transition (probability that one more wavelength becomes active) is somewhat more complicated, and depends on where the tunability is. If only the transmitters are tunable, a transition from k to $k+1$ occurs within Δt with probability

$$(N-k)\, p\, \frac{\Lambda - k}{\Lambda}\, \Delta t \tag{13.1}$$

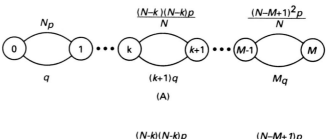

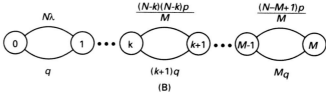

Figure 13-6. Markov chain representation of the number k of the total of Λ wavelengths in use.

because this happens only under the condition that a connection request is generated at one of the $(N-k)$ free transmitters *and* is addressed to one of the free receivers, i.e., a proportion $(\Lambda - k)/\Lambda$ of them. For the case of tunability at both transmitter and receiver, the proportion of receivers is larger, $(N-k)/N$, because all of the receivers can tune.

If only the receivers are tunable, the transition from k to $k+1$ occurs only under the condition that a connection request is generated at one of the free transmitters, of which the fraction is $(\Lambda - k)/\Lambda$, and is addressed to one of the $(N-k)$ free receivers. This again gives Equation 13.1 for the probability of a transition from k to $k+1$. Thus, the tunable-transmitter and tunable-receiver cases have identical behavior, and the case in which both ends are tunable has a higher probability of state transitions to the right (higher utilization of the Λ available wavelengths) than for the case where only the transmitters or only the receivers are tunable, because $(N-k)/N \geq (\Lambda - k)/\Lambda$ for $\Lambda \leq N$.

The advantage was investigated quantitatively [5] for the special case that p is a Poisson arrival process and $1/q$ is exponentially distributed. The Markov chain for this situation is called a *birth-death process*, and the solution for $\bar{k}$, representing here the average number of wavelengths in use, is well known. For some sample cases it was found that for very low and very high offered load p the case of both ends being tunable had a negligible throughput advantage over the case where only transmitters or only receivers are tunable, but at the value $p = q$ (short connections) the advantage of the former case amounted to a roughly 50-percent improvement in throughput.

Circular search protocol

Given a requirement that the transmitter of a node ℓ is obliged to stay fixed-tuned to λ_ℓ, setting up the full-duplex access path to some node m amounts to somehow notifying node m to tune to λ_ℓ, while the receiver of ℓ tunes to λ_m.

For circuit switching, the problem is fairly simple, because the applications served by this class of access path usually permit access times to be quite long– milliseconds, which is within the capability of today's tunable receivers and greater than the propagation latency of a MAN. As we shall see in the next gew sections, this is not true for packet switching which requires very rapid access time.

A particularly simple protocol, *circular search*, depicted in Figure 13-7, is used in the Rainbow wavelength-division MAN to be described in Chapter 16 [6, 7]. Let TX_i ($i = 1, \dots ,N$) denote the transmitter of node i, RX_i the receiver, $ADDR_i$ a bit pattern unique to node i, and $SOT_{i,j}$ a verb issued by i indicating start of data transmission to j. Let Λ denote the wavelength range spanned by all the λ_is. Assume that node ℓ wants to talk to m, but that m is temporarily busy communicating with k. The protocol is as follows:

At node ℓ: Listen for successful search completion

1. RX_ℓ tunes to listen at λ_m for the $SOT_{m,\ell}$ that indicates m is transmitting traffic to ℓ.

2. TX_ℓ transmits (on λ_ℓ), the "poll" consisting of the pair of fields $ADDR_m$, $ADDR_\ell$ requesting from m an access path with m and telling m what address ℓ wants to be known by.

3. When and if RX_ℓ hears the response $SOT_{m,\ell}$, it immediately begins to receive the ensuing traffic and it issues its own $SOT_{\ell,m}$ followed by traffic to m.

4. Periodically node ℓ times out (starts a timer) and executes the search described below until the timer expires, whereupon it resumes at step 1.

At node m: Circular search

1. As soon as node m becomes free from talking with k, it begins to sweep its tunable receiver from λ_k in a direction of increasing λ until it reaches the end of the range Λ over which any possible λ_is might exist.

2. As the tunable receiver passes over each λ_i it acquires bit clock and looks for any poll with $ADDR_m$ as its first field in the pair.

3. If it fails to find it, it passes on to λ_{i+1}.

4. If it does find some λ_ℓ that is sending the correct $ADDR_m$ as the first member of the pair, it stops the tuning, locks the tunable RX_m to that received wavelength, and sends $SOT_{m,\ell}$, using $ADDR_\ell$ to address it, followed by traffic for ℓ.

5. If the linear sweep carries RX_m to the end of the range Λ without encountering any node sending $ADDR_m$ as the first field in the poll, it restarts the wavelength sweep in the reverse direction.

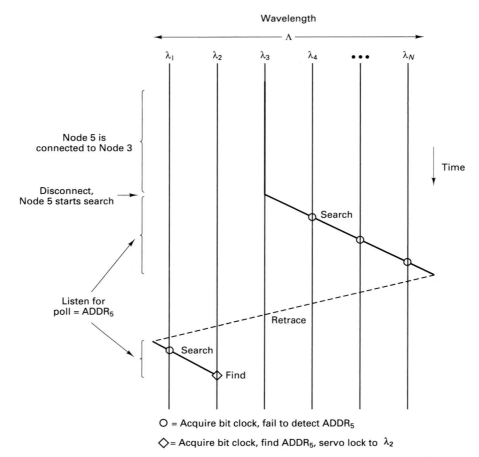

Figure 13-7. Time-wavelength diagram for circular search protocol. Node 2 establishes a connection with node 5.

The timeout mentioned in step 4 for node ℓ is required to avoid a deadlock situation that might occur, for example, if node A is listening to node B for a positive response while B is listening to C and C is listening to A. This circular situation is broken by forcing the receiver of every node to occasionally stop participating in the attempt to send traffic and detect whether some other node is trying to send to it.

The circular-search protocol provides very good modularity and growth flexibility. When a new node wants to join the network, no prior coordination with existing nodes is required, except that a new node should not use either a λ or an *ADDR* that is already in use. It can learn what these are by listening. The new node must only be sure to employ a new λ that lies within Λ and a new *ADDR* that is syntactically correct. Network control for one path is completely decoupled from that of another; any number of circular searches may be going on simultaneously in the network, so there is no call queuing as with classical circuit switching, which employs a shared controller.

At the bit rates used in the Rainbow network, acquisition of bit sync and poll recognition occur rapidly enough that the sweep of the tunable receiver is simply a linear ramp; the receiver does not have to dwell on each successive wavelength being searched.

13.5 Packet Switching

Lessons from first- and second-generation LANs

We now turn to packet switching. The first order of business is quickly to go over some first- and second-generation packet-switch MAC protocols looking for inferences to be drawn on how to design third-generation versions.

As mentioned in the last chapter, the access protocols most widely used in first- and second-generation LANs were token rings and CSMA/CD (carrier-sense multiaccess with collision detect). These *random-access* protocols were all developed because of the limitations of even older *scheduled-access* protocols in which the use of a portion of the capacity had to be scheduled in advance, the very opposite of the tell-and-go capability. The first- and second-generation random-access protocols proved very appropriate and successful in accomplishing what they set out to do: to provide a convenient and economical way of connecting a great many users having bursty traffic at rates of a few Mb/s at most over distances very short compared to a packet duration. They are quite inappropriate for third-generation optical networks because of the large ratio of propagation time to packet length and also for other reasons, as we shall now see.

We shall make a cursory examination of the delay and throughput. The simple idealizations we shall use for CSMA/CD and token rings are quite inexact, but they will serve to make the basic points. More exact analyses are available in many texts and references, for example [8, 9].

CSMA/CD

Figure 13-8 shows the time-space diagram for a simplistic model of CSMA/CD. Time reads vertically downward as before and distance reads horizontally, with the N nodes being equally spaced along the bus of length L. The diagram illustrates transmission from node 1. Each node wanting to transmit listens first to see if anyone else is transmitting ("carrier sense"), and withholds transmission if the channel is busy. Once the channel is sensed idle, a node with a packet ready to send first sends a probe ("contention") signal for the duration of one *contention slot*. While it is transmitting, it is also listening for a signal from any other node arriving at any time during the contention interval. If this *collision detection* turns out negative, the node then seizes the

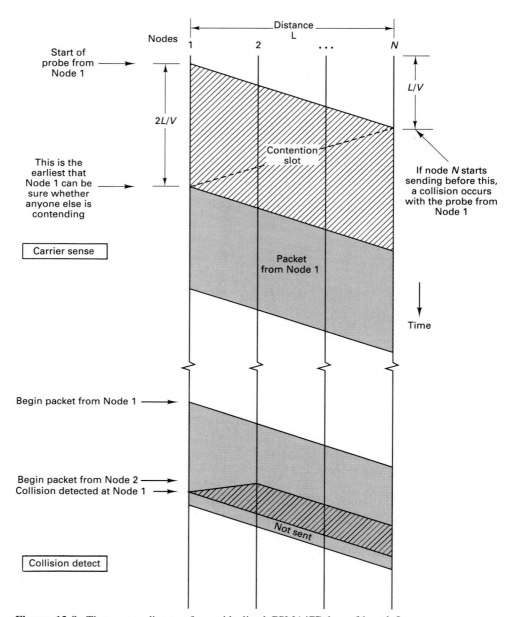

Figure 13-8. Time-space diagram for an idealized CSMA/CD bus of length L.

channel by immediately proceeding to send its packet, knowing that it now owns the channel, since nobody else is contending.

The length of the contention slot, and therefore the delay before a packet transmission can be attempted, is, unfortunately, very much affected by the bus length. The length of the contention slot is a systemwide parameter, and it must be set equal to or greater than the time required to be sure that no other node is transmitting. The upper part of Figure 13-8 shows that node 1 would have to wait an interval $2L/v$ seconds long, v being the velocity. This interval consists of two parts. There are L/v seconds after node 1 starts to probe the channel before the last node N hears the leading edge of the probe. This is followed by another L/v seconds for node 1 to make sure that N had not sent anything before N aborted because of the arrival at node N of the probe from node 1.

If the probe is successful, packet transmission starts immediately. If, during transmission of the packet, it is found that for some reason there are any other transmissions on the medium during that packet (another collision), the transmission fails and the packet is aborted, as shown in the bottom part of Figure 13-8.

The approximate analysis of CSMA/CD [10] assumes that each node is equally likely to send to any other node, that at the beginning of each slot each node is ready to send a packet with probability p (*Bernouilli trial*), and that if the slot is sensed to contain no energy from other transmitters, the node will in fact send the packet. The packet length P is assumed constant. Furthermore, the assumption is made that the system is memoryless in the sense that when a node fails to successfully send a packet because it lost a contention, the probability that it wants to send a packet in any later slot is still p. (In real life, of course, it remembers whether it got rid of a packet; the statistical correlation between packet attempts is one of the things that complicates an exact analysis).

The probability that a given node successfully seizes the channel at the end of any contention slot is $p(1-p)^{N-1}$. The probability that *some* node seizes the channel at the end of any contention slot is N times this,

$$S = Np(1-p)^{N-1} \tag{13.2}$$

The probability that a contention slot experiences no successful seizure by anybody is $(1-S)$, and the probability that there will be a sequence of $(j-1)$ failures followed by a success (i.e., the probability that the string of attempts will be j contention slots long) is $(1-S)^{j-1} \times S$, so that the mean length of a string of contentions is

$$\sum_{j=0}^{\infty} jS(1-S)^{j-1} = -S\frac{d}{dS}\sum_{j=0}^{\infty}(1-S)^j = -S\frac{d}{dS}\left(\frac{1}{S}\right) = \frac{1}{S} \tag{13.3}$$

contention slots, and the mean delay to complete the delivery of a packet between two nodes is therefore

$$\overline{D} = \frac{2L}{vS} + P + \frac{L}{2v} \tag{13.4}$$

The first term is the mean duration of the succession of contention slots that the node must wait before sending its packet. P is the packet duration. The last term expresses the fact that the mean spacing of the nodes between which the packet must travel is $L/2$.

We define the normalized throughput T as the fraction of the time that useful packets are being sent anywhere on the bus

$$T = \frac{P}{P + 2L/vS} \tag{13.5}$$

As the probability p that a node wants to send increases from zero, the probability S that after a given slot someone will be successful rises at first, and then falls as collisions set in and begin to dominate the traffic on the the channel. From Equation 13.2 the maximum value of S occurs at $p = 1/N$ and is $1/e$ for sufficiently large N. For this optimum value of p,

$$T_{max} = \frac{P}{P + 2eL/v} \tag{13.6}$$

and the mean delay is

$$\overline{D} = \frac{2Le}{v} + P + \frac{L}{2v} \tag{13.7}$$

Token ring

Figure 13-9 shows an an idealized token ring, with ℓ being the distance between equally-spaced nodes. To provide a fair comparison with the CSMA/CD bus, we set the circumference of the ring at L. We use the same simplistic traffic model, namely, when the token arrives at any node, that node is ready to put a single packet of duration P on the ring with probability p. Both the duration of the token and the time to process it are assumed to be negligible fractions of both the packet length P and the travel time ℓ/v between nodes.

The figure shows snapshots of the position of a single packet at various times t. The **T** at the trailing edge of the packet is the token. At the lower right at $t = 0$, the packet is about to be transmitted by a node that has just been given the token at the trailing edge of some preceding packet, this preceding packet having been consumed by this node. At $t = P$, the entire token has left the node. Averaged over a number of trials, the addressee will be halfway around the ring, so the average station is a distance $L/2$ away. The packet enters this half-way node at $t = L/2v$, and releases the token at $t = L/2v + P$. At $t = L/2/v + P + \ell/v$ the token reaches the next node downstream. The throughput on the loop will go up with p, the probability of this next packet being ready, and for p close to unity, this next node will always have a packet

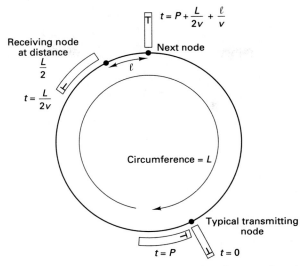

Figure 13-9. Token ring, showing the position of a typical packet at various instants of time.

ready to send. Under this condition the next packet transmission on the ring starts at $t = L/2v + P + \ell/v$.

From this reasoning we can conclude that as p approaches unity, the throughput is at a maximum Defining throughput T again as the ratio of packet duration to total time between leading edges of successive packet transmissions, we have

$$T_{max} = \frac{P}{P + \dfrac{L}{2v} + \dfrac{\ell}{v}} \tag{13.8}$$

The average delay between leading edge of a packet and the completion of the delivery of its trailing edge is

$$\overline{D} = P + \frac{L}{2v} \tag{13.9}$$

independently of N.

We have assumed that the ring has *early token release*, as in FDDI [11], in which the transmitting node releases the token immediately after completing transmission of the packet. This is in contrast to the IEEE 802.5 first-generation token-ring protocol [12], in which the packet and token must propagate completely around the ring and back to the transmitter before the token can be released.

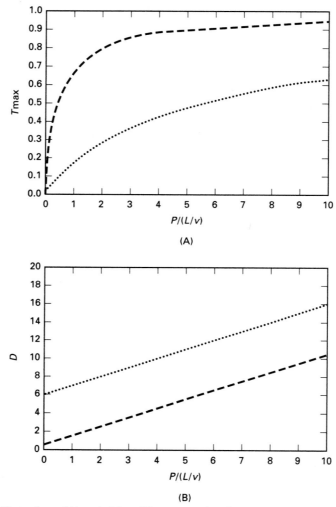

Figure 13-10. Throughput (A) and delay (B) versus ratio of packet length to propagation time across network for idealized first-generation LAN (CSMA/CD, dotted line) and second-generation LAN (token ring, dashed line).

Inferences

We have briefly and approximately analyzed a first-generation protocol (a crude representation of Ethernet) and a second-generation protocol (a crude representation of FDDI). Figure 13-10 compares the way in which throughput drops and delay builds up, respectively, as the physical size L of these networks increases. From analyses like this we can draw several object lessons useful for designing multiaccess protocols for third-generation lightwave networks.

- **Concurrency is missing.** In Figures 13-8, a snapshot taken at any one time by drawing a horizontal line shows that there is at most one packet in flight at any one time. The same thing can be inferred from Figure 13-9 for the token-ring case. There is no *concurrency*, which can be defined as the use, when a packet is in transit, of other parts of the medium that would otherwise be idle. Several second-generation lightwave networks have been devised that try to deal with this problem. For example, DQDB (Distributed Queue Dual Bus, the emerging IEEE 802.6 standard) [13] and also the Cyclic Reservation Multiple Access (CRMA) gigabit LAN [14] both take advantage of the topology of busses to allow different point-to-point transmissions to be in progress concurrently on physically disjoint portions of the bus or busses. Metaring [15] does the same thing with physically disjoint portions of a ring. The average number of concurrent packet flows that can take place with these systems ranges from less than two to about four, much less than N, the large number that Figure 13-1 suggests are possible with the WDMA or CDMA third-generation systems we shall be discussing.

- **Avoid unwise use of propagation time.** For both the contention bus and the token ring, it is rarely practical to route the links through the succession of nodes in a shortest-path manner. Instead, it is standard practice to bring the connection to each node back to a central wiring closet, or hub, so that failed stations may be easily bypassed, to provide modularity, and to improve problem determination capability. While providing these advantages, this expansion of propagation distance nevertheless impacts performance severely at high bit rates and large network extent, as Figure 13-10 shows.

- **Avoid introducing collisions**. It is bad enough to have to recover from random errors introduced by nature in the form of noise without exacerbating the problem by introducing man-made interferences having nothing to do with natural errors. In CSMA/CD systems of modest bitrate, for bursty traffic at low duty factors, this is a price worth paying for convenience of access, easy reconfigurability, and economy. However, in other situations it may be worth going to considerable trouble to avoid such self-inflicted (traffic-produced) randomness. An example of the pains sometimes taken to avoid collisions completely is packet switching by means of the multihop physical-route topology.

13.6 Survey of Third-Generation Packet-Switching Protocols

There is a rapidly-growing literature on packet-switch MAC protocols for all-optical networks, preceding the wide availability of the fast-switching component technology necessary for their realization. Several comparative surveys of this literature exist [16-18].

The circular search protocol described earlier for circuit switching uses what telephone people call *in-band signalling*. In this form of signalling, the information that sets up the access path travels at the same wavelength, time slot, or CDMA signal space as the information bitstream. Partly because of the copious bandwidth available with WDMA, it is attractive to set aside one or more additional wavelengths as *control channels* or *order wire* channels to be used for setting up and taking down all access paths. All of the packet-switch MAC protocols suggested to date do this.

Table 13-1.   Comparison of third-generation MAC protocols. (From [17]).

Scheme	Proc-essing	"Tell-and-go"?	Through-put	Sync needed?	Tunable elements per node	Wave-lengths per network
Habbab [19]	High	Yes	Low	No	2	≥ 2
Mehravari [20]	High	No	Low	No	2	≥ 2
Chen [21]	High	Yes	Moderate	Yes	1	$N+1$
Chen [22]	High	No	High	Yes	1	$N+1$
Chipalkatti [23]	Very high	No	High	Yes	1	$N+1$
Lu [24]	Very high	No	Moderate	Yes	2	≥ 2
Sudhakar [25]	High	Yes	Low	Yes	2	≥ 2
Li [26]	Low	Yes	Moderate	No	2	$2N$
Humblet [17]	Low	Yes	High	Yes	2	$2N$

We reproduce here the highlights of the survey presented in [16]. The protocols are listed according to first author in Table 13-1 where they are compared according to several parameters:

- **Processing requirements**. In Chapter 1 we mentioned the *electronic bottleneck* that occurs when every node has to handle the traffic of all the nodes, both payload data and protocol overhead. Analogously, there can be an *electronic processing bottleneck* in MAC protocols. This occurs when, even though a node does not have to see the aggregated user traffic, it still has to see a heavy overhead of aggregated protocol bits from all nodes.

- **Tell-and-go** was discussed in Section 13.2.

- **Network throughput** was discussed in the previous section.

- **Synchronization**. Many protocols use "slotted" time, and therefore each node needs to know the successive ticks of some clock that defines the beginning of each slot.

- **Number of tunable elements per node**. Tunable lasers and tunable filters are lumped together in this column of the table.

- **Number of wavelengths needed** Λ.

The protocols given in [19, 20] use either Aloha or CSMA/CD on a separate control wavelength and also use such contention schemes for sending the data packets. As we saw in the previous section, any such use of contention protocols becomes very inefficient when the packet size is small compared to the propagation time across the network. Moreover, collisions can occur on both control and data channels. Each node must look at all the control-channel information, so the processing cost is high. The protocol of [19] provides tell-and-go capability, since a node can send a packet as soon as it has one ready, whereas that of [20] must wait until it has determined whether its signalling information on the control channel was successful. This improves throughput but increases delay.

The next protocol on the list, *dynamic time WDMA* [21] is the one we have chosen for a detailed discussion in the next section. It has good delay and throughput characteristics, but requires considerable processing. The delay performance is particularly interesting, lying for most packets between one and two multiples of the propagation time.

The next two, [22, 23], are variants of dynamic time WDMA. They use the control channel to reserve capacity and thus gain in efficiency, but at the expense of the processing required. The delay is more than one to two multiples of the propagation time.

The scheme described in [24] improves on [19, 20] by avoiding collisions on the data channel. It does so by using the control channel to reserve data channel slots. However, this still leaves the possibility of collisions on the control channel. The variant of [25] aims at improving on this somewhat on this by transmitting on the data channel as soon as the control channel attempt is initated, thus reintroducing the possibility of collisions on both.

Following a suggestion of Humblet, the two protocols described in [17, 26] both use 2N wavelengths, N for a fixed-receiver control channel per node, and another N for a fixed-transmitter data channel per node. Each node has two tunable devices, a laser diode for selecting the control channel of the desired receiver and a receiver for selecting the data channel of the desired transmitter. The large number of wavelengths consumed in these two protocols buys considerable improvement in throughput, particularly in the second protocol which exchanges a slotted network-wide synchronization function for considerable improvement in flexibility and enhanced throughput.

It is clear that we can make a few rough generalizations from the data of Table 13-1. Certainly sticking with Aloha and CSMA/CD for gigabit optical networks of any size invites poor performance. As long as one goes to the expense of providing a dedicated wavelength to form a control channel, there are better ways of using this resource. Then if there are so many wavelengths available that each node can have its own control channel, performance and flexibility increases still further. It seems that one can trade added control channels for processing complexity.

We now discuss one protocol in which a single control channel is used quite resourcefully. It will serve to illustrate what can be done.

13.7 Dynamic Time WDMA

Dynamic time-division WDMA [21] is a protocol for packet switching, circuit switching, or hybrid (packet and circuit) switching in the situation that the transmitters are tunable and the receivers fixed-tuned. It is a *tell-and-go* protocol, based on two ideas, the first being that with fixed-tuned transmitters and agile receivers, many transmitters can be simultaneously active and no collisions will occur, because the transmissions will all be on different wavelengths. The second idea is to use the extra $(N+1)$st control channel wavelength for the transmitters to preannounce which receiver each is about to send to, so that a receiver can decide which transmitter it wants to hear. If the receiver follows a fixed rule in making the choice of whom to listen to, and all the nodes know what that rule is, then the successful transmitter will know that it has been successful and the unsuccessful ones will know that they must retry later.

It is assumed that the network is based on a central star coupler. Figure 13-11 shows the waveforms observed at the hub at all $N+1$ wavelengths, the N station wavelengths carrying data and the one "common wavelength" signalling channel. On all wavelengths, time is broken up into equal-length slots. On the signalling wavelength, each of these slots is further subdivided into *minislots*, exactly N in number, with the ith minislot belonging to transmitter i $(i = 1, \ldots, N)$. The method by which all the slot boundaries are time-aligned is that, when each node joins the network, it listens to its own transmission at its own receive wavelength to establish the exact delay to the hub.

Consider packet switching first. Any node i having a packet to transmit to a node j declares that fact when its minislot occurs and proceeds to send its packet in the next slot. It is now only necessary for the receiver to tune to the wavelength of the transmitter it wants to receive the packet from by the time this next slot begins. A *tuning time* is set aside at the end of each slot for the receivers to do this. The slot length, minislot number and format, tuning-time, and decision rule are all system parameters set for the entire network.

As long as the receiver uses the systematic rule to decide which transmitter to listen to and as long as all the transmitters know how the decision came out, the successful transmitter can proceed to queue up the next packet and the losers can proceed to contend again in the next slot. Since every node has received the sequence of N minislots, each has all this information.

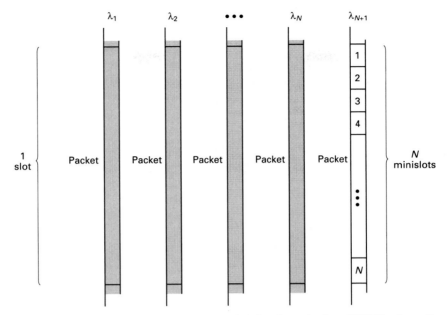

Figure 13-11. Waveforms seen at hub (star coupler) for dynamic time WDMA when all nodes are transmitting.

For example, suppose the rule is that the winner is to be that node out of those wanting to send to a given receiver whose packet will be the oldest when it arrives at the hub. This time will be the age of the packet at the sending node (the time it has been in queue) plus the known travel time to the hub. Each candidate transmitter can convey that age information in an appropriate subfield of its minislot. This algorithm guarantees a high degree of fairness, since it avoids favoring the nodes nearest the hub.

As we shall now see, the dynamic time WDMA protocol results in a high-concurrency situation in which in each slot time at least one packet is delivered, without collision, by *some* one of the transmitters that have traffic intended for each receiver. In one slot time, many receivers are simultaneously receiving packets while, during the same slot, the minislots on the signalling wavelength are being monitored and the winners being computed so that the very next slot can be used for packet transmission. The contention resolution and the packet transmission are thus time-overlapped or *pipelined*.

Each minislot contains not only the address of the intended receiver and the delay parameter, but also a *mode* bit that tells whether the receiver is in a circuit- or packet-switched connection. If it is in circuit-switched mode with some transmitter as the result of some earlier contention, all the other transmitters will know this and therefore know that their last packets failed, as will all future ones aimed at that receiver. Clearly all nodes could operate in this way, in which case we would have a pure circuit-switched network. Or, hybrid switching can be implemented by having some receivers in packet mode and some in circuit mode.

Figure 13-12 shows an example of two nodes, 1 and 3 contending to send a packet to node 2, and shows 1 having to resend its packet because node 3 won the contention, node 2 having heard both requests but having chosen node 3, perhaps because node 3 is closer to the hub than node 1. The way the choice by 2 came out is known at node 1 by time t_1 and at node 3 by time t_2. Node 1 then decides to retransmit the same packet as soon as possible and this time node 1 is successful. Meanwhile, two slots have elapsed during which node 1 successfully transmitted to nodes 4 and 5 and node 3 successfully transmitted to nodes 5 and 4.

The performance of this protocol has resisted a closed-form analysis because of the usual difficulty of handling correlations between successive attempts, but the same simple idealization we have been using for other protocols allows an analytical determination of the *maximum* throughput per wavelength. Imagine that the probability p that a packet will be ready to send at a given node is unity. At a given receiver the probability that a packet from a particular transmitter is directed to it is $1/(N-1)$. The probability that the packet from that particular transmitter goes somewhere else is $1 - 1/(N-1)$. The probability that this is true for all the $(N-1)$ transmitters is

$$[1 - 1/(N-1)]^{N-1} \tag{13.10}$$

and therefore the probability that there is a packet directed to this receiver is

$$\{1 - [1 - 1/(N-1)]^{N-1}\} \tag{13.11}$$

The total throughput per slot for the entire network is N times this,

$$N\{1 - [1 - 1/(N-1)]^{N-1}\} \tag{13.12}$$

which is $N(1 - e^{-1})$ for sufficiently large N. Therefore the maximum throughput per node per slot is $0.63\,N$ packets per slot.

Although this crude analysis only gives the asymptotic throughput for the offered load $p = 1$, simulations have provided results for smaller values of p. Figure 13-13 shows with curve (A) the throughput versus p for $N = 40$ nodes and transmitter queue length equivalent to three times the distance to the hub (which was 5 slots in the case simulated). The simulation asymptotic value for throughput was 0.60, rather than the 0.63 of the simple analytic model, the difference being due to the correlated packet reappearances. The simulations also showed that making the queue length more than three times the travel time to the hub did not increase throughput. Figure 13-13 shows with curve (B) the corresponding curve of delay versus offered load.

It is seen that, unlike many other protocols, with dynamic time WDMA some node gets its packet to the destination in one propagation time plus a small overhead, which is at most one packet length (the set of minislots). The contention losers take

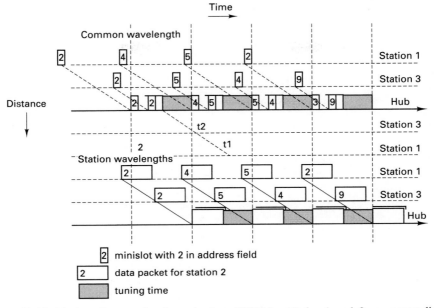

Time

Common wavelength

| | | | |
| Station 1 |
| Station 3 |
| Hub |
| Station 3 |
| Station 1 |

t2

t1

2

Station wavelengths

| | | | |
| Station 1 |
| Station 3 |
| Hub |

Distance

2   minislot with 2 in address field

2   data packet for station 2

tuning time

Figure 13-12. Timing diagram for dynamic time WDMA. Nodes 1 and 3 are contending to send to 2 and 1 loses the contention.

longer, but the mean delay was found to be never worse than about 17 packet durations for the example simulated.

The principal advantage of dynamic time WDMA is that it allows packet switching to be performed with a very bare minimum of packet delay for the contention winners, namely about one propagation time across the network. The disadvantages include the fact that an extra laser and fixed-tuned receiver must be provided at each node to support the signalling channel, and that there is a 3-dB energy loss at both transmitter and receiver if simple wavelength-flat couplers are used to merge and split the signalling wavelength from the data wavelength. Also, scalability is limited, since the maximum number of nodes N is fixed at the number of minislots, and the slot assignments must be coordinated.

13.8 Multihop Packet Switching

In the preceding chapter we mentioned imposing a carefully chosen multihop physical-path topology on a given physical-network topology in such a way that the former constrained a receiver at each node to receive traffic from one and only one transmitter. This means that for a wide variety of physical-network topologies (including a star or bus) such a physical-path topology can be employed for collisionless packet switching by providing internal electronic switching within each node and a buffer at each output, as shown in Figure 13-14. Each packet is routed from node to node

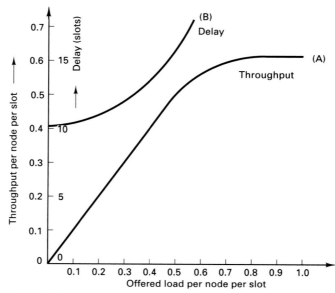

Figure 13-13. Dynamic time WDMA performance (A) Throughput versus offered load. (B) Delay versus offered load.

across the multihop physical-path topology until it arrives at the proper receiving node. There is no "multiple access" to a common medium and the MAC protocol layer is not involved; network control provides initial setup of each point-to-point internode physical connection. In each node the network layer, the third or routing layer, routes an incoming packet to the proper output port according to the usual routing tables set up by networkwide interaction of the control points in the nodes. The multihop network acts like a conventional packet-switched WAN, as in Figure 12-7(D). The switching can either be of the conventional electronic TDM form or advantage could be taken of the rapid tuning speed available from RF components, as mentioned in Section 9.7. Such a system has been partially prototyped [27].

The multihop scheme could be called a third-generation fiber optic network because, whether or not it uses wavelength division, it exploits the very large bandwidth capability of fiber optic technology rather than substituting fiber for copper in some existing scheme. On the other hand, we would not call a multihop network "all-optical," since there are many optical-electronic conversions along each path. The multihop design uses in-band signalling; there is no dedicated wavelength or control channel.

Multihop performance

The price paid for complete lack of collisions, and the fact that no tunability is required, is not only the possibility of an electronic bottleneck in the nodes, but a large delay, which would appear to be wasteful if the physical topology is already of a

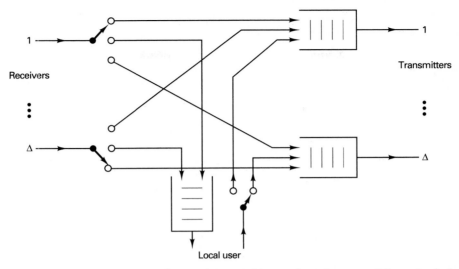

Figure 13-14. Block diagram of a packet switching node using a multihop physical-path topology. Switches are thrown on the basis of address information in the packet header. Packet buffers are also shown.

one-hop character. The LAN or MAN that was one-hop at the physical-network level has been made multihop at the physical-path level by using $N\Delta$ wavelengths (in the wavelength-division case for which multihop networks were invented). As with any packet-switched network, a buffer is required at the transmitting end that is large enough to accommodate each packet in flight to the receiving end for at least one round trip through the multiple hops and back again, until its receipt can be confirmed.

The mean number of hops $\overline{H}$ was discussed in Chapter 11, for both the shufflenet and deBruijn topologies of degree (number of ports per node) Δ, and any queuing delay at each node can only add to this. Therefore the mean delay in sending across a multihop network is at least

$$\overline{D} = \overline{H}\,\frac{L}{v} \tag{13.13}$$

Calculating maximum throughput is more complicated. Clearly throughput will increase with offered load, so let us examine what happens when p, the probability that at any instant a given node has a packet to transmit, increases to unity, as before. What now happens is that, because each edge on the graph must carry traffic on behalf of more than one pairwise connection, some of the wavelengths (edges of the graph) must carry many times more traffic than they would if each were dedicated to carrying traffic from only one transmitter. How bad this *edge loading* can get has been investigated [28] for deBruijn graphs but is largely unknown at present for the shufflenet graph.

Edge loading is calculated assuming that each node has a stream of packets at the per-node bit rate that are addressed, one-by-one, to any of the other $N - 1$ other nodes uniformly. Sooner or later there is a worst-case situation where a large number of these packets must pass through that edge that participates in the largest number of routes. The number of packets that pass simultaneously across that worst-case edge at that time is L_{max}, the maximum edge loading. For deBruijn multihop networks with the routing adjusted so that the edge loading is evened out across the edges,

$$L_{max} = H_{max} \, \Delta^{H_{max} - 1} \tag{13.14}$$

This number can become quite large, leading to the requirement that certain links involve transmitting and receiving electronic components that run at L_{max} times the bit rate of the transmission from that node to one other node. Thus, either the bit rate of the entire network must be degraded by a factor $L_{max}/(N - 1)$, or the number of nodes N must be reduced by the same factor.

Modified Aloha as an alternative to multihop

If the only reason for using the multihop option in the first place is to support packet switching, we might ask if there isn't a better alternative. Might there not be another packet-switching option that does not use one or more signalling channels, avoids the edge-loading problem, and has no worse a delay than that expressed by $\overline{H}$ and H_{max} of the multihop network? Specifically, we might expect that a single-hop packet network that *allowed collisions* but recovered from them by time-consuming retransmissions might nonetheless have delay performance superior to that of the multihop network which goes to great trouble to avoid them altogether by indirect routing.

One approach to this is the following. Assuming that tunable transmitters are available, we could use an Aloha protocol to send a packet to the desired fixed-tuned receiver and retry upon collision.

Since collisions occur per receiver, not per network, the scheme might have advantages. In [29] the shufflenet was compared with a *modified Aloha* protocol, where "modified" meant that the algorithm by which a node decides when it will retransmit a collided packet was a special one, the "pseudo-Bayesian stabilized algorithm" [8, 30], which is one of the few known that allow Aloha protocols to work at high offered load. Most other variations of Aloha exhibit instabilities, throughput reduction, and unbounded delays whose probability increases as the offered load increases. This is because, unlike CSMA/CD, there is no carrier sense and no preparatory contention interval that can be used to cut short a transmission that will otherwise collide.

It was found that for $\Delta = 2$, modified Aloha always had a smaller delay, but that when the number of ports Δ was doubled, the shufflenet hop count decreased so dramatically that at high offered loads it was better than modified Aloha.

13.9 Optical Code-Division Multiaccess

Motivation

It would be nice to have some way of accommodating collisions, because then the channel could accommodate many concurrent connections. Recall that we defined a collision as an error-producing time overlap of two signals on the shared communication medium. Spread spectrum [31] is a communication technique that allows *non-error-producing time overlaps*. Originally developed for its antijamming properties, its use in multiaccess systems, where it is called *code-division multiaccess (CDMA)*, has proved to be much the more important and enduring application.

The basic idea is shown in Figure 13-1(C). Many active nodes transmit simultaneously and in the same frequency band, and are therefore all technically in collision with one another. However, in CDMA these collisions are rendered as harmless as possible by choosing the signal waveforms to be as mutually noninterfering as possible. This requires that their bandwidth be several orders of magnitude greater than the information bandwidth $1/T$ of the digital modulation, T being a bit duration. CDMA waveforms are commonly called *pseudonoises* or *CDMA codes*. For optical CDMA [32, 33], pseudonoises can be generated purely optically by exciting with a short pulse a parallel set of various lengths of fiber, whose variously delayed outputs are then combined additively in a coupler to form the composite optical signal.

In general, whether with conventional CDMA or optical CDMA, it will prove impossible to choose the pseudonoises to be completely noninterfering. Also, with optical CDMA at 1 Gb/s, we are dealing with such extremely high bandwidths that dispersion and other factors become a serious problem.

In view of such difficulties, we might well ask what the advantage of CDMA might be for optical networking. The answer is that CDMA is a true *tell-and-go* protocol, which is incidentally an important reason for its use today in cellular telephony. When one node wants to send traffic to another node, prior coordination is required only with that node. A CDMA network is also highly scalable and modular. We may keep adding connections, and all that happens is that there is a steady buildup of a background noise that consists of the sum of the other CDMA signals plus any natural noise in the system. The natural noise consists of those noise components arriving at the receiver from any of a number of sources—for example, shot noise, thermal noise, signal-spontaneous amplifier noise, and so forth, as discussed in Chapter 8.

The buildup of interference with number of nodes N is the key performance parameter determining the usefulness of optical CDMA, and so most of this section is occupied with quantifying receiver output signal-to-interference ratio as a function of N.

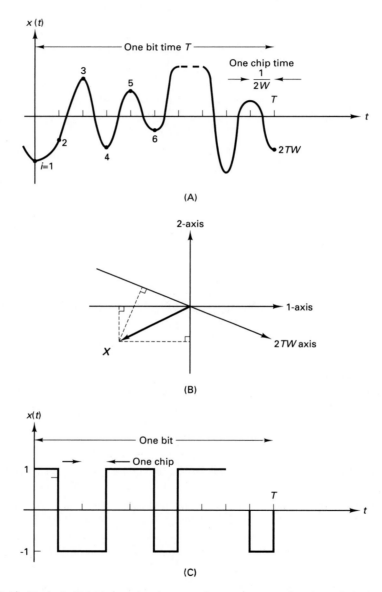

Figure 13-15. Typical CDMA pseudonoise waveform. (A) As a function of time. (B) As a vector in $2TW$-space. (C) A BPSK pseudonoise.

Vector representation of signals and noise

Consider a typical pseudonoise waveform of bandwidth W Hz lasting T seconds, as shown in Figure 13-15(A). If the product TW is large enough, we can consider all the energy to lie completely within the duration T and the bandwidth W. The sampling theorem of bandwidth-limited signals then says that this waveform is completely defined by $2TW$ numbers, each representing one sample every $1/2W$ seconds (the *Nyquist rate*).

$$X(t) = \sum_{1}^{2TW} x_i \, \psi(t + \frac{i}{2W}) \qquad (13.15)$$

where $\psi(t)$ is a unit amplitude pulse, all of whose energy is confined to W Hz and most of whose energy is confined to $1/W$ seconds.

In spread-spectrum nomenclature, each interval $1/W$ seconds long is called a *chip*. The number of chips per bit ($2TW$) is called the *spreading factor*, because it is the factor by which the bandwidth of the signal is artificially increased over the intrinsic bit rate ($1/T$). It is also called the *processing gain*, since it will turn out (Equation (13-24)) that under many conditions the output signal-to-noise ratio is $2TW$ times the input signal-to-noise ratio.

Output SNR equation

To see how CDMA works, it is most useful to start by representing a typical pseudonoise $X(t)$ (Figure 13-15(A)) as a vector in *signal space*, in this case a $2TW$-space, as depicted in Figure 13-15(B). In this scheme of signal representation, the value of the i th sample of $X(t)$, namely x_i, is represented as the projection of the vector X along the i th axis. In this representation the "vectors" are, of course, mathematical not physical vectors.

It helps in understanding the vector representation to note that it is a generalization to $2TW$ dimensions of the familiar two-dimensional phasor representation of a sinusoid (Section 3.3). Given the sinusoid's frequency, two additional numbers that completely characterize it are the length and polar angle of the phasor. Alternatively we could specify the in-phase and quadrature components. In the same way, the actual center frequency of a pseudonoise is suppressed in Figure 13-15, and the corresponding vector representation as $2TW$ numbers then completes the characterization of the pseudonoise waveform. The squared length of the vector X that represents the signal $X(t)$ is

$$X \bullet X = |X|^2 = \sum_{i=1}^{2TW} x_i^2 \quad \text{which, for sufficiently large } 2TW,$$

$$(13.16)$$

$$= \int_0^T X^2(t)\,dt = \mathcal{E}_x$$

the energy of the signal $X(t)$, assuming it to be a voltage or current applied to one ohm.

A CDMA receiver uses *correlation* of two waveforms, defined as the integration over one bit time T of their product. Specifically, the *crosscorrelation function* between two voltage functions of time $X(t)$ and $Y(t)$ having a relative delay τ (i.e., whose time samples are indexed by i and $(i + 2\tau TW)$, respectively) is given by

$$\phi_{xy}(\tau) = \int_0^T X(t)Y(t+\tau)\,dt \quad \text{which, for sufficiently large, } 2TW$$

$$= \sum_{i=1}^{2TW} x_i\, y_{(i+2\tau TW)}$$

$$(13.17)$$

The value at $\tau = 0$ is the *crosscorrelation, $\phi_{x\,y}(0)$,* and is represented in the vector diagram as the projection of one of the signal vectors upon the other

$$\phi_{xy}(0) = \sum_{i=1}^{2TW} x_i\, y_i$$

$$(13.18)$$

from which it is clear that the energy is the *autocorrelation*

$$\mathcal{E}_x = \phi_{xx}(0)$$

$$(13.19)$$

Given all this, it is now a simple matter to explain how a CDMA node uses crosscorrelation to receive a bit from a desired transmitting node and discriminate against signals from all the other nodes. We use the same optimum-receiver notions as those presented in Chapter 8. Both the transmitter and the receiver have stored copies of two waveforms $X_0(t)$ and $X_1(t)$, representing a binary "0" and "1," respectively. Everybody else has other waveform pairs. As shown in Figure 13-16(A), the receiver synchronizes its stored references $X_0(t)$ and $X_1(t)$ to the desired transmitter, multiplies each of these stored waveforms by the incoming signal $Y(t)$, integrates each product for the T seconds of the bit duration (Equation (13.17)), and declares "0" or

"1," depending on which integrated product (crosscorrelation) is the greater. $Y(t)$ consists of a mixture of random system noise and all the CDMA transmissions.

Figure 13-16(B) shows a vector representation of the action of such a CDMA receiver. There is a *decision surface*, in this case a plane of all points equidistant from the X_0 and X_1 vector ends. The receiver declares all observed values of Y that lie on the X_0 side of the plane to signify a "0," while those on the other side are taken to mean a "1." Errors will be committed when noise plus the interference from the other CDMA signals causes a crossover onto the wrong side of the decision surface.

It is easily seen that we want not only to make $X_0(t)$ as different as possible from $X_1(t)$, but also wants to make both of them as different as possible from all the possible received signals Y that might come from other nodes on the network or from noise. Clearly, given an X_0, the best choice for X_1 is for its vector representation to be antipodal to that of X_0,

$$X_1(t) = - X_0(t) \qquad (13.20)$$

In order to make the energy per bit (squared length of the vector) as large as possible under a constraint on *peak* power (a quite common engineering constraint in both electronic and optical transmitters), the best choice for $X_0(t)$ would be a constant-amplitude *binary phase-shift keying (BPSK)* CDMA waveform such as the one shown in Figure 13-15(C). Then $X_1(t)$ would be its phase-reversed mate.

Suppose the transmitter is sending a 1. Assume that the desired signal and all the $N - 1$ signals from other nodes arrive at the receiver with equal energy $\mathcal{E}_S$. Also assume, for simplicity, that the system noise and the interfering signals are all random noises and can thus be lumped together as a composite noise. Then it is a fairly simple matter to calculate the ratio of output signal energy to output noise energy in a CDMA receiver.

The pseudonoises are actually causal signals, i.e., nonrandom, but the noise is truly random. To get an answer, we shall lump the interfering signals and the system noise together and imagine that the system is observed for a very large number of trials, each with a different total-noise waveform sample, taken from a population of noise-waveform samples with the same statistics. By then averaging over these recurrences (*ensemble average*), the desired answer can be obtained.

If the random system noise has an energy (variance) over one bit time of $\mathcal{E}_n$, then we can consider the total interference energy in one bit time at the receiver input to be the sum of these energies

$$\mathcal{E}_{ni} = \mathcal{E}_n + (N - 1)\,\mathcal{E}_s \qquad (13.21)$$

(n for noise, s for signal, and i for input). The process of crosscorrelating this composite mixture of system noise and interference from other nodes against one of the two reference pseudonoises, say $X_1(t)$, can be thought of as adding up all the $2TW$ products gotten by multiplying the composite interference, sample by sample, against the

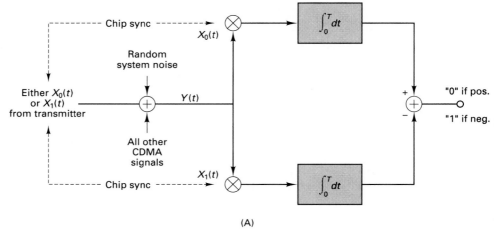

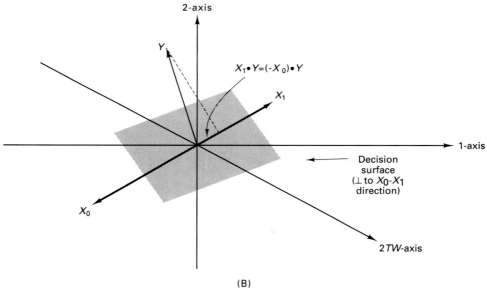

Figure 13-16. CDMA receiver. (A) Block diagram.   (B) Equivalent decision process in 2*TW*-space.

corresponding (time-aligned) samples of $X_1(t)$, namely x_i.   For convenience, assume that the reference $X_1(t)$ has unit energy $\mathcal{E}_x = 1$.   Then, as illustrated in Figure 13-17(A), each of the 2*TW* small intervals contains an average energy $\mathcal{E}_n/2TW$ and randomized amplitude and phase.   The signal is similarly constituted, and, averaged over many occurrences, the correlator output total noise energy $\mathcal{E}_{no}$ will be the sum of the squares of all 2*TW* of the little product vectors

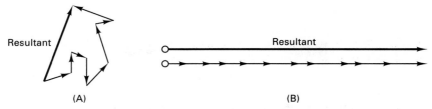

Figure 13-17. The phasor representation of the CDMA receiver correlation process (A) Local reference against received noise (including interfering transmissions). (B) Local reference against incoming replica (assuming the two are synchronized to within less than a chip).

$$\mathcal{E}_{no} = \left\langle \sum_{i=1}^{2TW} (\mathcal{E}_n/2TW) \right\rangle + (N-1) \sum_{i=1}^{2TW} \mathcal{E}_s/2TW = \mathcal{E}_n + (N-1)\mathcal{E}_s \qquad (13.22)$$

(o for output). As in Chapter 8, the $\langle\ \rangle$ indicates the ensemble average.

On the other hand, when the desired incoming X_1 from the transmitter will cross-correlate with the reference, all the little products are in phase (Figure 13-17(B)) and will produce an output energy which is the square of the sum

$$\mathcal{E}_{so} = \left[\sum_{i=1}^{2TW} x_i \right]^2 = 2TW\,\mathcal{E}_{si} \qquad (13.23)$$

Taking the ratio, the signal-to-noise (energy) ratio at the receiver output is

$$\rho_o = \frac{\mathcal{E}_{so}}{\mathcal{E}_{no}} = 2TW\,\frac{\mathcal{E}_{si}}{\mathcal{E}_{ni}} = 2TW\,\frac{1}{\mathcal{E}_n/\mathcal{E}_s + (N-1)} \qquad (13.24)$$

The second "equals" sign in this expression captures the whole point of CDMA, that the output signal-to-noise ratio is $2TW$ times the input signal-to-noise ratio. We can build a CDMA network with a desired number of nodes N that will achieve a desired bit error rate (given a low enough system noise) by making the processing gain $2TW$ suitably large. Given the desired value of ρ_o, and the system background noise level, we can determine the number of nodes N the CDMA network can support. In the case in which the system noise and the interferences from other nodes can all be approximated as gaussian processes, and $X_0(t) = -X_1(t)$, this expression can be substituted into the Q-function of Figure 8-7 to get the probability of bit error. For example, assuming the system noise to be of negligible power relative to a signal, a design objective of 10^{-9} for bit error rate will dictate that $2TW \geq 36\,(N-1)$.

Practical CDMA systems using radio transmission and advanced components have achieved processing gains $2TW$ of thousands to tens of thousands, for example by clever use of surface acoustic wave (SAW) pseudonoise waveform generators and matched filters [34].

Notice that, whereas a conventional digital receiver must maintain bit synchronization to a small fraction of the bit time T, with CDMA the synchronization tolerance is much more severe. Timing must be maintained to within a small fraction (say, one-tenth) of $1/W$, which can be less than T by three or four orders of magnitude. Nevertheless, in electronic (as distinct from optical) CDMA systems, a number of practical schemes exist for acquiring and maintaining sync [31].

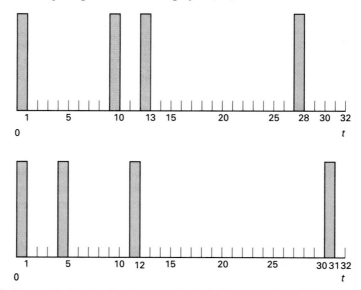

Figure 13-18. Two optical codes that have good correlation properties relative to each other.

Choice of waveforms

Having once chosen the 0 and 1 waveforms for one node as an antipodal pair, knowing that those for all the other $N-1$ other nodes in the network should be antipodal pairs doesn't help very much in reducing the mutual interference. The question is how to pick the pair of X s for each node *with respect to the pairs at all the other nodes*. Ideally, we would like to make them mutually orthogonal (zero correlation) under all relative time shifts τ, or, even better, negatively correlated with each other, as with BPSK. This problem has occupied designers of CDMA systems for years. A number of good solutions have been found only for the idealized situation in which the waveforms can take on negative as well as positive values (as in Figure 13-15), and furthermore in which the bits from all stations arrive time-aligned [35].

However, in real networks the interfering bit arrivals will not be time-aligned, and even more importantly, in real optical networks the chips will be *unipolar*, consisting of $+1$ s and 0 s, not *bipolar*, having $+1$ s and -1 s. Pseudonoise waveforms consisting entirely of 1s and 0s are called *optical codes* for this reason, and the state of the search for good optical codes, as summarized in [36, 37], has only yielded some bounds and some good rules for guiding enumerative computer searches for good codes, but no closed-form method of synthesizing optimal optical codes.

Figure 13-18 shows an example of two optical CDMA codes that have good properties with respect to each other. It is seen that there is no value of time shift for which the crosscorrelation would be greater than 1, whereas the autocorrelation would be 4. Good optical codes are very sparse in 1s, which means that to develop a certain energy per bit, either the peak power level or the number of chips per bit (or both) must be larger than for the traditional electronic CDMA systems that use waveforms in which every chip contains energy. Thus, while optical codes can be designed that

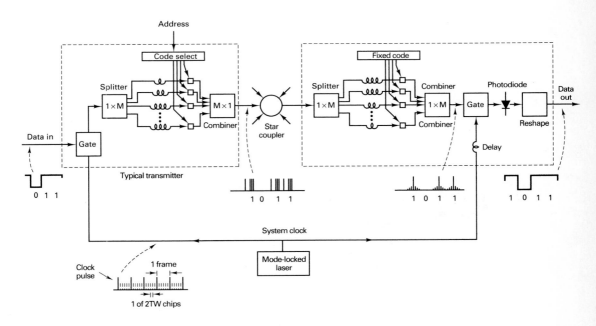

Figure 13-19. A CDMA optical network. (From [32]).

have few coincidences of 1s between the desired signal and the many interfering signals, the link budget suffers drastically. According to Equation (13.24), if there were no system noise in the system this could be tolerated, but in a real network with losses in star couplers or taps and with propagation over useful distances, this sparseness constraint on the signalling waveforms is a significant problem with optical CDMA in its current state of evolution. Also, since a CDMA chip is $2TW$ times as short as a bit, dispersion in the fiber can lead to smearing of adjacent chips in phys-

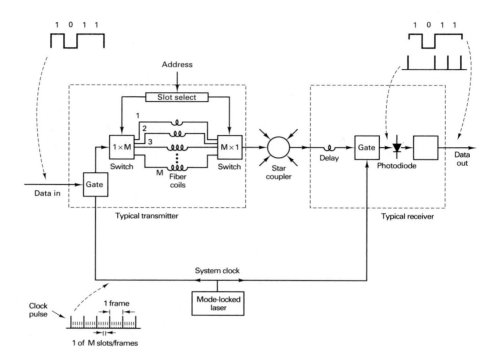

Figure 13-20. A TDMA optical network. (From [38]).

ically large networks working at any wavelength other than the zero-dispersion wavelength, and even there the higher orders of dispersion will be harmful (Section 3.6).

The current state of the component art for optical CDMA is limited to sending and receiving unipolar signals, but the search continues for clever CDMA technologies that employ bipolar signals [39].

Figure 13-19(A) shows an experimental CDMA network [32]. As usual, all nodes are connected together using a star coupler. Periodic clock pulses are distributed to each node from a central mode-locked laser source source. Each clock pulse, recurring every T seconds and lasting $\leq 1/W$ seconds, is converted by means of a *tapped delay line* (*transversal filter*) into either $X_0(t)$ or $X_1(t)$ consisting of 0s and 1s. At the receiver the two correlations are performed by sampling the outputs of a pair of *matched filters*, each made in the form of a tapped delay line, and having an impulse response $X_0(-t)$ and $X_1(-t)$, respectively. The clock pulse synchronizes the sampling.

13.10 Optical TDMA

Once we have take the trouble to distribute the clock pulse to each node and provide the corresponding receiver synchronization, we can build a fixed-assignment purely optical time-division multiaccess system, as in Figure 13-19(B), that is somewhat simpler than the CDMA system of Figure 13-19(A) [38].   What was formerly a CDMA chip now becomes a TDMA *slot*, and each slot is assigned to a different node. Node 1 sends a single bit ("0" or "1") in slot 1, node 2 in slot 2, and so forth. For a node to receive a bitstream from transmitter i it is only necessary to look in the i th time slot after the clock pulse that indicates the beginning of the bit time T seconds long.

The number of nodes that can be accommodated is $2TW$, a much greater number than for CDMA, but the modularity and "tell-and-go" flexibility of CDMA are lost, since the slot assignments must be coordinated across all nodes.

13.11 What We Have Learned

In this chapter, we gained a quantitative idea of the performance loss in trying to use first- and second-generation LAN and MAN multiaccess protocols for third-generation solutions. Either optical WDMA, CDMA, or fixed TDMA is required to exploit the bandwidth available in fibers for networks of any interesting physical size running at any of the interesting bitrates mentioned in Chapter 2.

We saw that both TDMA and CDMA pose such severe problems of synchronization and vulnerability to dispersion effects that WDMA is clearly the preferred direction.   All the third-generation circuit- and packet-switch multiaccess protocols that have been addressed in the literature take this approach.

As the reader can see, a definitive selection of the best possible packet-switch MAC protocols for third-generation networks has yet to be made. A number of clever protocols have been generated on an ad hoc basis, and there are the beginnings of some understanding about whether optimally to place the tunability at transmitter, receiver or both. However, this understanding is fairly superficial; also, those packet-switch protocols that have been proposed suffer from large delays due to contention or require either a large amount of protocol processing or a fairly wasteful use of the number of wavelengths available.

The area of protocols for third-generation lightwave networks is a very new one. As more and more networks are actually built and standards begin to be developed, there are certain to be many more additions to the repertoire that has been introduced in this chapter.

13.12 Problems

13.1 Discuss whether you would rather have a circuit-switch call blocked or suffer a packet loss.

13.2 Why is the finite-state machine of Figure 13-6 finite?

13.3 Discuss whether the analysis of "tunability at transmitter or receiver or both" in Section 13.4 is valid for packet switching.

13.4 How would you add concurrency (spatial reuse) to the simple model of FDDI in Section 13.5? That is, how could you change the protocol so that there is more than one packet on the ring at any given instant of time?

13.5 Discuss whether it is a fair comparison to make the length of a CSMA/CD bus the same as the circumference of a token ring as was done in Section 13.5.

13.6 If the two numbers representing a sinusoid (a form of bandwidth-limited signal) are amplitude and phase (they could just as easily have been in-phase and quadrature amplitudes, for example), then what are the two numbers for a ($\sin x/x$)-shaped pulse, which is also a band-limited waveform?

13.7 (*Regular simplex* signal set) The optimum pair of signals was said to be an antipodal pair in the case of a binary signalling alphabet (*m*-ary alphabet with $m = 2$). This signalling set spanned only one dimension, even though the signal space had $2TW$ dimensions. What do you think the optimum signal set is for $m = 3$, and how many dimensions does it span? How about $m = 4$?

13.8 Show that, in the absence of interference from other nodes, CDMA (spread spectrum) does not offer the improvement against natural noise that it does against bandwidth-limited noise of finite power. (Natural noise is characterized by a noise spectral power density, not a total power; that is, as the bandwidth of the receiver is opened up, more noise power enters).

13.9 Suppose you want to build a network of 243 nodes at 1.5 μ wavelength, spanning a maximum internode distance of 50 km. What is the maximum bit rate you could use for (A) TDMA. (B) CDMA with an error probability of 10^{-9}, assuming no natural noise in the system. (C) Multihop using the deBruijn physical-routing topology with $\Delta = 6$ and $k = 3$, assuming that the per-node electronics can run no faster than 1 Gb/s.

13.13 References

1.   D. D. Clark, "Abstraction and sharing," *Conf. Record, IEEE LEOS Topical Meeting on Optical Multiaccess Networks, Monterey, CA*, July, 1990.

2.   P. E. Green, ed., *Network Interconnection and Protocol Conversion*, IEEE Press, 1988.

3. P. E. Green and K. Naemura, R. C. Williamson, ed., "Heterogeneous computer network interconnection," *IEEE Jour. Selected Areas in Comm.*, vol. 8, no. 1, pp. 1-159, January, 1990.

4. P. E. Green, "Protocol conversion," *IEEE Trans. on Commun.*, vol. 34, no. 3, pp. 257-268, March, 1986.

5. R. Ramaswami and R. Pankaj, "Tunability needed in multi-channel networks: Transmitters, receivers, or both?," *IBM Research Report RC-16237*, October, 1990.

6. N. R. Dono, P. E. Green, K. Liu, R. Ramaswami, and F. F. Tong, "Wavelength division multiple access networks for computer communication," *IEEE Jour. Sel. Areas in Comm.*, vol. 8, no. 6, 1990.

7. F. J. Janniello, R. Ramaswami, and D. G. Steinberg, "A prototype circuit-switched multi-wavelength optical metropolitan-area network," *Conf. Record, IEEE Intern. Commun. Conference*, 1992.

8. D. Bertsekas and R. G. Gallager, *Data Networks - Second Edition*, Prentice Hall, 1987.

9. J. D. Spragins, J. L. Hammond, and K. Pawlikowski, *Telecommunications: Protocols and Design*, Addison-Wesley, 1991.

10. R. M. Metcalfe and D. R. Boggs, "Ethernet: Distributed packet switching for local computer networks," *Comm. ACM*, vol. 19, no. 7, pp. 395-404, 1976.

11. F. E. Ross, Fiber Digital Data Interface - Token Ring Media Access Control, American National Standards Institute X3.139, 1987.

12. J. L. Hammond and P. J. O'Reilly, *Performance Analysis of Local Computer Networks*, Addison-Wesley, 1986.

13. "Draft IEEE Standard 802.6, Distributed Queue Dual Bus (DQDB) Metropolitan Area Network (MAN)," *IEEE Computer Society*, June, 1988.

14. H. R. Mueller, M. M. Nassehi, J. W. Wong, E. A. Zurfluh, W. Bux, and P. Zafiropulo, "DQMA and CRMA: New access schemes for Gbit/s LANs and MANs," *Proc. IEEE Infocom*, pp. 185-191, 1990.

15. I. Cidon and Y. Ofek, "Metaring, a full-duplex ring with fairness and spatial reuse," *Conf. Record, IEEE INFOCOM*, pp. 969-981, 1990.

16. R. Ramaswami, "Multi-wavelength lightwave networks," *IEEE Commun. Magazine*, vol. 30, 1992.

17. P. A. Humblet, R. Ramaswami, and K. N. Sivarajan, "An efficient communication protocol for high-speed packet-switched multichannel networks," *Submitted to IEEE Jour. Selected Areas in Commun.*, March, 1992.

18. B. Mukherjee, "Architecture and protocols for WDM networks," *Univ. of Calif., Davis, Research Report CSE-91-32*, 1992.

19. I. M. Habbab, M. Kevehrad, and C. E. Sundberg, "Protocols for very high speed optical fiber local area networks using a passive star topology," *IEEE/OSA Jour. Lightwave Tech.*, vol. 5, no. 12, pp. 1782-1794, 1987.

20. N. Mehravari, "Performance and protocol improvements for very high speed optical fiber local area networks using a passive star topology," *IEEE/OSA Jour. Lightwave Tech.*, vol. 8, no. 4, pp. 520-530, 1990.

21. M. S. Chen, N. R. Dono, and R. Ramaswami, "A dynamic time division multi-access protocol for WDMA networks," *IEEE Jour. Selected Areas in Commun.*, vol. 8, no. 6, pp. 1048-1057, August, 1990.

22. M. S. Chen and T. S. Yum, "A conflict-free protocol for optical WDMA networks," *Conf. Record, IEEE Globecom*, pp. 1276-1281, 1991.

23. R. Chipalkatti, Z. Zhang, and A. S. Acampora, "High-speed communication protocols for optical star networks using WDM," *Conf. Record, IEEE Infocom*, 1992.

24. J. Lu and L. Kleinrock, "A wavelength division multiple access protocol for high-speed local area netowrks with a passive star topology," *Submitted to Performance Evaluation*, 1991.

25. G. N. Sudhakar, N. D. Georganas, and M. Kavehrad, "Slotted aloha and reservation aloha protocols for very high-speed optical fiber local area netowrks using passive star topology," *IEEE/OSA Jour. Lightwave Tech.*, vol. 9, no. 10, pp. 1411-1422, 1991.

26. C. S. Li, M. S. Chen, and F. K. Tong, "Architecture of a passive optical packet-switched wide-area network (POPSWAN) using WDMA," *IBM Research Report*, 1992.

27. M. M. Choy, S. M. Altieri, and K. Sivarajan, "A 200 Mb/s packet-switched WDM-SCM network using fast RF tuning," *Submitted to IEEE Journal of Lightwave Technology*, March, 1992.

28. K. Sivarajan and R. Ramaswami, "Multihop lightwave networks based on deBruijn graphs," *Submitted to IEEE Trans. on Comm.*, vol. 39, 1991.

29. R. Pankaj, "Comparison between single hop and multihop schemes on a passive star coupler based lightwave network," *Private communication*, July, 1990.

30. R. L. Rivest, "Network control by Bayesian broadcast," *MIT Lab. for Computer Sci., Tech. Report TM-285*, 1985.

31. M. K. Simon, J. K. Omura, R. A. Scholtz, and B. K. Levitt, *Spread spectrum communications (3 vols.)*, Computer Science Press, 1985.

32. P. R. Prucnal, M. A. Santoro, and T. R. Fan, "Spread spectrum fiber-optic local area network using optical processing," *IEEE/OSA Jour. Lightwave Tech.*, vol. 4, no. 5, pp. 547-554, 1986.

33. J. A. Salehi, "Code-division multiple-access techniques in optical fiber networks–Part I: Fundamental principles," *IEEE Trans. on Commun.*, vol. 37, no. 8, pp. 824-833, 1990.

34. C. K. Campbell, "Applications of surface acoustic and shallow bulk acoustic wave devices," *Proc. IEEE*, vol. 77, no. 10, pp. 1453-1484, October, 1989.

35. J. G. Proakis, *Digital Communications*, McGraw-Hill, chap. 8.3, 1983.

36. R. Gagliardi, J. Robbins, and H. Taylor, "Acquisition sequences in PPM communications," *IEEE Trans. on Info. Theory*, vol. 33, no. 5, pp. 738-744, 1987.

37. F. R. Chung, J. A. Salehi, and V. K. Wei, "Optical orthogonal codes: Design, analysis and applications," *IEEE Trans. on Info. Theory*, vol. 35, no. 3, pp. 595-604, 1989.

38. P. R. Prucnal, M. A. Santoro, and S. K. Sehgal, "TDMA fibre-optic network with optical processing," *Elect. Ltrs.*, vol. 22, no. 23, pp. 1218-1219, November, 1986.

39. J. A. Salehi, A. M. Weiner, and J. P. Heritage, "Coherent ultrashort light pulse code-division multiple access communicaton systems," *IEEE/OSA Jour. Lightwave Tech.*, vol. 8, no. 3, pp. 478-491, 1990.

Part IV
REALIZATION

CHAPTER 14

Operating
Third-Generation Links

14.1 Overview

The reader will recall that at the outset of this volume, a topological distinction was made between links, multipoints, and networks, and a second distinction was made on the basis of generations of such systems. We defined the first-generation systems as those not using fiber, the second generation as those that substituted fiber for copper without changing the architecture, and the third generation, the subject of this book, as those that exploited some of the unique properties of fiber technology.

Having spent many chapters discussing building blocks and then architecture, we now close this volume with three brief chapters illustrating how the different building blocks have been assembled into real third-generation systems. These chapters deal, respectively, with links, multipoints, and networks. In each area, a number of systems are listed, and then a particularly interesting one is discussed as a case study.

In all three areas, particularly links, a great many laboratory experiments have explored basic concepts. As long as the link experiments were strictly laboratory efforts, no attempt will be made to cover them here, since so many have dealt with ideas not yet ready for the test of practical field usage. A good example would be soliton links. Such links have been operated in the laboratory using repeated passes through a single spool of fiber to simulate thousands of kilometers of propagation distance. However, since real thousand-kilometer soliton links carrying real bit traffic have yet to be built and tested outside the research laboratory, we shall not cover soliton-based systems here.

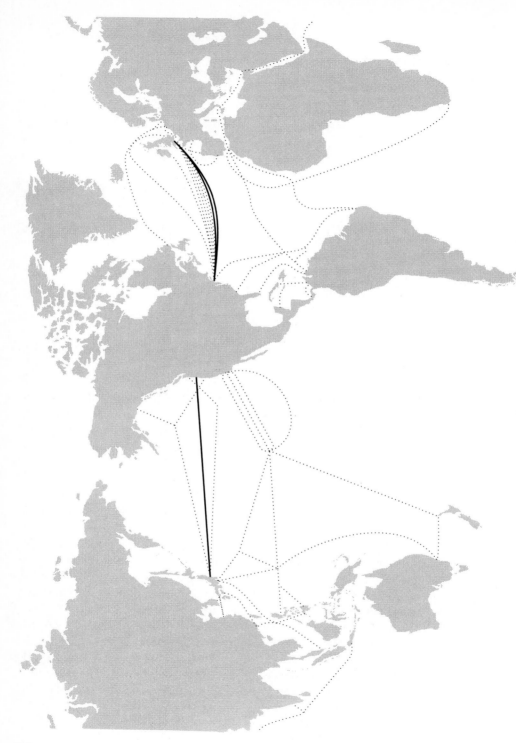

Figure 14-1. Undersea fiber optic links. Heavy lines: those in service as of Year End 1991. Light lines: those planned. (Based on data from AT&T and British Telecom).

What will be emphasized, instead, are those ideas that have either seen service usage or at least have been used to carry real user traffic in real "field trials" or "operating prototypes." In other words, wherever possible the criterion will be whether the system ideas got out of the laboratory or not.

In the link area, the interesting things that are happening involve intercity and intercontinental links. Figure 1-1 pretty well captures what the issues are; to maximize the total bitrate carried and also to maximize the distance between conversions from photonic form back to electronic form.

In recent developments of point-to-point links, one may discern several currents of third-generation system thinking. The use of coherent optical detection is certainly one of these themes, as is the use of photonic amplification. Wavelength-division multiplexing is still another. We give a sampling of such realizations in this chapter.

14.2 Second-Generation Link Architectures

For some years there have been several standard hierarchies of successively higher-speed time-division transmission standards, all based on 125-microsecond framing. Recently all these have been defined assuming that the highest hierarchical level runs on fiber optic transmission facilities.

The 125-microsecond frame rate was dictated by a choice of sampling rate made thirty years ago for voice, which dominated and still dominates the design of all transmission facilities offered by the world's carriers. Rates such as T1 (1.544 Mb/s in the United States, 2.048 elsewhere) and T3 (44.736 Mb/s in the United States and 34.368 elsewhere) have been typical. T1 and T3 are also known as DS-1 and DS-3, respectively, DS standing for "Digital Service." Higher bitrates always existed in the interior of the telephone network, but, until recently, these rates were limited to the capability of microwave transmission facilities. Now the maximum internal rates are on the order of several Gb/s, and T1 and also T3 are become widespread in the United States as leased subscriber facilities provided to businesses.

Thus, with the widespread installation of truly practical fiber links, beginning about ten years ago, a push toward very high bitrates began, still keeping the 125-microsecond framing. Table 14-1 presents a broad synopsis of the two standards that have grown up, side by side [1, 2]. The Standard Optical Network (SONET) convention of the United States has evolved into the Synchronous Digital Hierarchy (SDH), being developed into an international standard by CCITT. In the table, *OC* means "Optical Carrier" and *STM* means "Synchronous Transport Module."

In Chapter 16, we shall mention some of the multiaccess structures being implemented on top of SONET and SDH.

Perhaps the most spectacular second-generation link has been the fiber-optic undersea cable [3]. Figure 14-1 shows installed and planned undersea fiber facilities. TAT-8, installed in 1988 runs at 280 Mb/s to carry 4200 64-Kb/s voice circuits. There is a repeater every 70 kilometers, and each repeater has two lasers with a mechanical switchover if one of them fails. If installation plans proceed as scheduled, the present two transatlantic cables will become nine by 1995, representing a 30-fold increase in capacity. Similar interconnections are going into service in the Pacific.

Table 14-1. SONET and SDH bitrates

SONET	SDH	Mb/s
OC-1		51.840
OC-3	STM-1	155.520
OC-9		466.560
OC-12	STM-4	622.080
OC-18		933.120
OC-24		1,244.16
OC-36		1,866.24
OC-48	STM-16	2,488.32

14.3 Coherent Links

As mentioned in Chapter 8, as long as one is not attempting to send multiple bitstreams at multiple wavelengths, the limited tuning range of today's tunable lasers is ample to cover system drifts, and partly because of this, single-wavelength coherent links have in the last two years reached the stage of significant field trials. External-cavity lasers were a sufficient solution for the local oscillators, and therefore much of the development effort centered on another big problem, handling the slow drift of received state of polarization (Section 3.13).

A historical review of various coherent link field tests has been given in [4]. The earliest field-tested system was reported by the Japanese KDD group [5], who in 1987 used polarization diversity to achieve 560-Mb/s transmission over a 95-kilometer round-trip loop of undersea cable. Very shortly thereafter, British Telecom Laboratory [6] began a series of field demonstrations involving telephone exchanges in England, Scotland, and the Channel Islands, with bitrates in the half-gigabit range using FSK and DPSK. AT&T conducted a series of field tests of 1.7- and 2.5-Gb/s coherent systems using FSK to achieve repeaterless distances of over 400 km [7]. Typical of later achievements in Japan was a 2.5-Gb/s transmission experiment using coherent reception over undersea cable distances up to 157 km [8]. Many of the field tests involving coherent detection used some form of photonic amplifier in the link, so that the improved receiver detectability limits achieved were sometimes not entirely due to coherent detection.

It will be useful to pick one of these systems and use it to illustrate how the ideas of the preceding chapters were applied. The classic 1988 British Telecom experiment serves as a good example. Figure 14-2 shows a block diagram of the system that was run for some months between Cambridge and Bedford, 44 kilometers away.

Four fiber strands in the installed plant between the exchanges were spliced to get a total length ranging from 44 to 176 km. Since DPSK was used, both transmitter and local-oscillator lasers were obliged to have very small linewidths, as discussed in Section 8.13. And, of course, the LO had to have some degree of tunability. For these reasons, an external-cavity laser essentially identical with that shown in Figure 5.28 was used for both these functions. External reflections, which would otherwise have enlarged the linewidth, were suppressed by a Faraday-rotation isolator (Section 3.13), and the signal was phase-modulated using a lithium niobate Mach-Zehnder phase modulator, as described in Section 7.3. The photonic amplifier (Chapter 6) was a travelling-wave laser-diode amplifier, which was replaced in subsequent British Telecom experiments with erbium-doped-fiber technology.

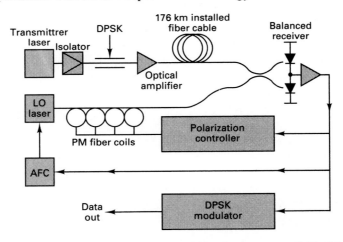

Figure 14-2. Block diagram of British Telecom's 1988 coherent system field trial.

The treatment of polarization (discussed in Section 8.12) was particularly interesting. Full tracking of the SOP was used, instead of polarization diversity or polarization scrambling, which, though convenient, both introduce a loss of detectability [9]. The SOP was measured, then calculations in a microcomputer determined in real time how much voltage to apply to each of a pair of piezoelectric cylinders, around which the fiber was wrapped. This portion of the fiber run was made of polarization-maintaining fiber so as to have a particularly high birefringence ($n_e - n_o$) (Section 3.13). The voltage-controlled expansion of these cylinders produced small changes in the eccentricity of the fiber core, which in turn introduced a change in birefringence. The two stretchers were arranged so that they produced trajectories on the Poincaré sphere that were great circles at 90° from each other, thus allowing any SOP point on the sphere to be reached by a suitable combination of the two voltages. As might be imagined, early versions of this system were quite bulky, but later commercialized versions were contained in a volume the size of a can of tennis balls, not including the microprocessor.

The usual balanced receiver structure was used (Section 8.13), and the DPSK demodulator at the intermediate frequency consisted simply of a delay line one bit duration long, followed by a comparator that determined which was most likely: that the phase between bit times had stayed constant or that it had been reversed.

Figure 14-3. Series of undersea fiber trunks linking west coast Italian and Sicilian cities, as of Year End 1991. The link connecting Pomezia, Formia and Napoli uses erbium doped fiber amplifiers.

Experiments such as this one have shown that, given enough attention to the messy details of laser linewidth, polarization control, cancellation of laser amplifier noise, and precise control of optical carrier frequency, coherent detection can be made to work in operational systems. Whether coherent reception will remain a candidate that is competitive with the simpler amplifier-plus-direct-detection option is not so obvious.

14.4 Direct-Detection Links Using Photonic Amplification

For a long time, it was thought that only coherent techniques were capable of pushing link performance anywhere near the quantum limits discussed in Chapter 8, but once it was realized that, with ASK modulation, direct-detection receivers that used optical preamplification at the receiver could do almost as well (Section 8.11), a number of field installations have exploited this system direction.

One of the most interesting and most advanced installations is the bidirectional link between the three Italian cities of Pomezia, Formia, and Napoli (Figure 14-3). This link forms part of the chain connecting many of the Italian mainland and Sicilian cities. This technique, which has been given the quaint designation *coastal stitching*, is being widely considered elsewhere for linking long chains of coastal cities.

In the case shown in Figure 14-3, there were formerly two bidirectional 565-Mb/s links running at 1.3-μ wavelength, one between Pomezia and Formia and the second between Formia and Napoli. At Formia, the signal had been electronically regenerated. In order to simplify the system and reduce costs, the Formia station was relegated to the role of a simple splice point to make the two links into one continuous link, whereupon the total length became 263 km, a distance too great to be spanned without the aid of amplification. Therefore, in the southbound direction, the arriving 1.3-μ signal at Pomezia was remodulated onto a 1.5-μ wavelength, optically amplified by about 20 dB, and upon reception at Napoli was passed through a predetection optical amplification having another 6 dB of gain, and then remodulated back to 1.3 μ. The same thing was done with the second fiber path running in the northbound direction.

Installations such as the one in Italy have very quickly proved that erbium-doped-fiber amplifiers can be economically and reliably used for installed systems. Current plans are to install fiber amplifiers by 1966 in those undersea systems in Figure 14-1 shown as solid lines [10].

14.5 Wavelength-Division Links

In a classic 1984 laboratory experiment, Olsson, et al., demonstrated convincingly that WDM offers a simple way of opening up very large capacities on already-installed fiber paths. They used ten wavelengths to send a total of 1370 Gb/s over a 68-km spool of fiber [11] This is plotted in Figure 1-1 as the point ×.

Since that time, there has been much discussion of using "dense WDM" not only for networks, but also for links. Most of the coherent links and amplifier-based links that have been field-tested in real systems have had as a stated objective the ultimate use of dense WDM (e.g., [4, 6]).

In spite of the promise of WDM for link applications, there has been little attempt to implement very large numbers N of independent WDM channels on an operational single link. Typical of direct-detection WDM links was an $N = 4$ field test at 1.7 Gb/s per wavelength conducted by ATT over a 70-km link [12].

The use of WDM has been resisted for extremely long links that require many cascaded optical amplifiers, partly because this compounds the unevenness of gain with wavelength illustrated in Figure 6-11. The principal spectral feature that causes the problem is the erbium resonance at 1532 nm. Clever passive optical filtering schemes are being devised that place a transfer-function minimum at that wavelength, and this evens out the gain spectrum considerably. Some of these schemes involve fabricating systematic changes of core cross-section along some length of the doped-fiber path [13, 14].

14.6 What We Have Learned

The fiber optic revolution affected the way links are built long before it did multi-points and networks. The common-carrier research community has brought out of the laboratory and into real field tests almost every idea we have covered in the earlier chapters of this book that could help increase the bitrate and the distance between photonic-electronic conversions.

Although the research literature is beginning to show a discernible broadening of interest away from links and toward multipoints and especially networks, it is still dominated by the emphasis on point-to-point link questions.

At the level of real systems, coherent detection, optical amplification, and wavelength-division multiplexing dominate the picture. We have not discussed in this book some of the link-related issues that once promised important improvements, since they either have proved impractical or are not yet sufficiently mature. Among the former ideas are dispersion-shifted fibers, post-detection equalization of fiber dispersion, Raman amplification and others. Among the latter are the quantum squeezing of light and ultra-high-speed digital transmission using solitons.

14.7 References

1.   R. Ballart and Y. C. Ching, "SONET: Now it's the Standard Optical Network," *IEEE Commun. Mag.*, vol. 27, no. 3, pp. 8-15, March, 1989.
2.   T. Miki and C. A. Siller, eds., "Special Issue on the Synchronous Digital Hierarchy," *IEEE Commun. Mag.*, vol. 28, no. 8, pp. 1-65, 1990.
3.   P. K. Runge and P. R. Trischitta, eds., *Undersea Lightwave Communication*, IEEE Press/IEEE Commun. Soc., New York, 1986.

4.   M. C. Brain, "Coherent transmission field demonstrations," *Conf. Record, Optical Fiber Commun. Conf.*, p. 29, 1991.

5.   S. Ryu, S. Yamamoto, Y. Namihara, K. Mochizuki, and H. Wakabayashi, "First sea trial of FSK heterodyne optical transmission system using polarization diversity," *Elect. Ltrs.*, vol. 24, no. 7, pp. 399-400, March, 1988.

6.   M. J. Creaner, R. C. Steele, G. R. Walker, N. G. Walker, J. Mellis, S. Al-Chalabi, I. Sturgess, M. Rutherford, J. Davidson, and M. Brain, "Field demonstration of 565 Mbit/s DPSK coherent transmission system over 176 km of installed fibre," *Elect. Ltrs.*, vol. 24, no. 22, pp. 1354-1356, 1988.

7.   Y. K. Park, "1.7 Gb/s coherent optical transmission field trial," *Conf. Record, Optical Fiber Commun. Conf.*, p. 30, 1991.

8.   T. Imai, Y. Hiyashi, N. Ohkawa, T. Sugie, Y. Ichihashi, and T. Ito, "Field demonstration of 2.5 Gbit/s coherent optical on transmission through installed submarine fibre cables," *Elect. Ltrs.*, vol. 26, no. 17, pp. 1407-1409, August, 1990.

9.   G. R. Walker and N. G. Walker, "Rugged all-fibre endless polarisation controller," *Elect. Ltrs.*, vol. 24, no. 22, pp. 1353-1354, 1988.

10. P. K. Rudge and J. M. Sipress, "Advances in international communications via undersea systems," *Conf. Proc. Telecom-91*, p. 7, October, 1991.

11. N. A. Olsson, J. Hegarty, R. A. Logan, L. F. Johnson, K. L. Walker, L. G. Cohen, B. L. Kasper, and J. C. Campbell, "68.3 km. transmission with 1.37 Tbitkm/s capacity using wavelength division multiplexing of ten single-frequency lasers at 1.5 micrometers," *Elect. Ltrs.*, vol. 21, pp. 105-106, January, 1985.

12. J. A. Nagel, D. A. Fishman, and S. M. Bahsoun, "Multigigabit capacity 1.5," *Conf. Record, Optical Fiber Commun. Conf.*, p. 28, 1991.

13. M. Wilkinson, A. Bebbington, S. A. Cassidy, and P. McKee, "D-fibre filter for erbium gain spectrum flattening," *Elect. Ltrs.*, vol. 28, no. 2, pp. 131-132, January, 1992.

14. Ampliphos Erbium Fiber Optical Amplifier, Pirelli Cable Corporation, 1991.

CHAPTER 15

Operating Third-Generation Multipoints

15.1 Overview

The physical topology we have been arbitrarily calling a "multipoint" consists of one central node and a collection of subsidiary nodes. Figure 15-1 shows several multipoint physical-network topologies.

In this chapter, we shall mention current activity of a second-generation type in making direct replacements of copper with fiber and then discuss current work that uses the more advanced possibilities of optical fiber that merit the designation "third generation."

15.2 Second-Generation Multipoints

The dominant application area for multipoints is distribution of entertainment material, principally television. At this time the technical and economic directions of the video and audio distribution industry are in a state of flux. A particularly compelling issue concerns competition between *telco* (telephone company, or common carrier) and cable television (CATV) industries over control of future deployment of fiber as a substitute for copper in distributing digital audio and analog or digital video, particularly high-definition television (HDTV). It has proved difficult to identify reasons that businesses have an early need for HDTV-level display on a broad scale, and so the emphasis of both telco and cable industries seems to be on bringing fiber to the home rather than to the office.

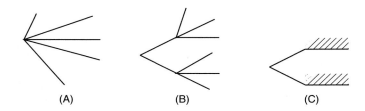

(A)   (B)   (C)

Figure 15-1. Four physical network topologies underlying a multipoint physical path topology (dotted lines). (A) Star, (B) Double star, as implemented by the common carriers, and (C) Star-herringbone, as implemented by cable television companies.

Some of the HDTV technical issues concern the choice of standards, the timing of HDTV's introduction at the consumer level and the question of delivery medium, coaxial cable, fiber-optic cable, and satellite direct broadcast. There seems to be general agreement that modulation formats, for video at least, will continue to be analog for today's TV standards, but that all video will eventually be in either compressed or uncompressed digital form.

As discussed in Chapter 9, both FM and digital video have the advantage over AM of requiring only 16-18 dB of carrier-to-noise ratio for studio-quality picture transmission, whereas the corresponding requirement for AM is 56 dB. Still, the universal use of AM in consumer TV receivers today means that the outermost part of the distribution system must use AM. This then often dictates that the entire multipoint video distribution system remain AM all the way back from the home or office to the source of program material.

The telco community, which has almost singlehandedly borne the burden of research and technology development in optical communication, starts with a pervasive installed base of 22-gauge copper. The bandwidth of this installed base is impossibly narrow, but the ubiquitous presence of this form of copper interconnection, plus worldwide coordination of voice communication standards and architectures forms a strong base on which to launch the future "fiber to the home" or "fiber to the office."

The cable television industry, on the other hand, though fragmented and dependent on others for technical advances, possesses a growing penetration of installed coaxial cable having up to 1-GHz bandwidth, experience with wideband analog distribution, and (probably most importantly) significant control over program material, or at least superior access to it.

Both industries are already hard at work on extending fiber from central points out toward individual residences and offices. In the telco case, the central points are the individual central exchanges. In the cable TV case, they are "headends," typically receive-only satellite earth stations. Numerous economic studies by both groups have shown that costs have yet to decrease to the point that fiber paths all the way to the subscriber are sufficiently economical. Replacing copper with fiber for entertainment distribution is extremely cost-sensitive, since it must compete with other technologies whose costs are also dropping rapidly, notably direct-broadcast satellite dishes and video rentals. On the other hand, the savings to be had by replacing copper with fiber out to distribution points that are usually within a small number of kilometers of a subscriber are already great enough to stimulate widespread installation of fiber for that part of the system.

Thus, one is hearing less about "fiber to the home" as an immediate possibility, and more about "fiber to the curb" (telephone companies [1]) and "the fiber backbone" (cable TV companies [2]).

The favored topology for telephone connections has been the double star of Figure 15-1(B). The outer layer of spokes consists of individual copper pairs. The inner star formerly consisted simply of bundles of pairs, one pair for each subscriber, but in recent years the spokes of the inner star have gradually been replaced by a small number of copper pairs. On each pair, a *subscriber loop multiplexor* is used, usually employing TDM to support a number of voice channels on one copper pair.

For cable TV companies, the star-herringbone or "tree-and-branch" of Figure 15-1(C) has been a favored physical topology in the United States, while in Canada [3] a hierarchy of rings seems to be the topology of choice. Until recently, coaxial cable has been the technology of choice throughout both hierarchical levels of Figure 15-1(C). At the center is the *headend*, involving such sources of program material as satellite earth stations, off-the-air antennas, videotape machines, and (occasionally) incoming trunks from other headends. The spokes out to the beginning of the herringbone arrangement are the *trunks*, and it is here that fiber optic technology has been most effective.

The distribution point at the end of a cable TV trunk has been given various names, usually being called simply a "node." It plays very roughly the same role as the subscriber loop multiplexor in telco voice installations. The "last kilometer" subsystem from node to individual residence drop-off points constitutes by far the major investment in the entire cable TV plant. Therefore the replacement of copper with fiber, which is becoming widespread, is limited to the central star for cable television, just as it has been with telco use of fiber.

Figure 15-2 shows a "before-and-after" view of the current wave of retrofit within the cable TV industry that goes under the name *fiber backbone* [4]. All coaxial cable trunks are replaced with fiber, and the conversion from photons back to electrons takes place at each of the nodes. The advantage of fiber trunking over the traditional coaxial cable trunking has been mainly in the lower loss rather in the wider bandwidth available. Coaxial trunks and distribution lines require many cascaded high-quality broadband RF amplifiers, typically one every 1000 to 2000 feet, eventually causing degradations of signal quality. Single-mode fiber has usually allowed this function to be done with no amplification.

A *supertrunk* is a connection between headends, sometimes over intercity distances. This allows different headends to exchange program material, each being able to add and drop program material tailored to separate audiences. Fiber links are now

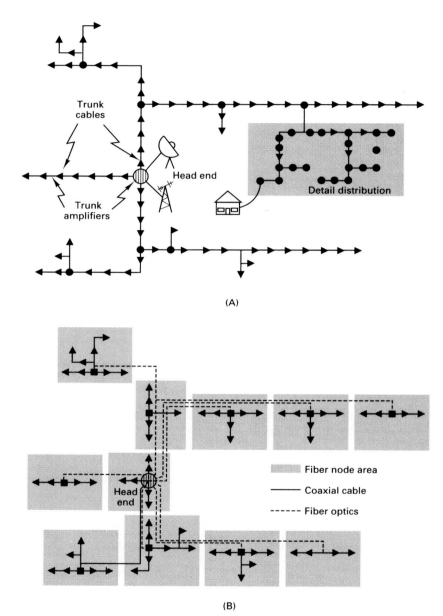

Figure 15-2. Typical fiber backbone cable television installation. (A) Preexisting all-coax distribution system. Each triangle is a separate wideband RF amplifier. (B) A star of fibers, usually without amplifiers, replaces the central star of coax with many RF amplifiers. Some RF amplifiers are reversed. (From [4]).

the preferred medium for CATV supertrunks [5]. Recently, GTE put into operational test service an 80-channel supertrunk system [6] in order to bring headend services to the community of Cerritos, California, where a "video-on-demand" system is being field-tested in a number of homes.

All fiber-optic trunks, supertrunks, and the occasional subscriber link use the SCM techniques (subcarrier multiplexing) that were described in Chapter 9. This has required laser diodes of uncommonly high linearity and low *RIN* [7].

The trend toward eventually having an all-fiber path from hub to subscriber will lead to third-generation "all-optical" multipoints in the same sense that "all-optical" networks are being developed, as discussed in the next chapter. Nippon Telephone and Telegraph (NTT) has announced plans to replace 90 percent of the copper in the local subscriber routes with fiber running Broadband Integrated Services Data Networking (BISDN) services by the year 2015 [9].

Those third-generation ideas that have been most widely used to date in either telco or cable TV multipoints involve splitting techniques (*passive optical networks, PONs*), the use of optical amplifiers to provide an all-optical, very long trunking system, SCM (the subject of Chapter 9) and, to a minor extent, WDM.

15.3 Passive Optical Networks (PONs)

The success of fiber-fed subscriber loop multiplexors and of fiber CATV trunking led recently to further extension of fiber technology. British Telecom Laboratories seems to have been the first to exploit the potential of passive splitters to allow an inexpensive means of extending the fiber beyond the subscriber loop multiplexor or node. BT has pioneered the use of the passive double-star configuration, under the title *passive optical networks* [8]. Several American vendors and continental European countries have recently begun to follow BT's lead to various degrees [10].

In the United Kingdom, 90 percent of the subscriber lines are less than 3 kilometers in length, whereas in the United States, the lengths are at least twice that figure. When the distance attenuation is low, it is possible to think of tolerating a large splitting loss in broadcasting the same signal to many subscribers in exchange for the savings on remote multiplex/demultiplex equipment. The PON concept involves replacing the active subscriber loop multiplexor using copper or fiber with an all-fiber system in which passive optical splitters form the boundary between the central star and the subsidiary stars.

A typical arrangement is shown in Figure 15-3. Bidirectional voice capability is provided by broadcasting and "broadgathering" as many as 294 64-Kb/s digital voice signals to and from a maximum of 128 fiber ends, each one terminating on the subscribers' premises. Thus, at the headend there is only one transmitter and one receiver. This *Telephony Passive Optical Network (TPON)* capability is all at one optical wavelength, while concurrently, at an optional set of other wavelengths, *Broadband PON (BPON)* is proposed. Commercially available optical filters, such as the Fiber Fabry Perot of Figure 4-9, can be used to prevent BPON channels from interfering with TPON, and to separate individual BPON channels from each other. It is expected that

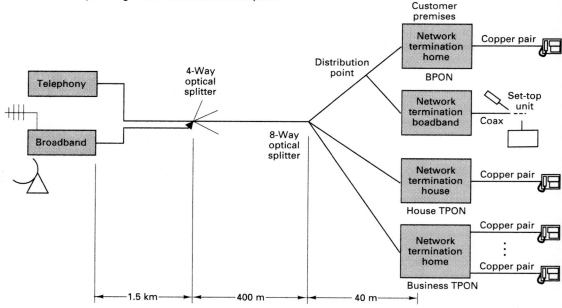

Figure 15-3. Typical Passive Optical Network (PON), as proposed and field tested by British Telecom Laboratories. (From [8]).

the BPON channels will be used for video and for services based on the transmission of ATM (Asynchronous Transfer Mode) switching—for example, Broadband Integrated Services Data Network (BISDN). Both ATM and BISDN are discussed briefly in the next chapter.

A series of field trials of TPON have been underway for several years, using a customer set of 400 business and residential customers. The services provided have included telephony, stereo audio, videotex, and (using multiple TPON channels) cable television.

15.4 Use of Photonic Amplification

The cable television industry is currently recognized as forming a large mass market for doped-fiber amplifiers, considerably greater than that generated by undersea cables and intercity telco trunks. All manufacturers of erbium doped-fiber amplifiers are vying for a share of this emerging market with erbium doped-fiber amplifiers of either the power-amplifier or line-amplifier type. Many American CATV operators have funded plans to place erbium doped-fiber amplifiers in service use as power amplifiers.

15.5 Multipoints Using Wavelength Division

While subcarrier multiplexing (SCM) onto a single laser transmission is a widely installed practice, as mentioned in Section 15.2, the use of WDM is in its infancy. This is not hard to understand, because the cost of the DFB lasers required in order to avoid crosstalk when doing wavelength demultiplexing is much higher than the equivalent fixed-tuned RF equipment. The SCM approach requires only one transmitting laser, and even though the laser diodes are especially costly, they are still cheaper than tens of lasers and filters. For SCM video service, the requirements are for high linearity (low CSO and CTB) and low internal noise (*RIN* below − 150 dBW) [7].

As we shall see in the next chapter, this preference is reversed in the case of networks, since transmitters are not collocated. SCM techniques have proved to be more expensive than WDM techniques for networks.

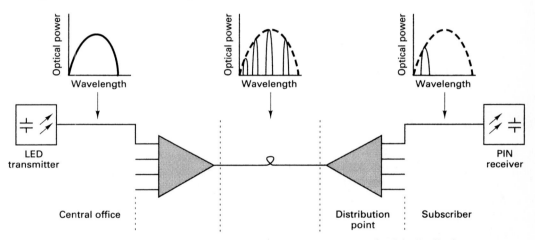

Figure 15-4. LED slicing as used by British Telecom for a low-cost 2 Mb/s distribution system. (From [11]).

15.6 Multipoints Using LED Slicing

Several years ago, British Telecom took the idea of LED slicing out of the laboratory and into a series of field tests [11] in London. LED slicing is discussed in Section 5.14.

The hub of the multipoint was at a telephone central office and the end points were 3.5 km away. The system is illustrated in Figure 15-4. Each of 8 LEDs was modulated with its own ASK bitstream and each was fed to a different input of an eight-way fixed grating MUX unit, such as those described in Section 4.12. In spite of the fact that each LED only developed 10 microwatts of power and the additional fact that only 0.01 of this power was passed by each grating, it was possible to send a 2-Mb/s signal over the system with 10^{-11} bit error rate.

This interesting experiment demonstrated a relatively simple and inexpensive approach to the distribution of fairly low data rates over modest distances. The cost reduction came principally in the use at the transmitter of the combination of N LEDs and one grating instead of the more expensive approach of providing N DFB laser diodes.

15.7 What We Have Learned

A high degree of architectural and technology sophistication does not seem to have permeated the multipoint form of real system realizations as much as it has for networks (next chapter) and especially for links (last chapter). This is probably due to the less technically aggressive character of the cable television industry when compared to the telco industry. As competition between the two increases, as exemplified by the PON trials, we may expect this to change.

Meanwhile, third-generation content in multipoint systems is dominated by SCM systems operating on a single wavelength channel.

15.8 References

1. G. T. Hawley and P. W. Shumate, Jr., eds., "Special Issue on the 21st Century Subscriber Loop," *IEEE Commun. Mag.*, vol. 29, no. 3, pp. 24-119, March, 1991.

2. J. A. Chiddix, "Fiber backbone trunking in cable television networks: An evolutionary adoption of new technology," *IEEE LCS Magazine*, vol. 1, no. 1, pp. 32-37, February, 1990.

3. G. M. Hart and N. F. Hamilton-Percy, "A broadband urban hybrid coaxial/fiber telecommunications network," *IEEE LCS Magazine*, vol. 1, no. 1, pp. 38-45, February, 1990.

4. J. A. Chiddix, H. Laor, D. M. Pangrac, L. D. Williamson, and R. W. Wolfe, "AM video on fiber in CATV systems: Need and implementation," *IEEE Jour. Selected Areas in Commun.*, vol. 8, no. 7, pp. 1229-1239, September, 1990.

5. J. A. Chiddix, "Optical fiber supertrunking–A performance report on a real world system," *IEEE Jour. Selected Areas in Commun.*, vol. 4, no. 5, pp. 758-769, August, 1986.

6. R. F. Kearns, V. S. Shukla, P. N. Baum, and L. W. Ulbricht, "An 80-channel high-performance video transport system over fiber using FM-SCM techniques for super-trunk applications," *Conf. Record, 40th Annual Meeting, Nat. Cable Telev.*, pp. 71-76, March, 1991.

7. R. Pidgeon, F. Little, and L. Thompson, "Developments in laser technology for CATV," *Conf. Record, Fiber Optics Plus, Soc. Cable Telev. Engrs.*, pp. 99-104, January, 1992.

8. T. R. Rowbotham, "Local loop developments in the U.K.," *IEEE Commun. Magazine*, vol. 29, no. 3, pp. 50-59, 1991.

9. T. Miki, I. Yamashita, and K. Okada, "Development concept on NTT's optical subscriber network system," *Conf. Record, IEEE Workshop on Passive Optical Networks for the Local Loop*, pp. 1.3/1 - 1.3/5, London, May, 1990.

10. K. Pyle, "A bandwidth transport, fiber to the curb system," *Conf. Record, Fiber Optics Plus, Soc. Cable Telev. Engrs.*, pp. 239-247, January, 1992.

11. A. R. Hunwicks, L. Bickers, and P. Rogerson, "A spectrally sliced, single-mode, optical transmission system installed in the U.K. local loop network," *Conf. Record, IEEE Globecom*, pp. 1303-1307, 1990.

CHAPTER 16

Operating Third-Generation Networks

16.1 Overview

In discussing real third-generation networks, it is important at the outset to recognize exactly what the differences are between the world of links and the world of networks. Fiber optic communication originated in the common-carrier laboratories of the world, and this origin has been the overriding influence in link transmission ever since. Chapter 14, therefore, was entirely an account of the use of third-generation optical communication ideas by carriers. Fiber optic communication's effect on any-to-any networks, which are the traditional domain of the data processing industry, came much later, and encountered a different set of priorities.

It is not always appreciated how extensive the differences are between the carriers' world of fiber links and the computer industry's world of fiber optic networks. Clearly there are many cultural differences between a community used to worldwide cooperation and standardization, and one used to rapid and competitive change. But in addition to cultural diferences there are many differences that are quite specifically technical. These are summarized in Table 16-1.

It is useful to keep these technical differences in mind in observing which technologies and architectures seem to be well-suited to one community and yet don't seem to catch on in the other.

In this chapter, as in the preceding two, we first mention the systems that the reader will encounter that belong to the second generation, and then proceed to third-generation networks. Because cost is such a factor with networks, third-generation designs have concentrated on wavelength division with tunable-filter receivers and direct detection.

Table 16-1.   Differences in Emphasis between Common-Carrier Links and Computer Networks

Parameter	Common-carrier emphasis	Computer network emphasis
Performance measure	Bitrate × distance	Any-any connectivity
Cost	In the cable	In the O/E modules
Target bitrates	Many Gb/s per link	≤ 1 Gb/s per node
Nominal BER	10^{-9}	10^{-15}
Topology	Interexchange: mesh. Local exchange: double star	Any-any LAN MAN or WAN
Switching	Circuit switching is traditional	Packet switching is traditional
Packaging	Hybrid packaging	Dense packaging
Safety	Class II	Class I
Skill level available	Craftsmen	Low or no operator skills
Ideal transmission medium	Dark fiber is controversial; offer SONET, SDH, ATM instead	Dark fiber is desirable

16.2 Second-Generation Networks

In Chapter 13 on protocols, we noted the existence of several MAN and LAN networks whose properties are driven by fiber technology, but which are essentially straightforward extensions of prefiber architectures. Notable among these are FDDI as a LAN, and DQDB as a MAN, both of which are rapidly being standardized. In the case of FDDI, the process is sufficiently mature that many products are already on the market. Having already touched on these networks in Chapter 13, we shall not elaborate further. The interested reader can get the details from such references as [1-4].

Several multiaccess structures are being built specifically to run on top of the fiber-based SONET and SDH link technologies described in Chapter 14. Two particularly important ones are discussed in [5]. The first of these is Asynchronous Transfer Mode (ATM) [6], in which each 125-microsecond frame is subdivided in such a way as to support the flow of very short (53-byte) packets, called *cells* This short length was chosen to satisfy the imperatives of voice communication, rather than computer communication. In packet-switched voice connections, the end-to-end delay can grow as the product of the number of intermediate nodes on the path times the packet duration. Making the packet duration only 53 bytes eases this problem, at the expense of greatly increased packet header processing.

Another respect in which ATM cells are not ideal for computer communication has to do with the frequency of packet loss, both due to buffer overflow and due to bit errors. The SONET standard [7] requires no better than 10^{-10} bit error rate.

Table 16-2.   U.S. National Gigabit Testbeds. (From [8]).

Name	Nodes at	Principal application
Aurora	Bellcore, Morristown; IBM Research; MIT; Univ. of Penn. (with Bell Atlantic, MCI, NYNEX)	Technology evolution
Blanca	Bell Labs, Holmdel; National Center for Super- computer Applications., Univ. of Illinois; Univ. of California, Berkeley; Univ. of Wisconsin	Technology evolution
Casa	Los Alamos Nat. Lab.; Caltech; Jet Propulsion Lab.; San Diego Supercom- puter Center. (with MCI, PacBell, US West)	Distributed supercomputing
Nectar	Carnegie Mellon Univ.; Pittsburgh Supercomputing Center	High-speed protocols, oper- ating systems and program- ming
Vistanet	Microelectronics Center of N.C.; Univ. of N.C., Chapel Hill. (with Bell South and GTE)	Communication standards testing; medical image proc- essing

Broadband Integrated Services Digital Networking (BISDN) [5] is the second form of multiaccess being supported on SONET and SDH. BISDN is built on the 155-Mb/s (OC-3/STM-1) or 622-Mb/s (OC-12/STM-4) class of links. ATM cells are the quanta of information exchanged by BISDN, which then adds higher-level functions that support voice, video, and other user-level services.

Meanwhile, the same United States Government-supported research community that developed ARPAnet, CSNET, and TCP/IP has been quite active. The most visible set of high-speed second-generation networks are the five that are parts of the "National Gigabit Testbed," being stimulated and organized by the United States Government [8]. Table 16-2 lists these. Most of the five use OC-12 and OC-48 links at

622 Mb/s and 2.4 Gb/s, respectively. These are provided by the participating interexchange and local exchange carriers.

16.3 WDMA Networks

As the reader will observe from the preceding chapter, work on links and multipoints has long proceeded on the basis that sending channels on different wavelengths and using tunable filters preceding direct detectors would eventually offer a particularly economic multipoint structure. Because of all this preliminary attention given the required components, especially tunable devices, it happened that, when the computer communication community discovered third-generation optical networking as a possibility, many of the building blocks were already moving out of the research laboratories and into commercial practice.

Wavelength-division multiaccess (WDMA) networks have been investigated and prototyped at the laboratory level by a number of groups [9, 10]. The programs at Bell Communication Research, AT&T Bell Laboratories, British Telecom Laboratories, Heinrich Hertz Institute, NTT Laboratories, Columbia University, and IBM Research have been particularly active ones. Bellcore's Lambdanet of 1987, described in Section 11-10 was the earliest WDMA network to be completely prototyped. However, it never underwent field tests.

16.4 The NTT 100-Wavelength Network

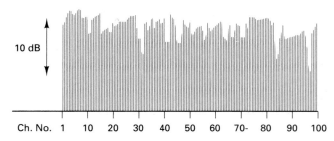

Figure 16-1. Observed spectrum at the star coupler output of NTT's 128-wavelength WDM network

The NTT Laboratory at Yokosuka has built the largest functioning WDM network to date [11], involving 100 wavelengths, each carrying 622 Mb/s. Each transmitter used a DFB laser, adjusted to an assigned wavelength. The wavelengths were spaced by 10 GHz (0.08 nm). All 100 signals were combined and then broadcast using a 128 × 128 single-mode star coupler, followed by an erbium doped-fiber amplifier at the receiver. Figure 16-1 shows the spectrum observed at one of the star-coupler outputs.

Special stabilization circuitry was provided for the laser transmitters. In the prototype, there was one receiver, at which channel selection was done by one monolithic 128-wavelength Mach-Zehnder interferometer chain, an earlier version of which is depicted in Figure 4-17. Millisecond tuning was accomplished using the small thermal heating pads shown shaded in Figure 4-17.

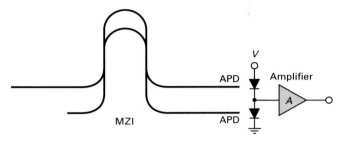

Figure 16-2. NTT's FSK demodulator, using a single Mach-Zehnder interferometer.

The FSK detector is particularly interesting. As shown in Figure 16-2, a Mach-Zehnder interferometer was used in such a way that the two outputs that feed the two photodetectors have nulls at the "0" and "1" frequencies, respectively. The spacing between the two frequencies was 2 GHz.

16.5 Columbia University's Teranet

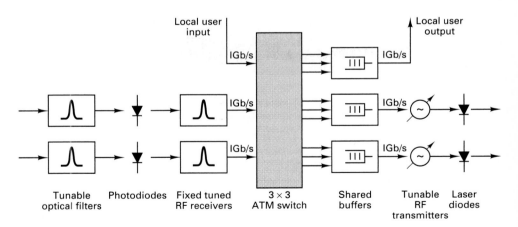

Figure 16-3. Block diagram of one of the network interface units of Columbia University's Teranet. There are $\Delta = 2$ input and output ports, plus one full-duplex user port, thus allowing multihop packet switching by means of the 3×3 switch, which handles ATM cells.

Teranet [12] is an 8-node network that uses the multihop principle of Sections 11-12 and 13-8. The number of ports per node $\Delta = 2$, and k, the order of the perfect shuffle, is also 2 (see Table 11-2). Thus the topology is exactly that shown in Figure 11-20(A). The purpose of Teranet is to provide fast packet switching without requiring wavelength tunability. Teranet is a key component of Columbia's ACORN project (Advanced Communications Organization for Research Networking) being pursued under the multidepartmental Center for Telecommunications Research.

Figure 16-3 shows a block diagram of each Teranet node. In addition to the two input network ports and two output network ports, there is a full-duplex user port, operating at a nominal 1.0 Gb/s bitrate (actually 1.25 Gb/s, because of the use of 4-out-of-5 line coding). The key to the packet-switching capability is the 3×3 switch. The packets handled by the system are ATM cells. The switch itself is non-blocking, and an output packet buffer capacity of 186 packets is provided to reduce the probability of overflow (and thus packet loss) to very low values. Each of the output buffers is shared between the three inputs from the switch.

While electronic switching is used for packet routing, wavelength switching using tunable Fabry-Perot filters is being built into the system so that it can be easily and quickly reconfigured. Such reconfiguration does not demand the microsecond tuning time that packet switching requires.

16.6 AT&T Bell Laboratory's Wavelength-Division Networks

Bell Laboratories has been pursuing two parallel network prototyping efforts, one using tunable optical filters and direct detection, the other using coherent detection.

The direct-detection system [13] originally supported four nodes at 45 Mb/s each, of which two were realized in the experiments. At the transmitter a DBR laser diode was frequency-shift keyed between two frequencies 144 MHz apart simply by modulating the bias current. At the receiver, a Fiber Fabry-Perot filter [14], of the type described in Section 4.4, was used both to isolate one of the wavelength channels and to perform the FSK demodulation function using the FSK/ASK scheme illustrated in Figure 7-4.

The direct detection system was later upgraded [15] to run at 600 Mb/s and then 1.2 Gb/s. These high bitrates necessitated the use of a special two-electrode DFB laser in order that the FM deviation obtained would be relatively uniform over a wide range of modulating frequencies. In the absence of such precautions, the deviation was bit-pattern dependent.

The coherent detection system [16] also supported four nodes, initially at 45 Mb/s. Three transmitters were implemented in the form of external-cavity tunable laser diodes. The optical frequencies were very close together (300 MHz), so the system was called an FDM system, not a WDM system. (As mentioned in Chapter 1, in this book we use "wavelength-division" to refer to both situations). One receiver was implemented, the local oscillator being an external-cavity laser diode and the polarization compensation being accomplished manually. Perhaps the most significant thing about this network has been that it has served as a test-bed for an increasingly sophisticated set of wavelength-stabilization techniques, leading eventually to the network-wide scheme shown in Figure 10-2.

16.7 British Telecom's Wavelength-Routing Network

The wavelength-routing principle of Section 11-7 is the basis of a network built by British Telecom to interconnect four telephone central offices in London [17]. These exchanges were all within a 45-kilometer radius of the central site at which each transmission and reception took place..

The channels were 12 nm apart in wavelength and each carried 622 Mb/s traffic. As shown in Figure 16-4, all the transmit and receive equipment was in one exchange at Ilford, the others serving simply as loop-back connection points. In effect the gratings were 3 × 3 units, so that each transmitted wavelength (upper left) was directed in a fixed route to one of the other nodes. To provide some variation in the route taken by a given bitstream, the wavelength-routing idea calls for tuning the transmit laser to a different wavelength. Since this was not possible with the lasers in use, some route-switching capability was provided by use of a 4 × 4 lithium-niobate photonic switch, shown as the block containing the "×" in the figure. The switch was used in a 1 × 3 mode. In this way, one of the bitstreams could be directed to one of three other nodes.

16.8 IBM's Rainbow-1 Network

IBM's Rainbow-1 network does circuit switching only. It was realized in commercial-grade computers and used in field trials in Westchester County, NY, in 1990 and in the Geneva, Switzerland, area in 1991. Figure 16-5 shows the latter configuration. Permitted data rates per node were 300 Mb/s maximum, and the maximum number of nodes supportable was 32. In the 1990 system there were four active wavelengths, corresponding to three nodes at one site (Hawthorne) and one at another (Yorktown), 25 kilometers away. The 1991 experiments used six wavelengths, four for 270-Mb/s digitized TV and two to carry 125-Mb/s FDDI bitstreams. Three computers were at the Telecom-91 exposition in Geneva and one at the Centre Europèen de Recherche Nucleaire (CERN), ten kilometers distant.

Two things are worth pointing out about this system. First, in the 1991 trials, the implementation included a completely realized multiaccess protocol, rather than a manual wavelength set-up procedure [18]. That is, connections were made by entering "log-on"-style requests at one computer, after which the protocol automatically set up a full-duplex connection with the desired other station by causing the latter to tune its

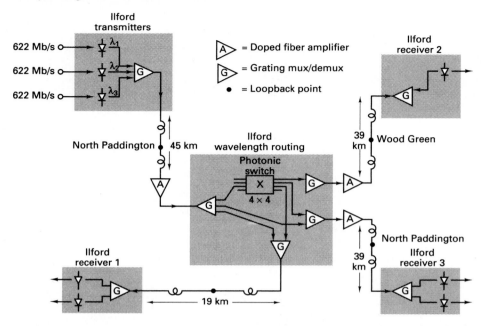

Figure 16-4. Block diagram of the British Telecom wavelength-routing and photonic-switching field test involving four London central offices.

tunable filter properly and by also causing the originating station to do the same. The multiaccess protocol used was the circular-search protocol described in Section 13.4. The slowness of this protocol (over 10 milliseconds) precluded any packet-switched service and limited this network to circuit-switched applications.

Second, each node took the form of a single plug-in printed-circuit card of a type widely used in the computer industry. This card could plug into the system bus (Microchannel) of a variety of workstations, controllers, and desktop computers. The fact that other communication cards are commercially available that also intercommunicate via the Microchannel means that any node pair in such a WDMA network not only can be used to support applications in the two nodes, but also can allow the network to act as a "bridge," or "network interconnection," between other networks.

In the 1991 Geneva field trial, there were five small computers in all, but two of them (one in Geneva and one at CERN) actually had two cards installed, so that each of these machines acted as two WDMA nodes. This allowed testing of the bridge function of a WDMA network, which was done by having one FDDI network at Geneva talking to another FDDI network at CERN over the Rainbow network.

16.9 What We Have Learned

It is clear that, in the last several years, technology has become available that has made third-generation networks entirely practical. This practicality is attested by the

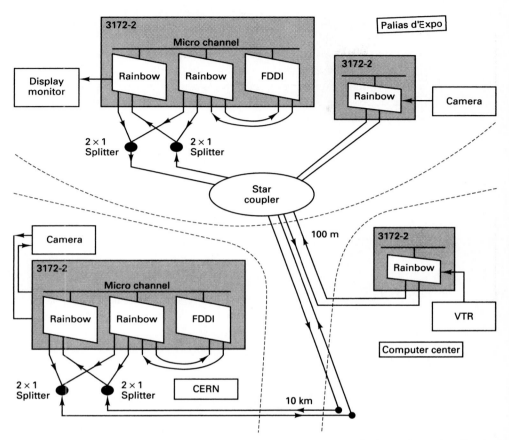

Figure 16-5. Block diagram of IBM's Rainbow-1 network, as field-tested at the Geneva Telecom-91 industrial exhibit.

fact that they have been implemented at the printed-circuit card level and employed for live user traffic.

It is equally clear, though, that until sub-microsecond tuning speeds become available for fully optical packet switching, such strategems as multihop indirect rerouting are required. Full packet-switching capability using pure WDMA should become available in the not-too-distant future, thanks to research efforts underway today on such fast-tunable components as index-tunable Fabry-Perot filters (Section 4.4), electrically tunable MZI chains (Section 4.6), switched grating filters (Section 4.12), and tunable laser diode structures (Sections 5.11, 5.12 and 5.13).

It is interesting to note that, in the network experiments described, most of the system complexity is in the electronics and most of the cost is in the photonics. The lasers, filters, photoreceivers, and photonic amplifiers are all structurally very simple, especially compared to the sophisticated memory, logic, and microprocessor chips that make up the electronics.

Given the same kind of cost-reduction evolution that electronic memory and logic have experienced, there is no reason to doubt that the key photonic components can ultimately be made very inexpensively. The technical ideas are already here. It is only necessary for the marketplace to drive the customary circular process of increased volumes producing lower costs and vice versa. As this plays out, third-generation lightwave networking components and architecture will provide a universally and economically available, easy-to-use platform upon which to build the applications of the future.

16.10 References

1. F. E. Ross, Fiber Digital Data Interface - Token Ring Media Access Control, American National Standards Institute X3.139, 1987.

2. "Draft IEEE Standard 802.6, Distributed Queue Dual Bus (DQDB) Metropolitan Area Network (MAN)," *IEEE Computer Society*, June, 1988.

3. I. Cidon and Y. Ofek, "Metaring, a full-duplex ring with fairness and spatial reuse," *Conf. Record, IEEE INFOCOM*, pp. 969-981, 1990.

4. H. R. Mueller, M. M. Nassehi, J. W. Wong, E. A. Zurfluh, W. Bux, and P. Zafiropulo, "DQMA and CRMA: New access schemes for Gbit/s LANs and MANs," *Proc. IEEE Infocom*, pp. 185-191, 1990.

5. G. Estes, ed., "Special Issue on Broadband Integrated Services Networks," *IEEE LTS Mag.*, vol. 2, no. 3, pp. 1-66, August, 1991.

6. M. dePrycker, *Asynchronous Transfer Mode–Solution for Broadband ISDN*, Prentice Hall, 1991.

7. Synchronous Opticl Network, American National Standards Institute, T1.106.

8. "Gigabit network testbeds," *IEEE Computer Magazine*, vol. 23, no. 9, pp. 77-80, 1990.

9. N. K. Cheung, K. Nosu, and G. Winzer, eds., "Special Issue on Dense Wavelength Division Multiplexing for High Capacity and Multiple Access Communications Systems," *IEEE Jour. Sel. Areas in Commun.*, vol. 8, no. 6, August, 1990.

10. C. A. Brackett, ed., "Special Issue on Lightwave Systems and Components," *IEEE Commun. Mag.*, vol. 27, no. 10, October, 1989.

11. H. Toba, K. Oda, K. Nakanishi, N. Shibata, K. Nosu, N. Takato, and K. Sato, "100-channel optical FDM transmission/distribution at 622 Mb/s over 50 km using a waveguide frequency selection switch," *Elect. Ltrs.*, vol. 26, no. 6, pp. 376-377, March, 1990.

12. R. Gidron and A. Temple, "TeraNet, a multihop multichannel ATM lightwave network," *Conf. Record, IEEE Intern. Commun. Conf.*, pp. 602-608, 1991.

13. I. P. Kaminow, P. P. Iannone, J. Stone, and L. W. Stulz, "FDMA-FSK star network with a tunable optical filter demultiplexer," *IEEE/OSA Jour. Lightwave Tech.*, vol. 6, no. 9, pp. 1406-1414, 1988.

14. J. Stone and L. W. Stulz, "Pigtailed high-finesse tunable fiber Fabry-Perot interferometers with large, medium and small free spectral range," *Electron. Lett.*, vol. 23, pp. 781-783, 1987.

15. A. E. Willner, I. P. Kaminow, M. Kuznetsov, J. Stone, and L. W. Stulz, "1.2 Gb/s closely-spaced FDMA-FSK direct-detection star network," *IEEE Photonics Tech. Ltrs.*, vol. 2, no. 3, pp. 223-226, 1990.

16.   B. S. Glance, J. Stone, K. J. Pollock, P. J. Fitzgerald, C. A. Burrus Jr., B. L. Kaspar, and L. W. Stulz, "Densely spaced FDM coherent star network with optical signals confined to equally spaced frequencies," *IEEE/OSA Jour. Lightwave Tech.*, vol. 6, no. 11, pp. 1770-1781, 1988.

17.   H. J. Westlake, P. J. Chidgey, G. R. Hill, P. Granestrand, L. Thylen, G. Grasso, and F. Meli, "Reconfigurable wavelength routed optical networks: A field demonstration," *Conf. Record, Eur. Conf. on Optical Commun.*, pp. 753-756, September, 1991.

18.   F. J. Janniello, R. Ramaswami, and D. G. Steinberg, "A prototype circuit-switched multi-wavelength optical metropolitan-area network," *Conf. Record, IEEE Intern. Commun. Conference*, 1992.

List of Symbols

Symbol	Chapter introduced	Definition	Units, if not dimensionless
a	3	Core radius	Meters
	5	Differential gain constant	Cm^2
A	8	Electrical amplifier gain	
	4	Mirror absorption loss	
α	3	Attenuation (imaginary part of propagation constant)	Meters^{-1}
	3, 11	Coupler power splitting ratio	
B	3, 8	Bandwidth	Hertz (cycles per second)
β	3	Real part of propagation constant	Radians per meter
	5	Linewidth enhancement factor	
	9	Frequency deviation ratio	
	11	Coupler excess loss	
c	3	Free-space propagation velocity	Meters per second
C	8	Capacitance	Farads

Symbol	Chapter introduced	Definition	Units, if not dimensionless
d	5	Laser active region depth	Meters
D	13	Delay	Seconds
$\Delta,\ \delta$	3	Differential operators	
δ	11	Tap divider ratio	
Δ	3	Fractional index difference	
	11	Topological degree of a node	
E	3	Electric field strength	Volts per meter
ϵ	3	Electric permittivity	Farads per meter
η	5, 8	Efficiency	
$\mathcal{E}$	5, 13	Energy	Joules
f	3	Frequency	Hertz
	3	Focal length	Meters
	4	Finesse	
F	4	Acoustic frequency	Hertz
	8	APD noise factor	
g	5	Gain per unit length	Cm^{-1}
G	6	Optical amplifier gain	
γ	3	Angular separation on Poincaré sphere	Radians
	8	Amplitude signal-to-noise ratio	
Γ	5	Spatial confinement factor of optical mode	
h	5	Planck's constant (6.63×10^{-34})	Joules $\times$ seconds
H	5	Magnetic field strength	Amperes per meter
	4	Amplitude transfer function	
	11	Hop count	

Symbol	Chapter introduced	Definition	Units, if not dimensionless
i, j, k ℓ, m, n ℓ, p, q		Indices	Integer
I	4	Pump current	Amperes
j	3	$\sqrt{-1}$	
J	3	Bessel function of the first kind	
k, K	3	Propagation constant	Radians per meter
$\mathbf{k}, \mathbf{K}$	4	Propagation vector	Radians per meter
k	11	Shufflenet parameter	
K	5	Modified Bessel function of the first kind	
	5	Boltzmann's constant (1.38×10^{-23})	Joules per °K
ℓ	13	Node separation	Kilometers
L	3, 5	Length	Meters
	11	Coupler loss	Decibels
	13	Network size	Kilometers
λ	3	Wavelength in free space	Meters per cycle
Λ	4, 5	Acoustic or grating wavelength	Meters per cycle
	11	Total number of wavelengths	
	13	Total wavelength range	Meters per cycle
m	4	Mass	Kilograms
	8	Mean	
	9	Modulation depth	
M	3	Number of permitted modes	
	8	APD gain	
μ	1		Microns
	3	Magnetic permittivity	Henrys per meter
n	3, 4	Refractive index	

Symbol	Chapter introduced	Definition	Units, if not dimensionless
n_{sp}	5	Spontaneous emission factor	
N	1	Number of nodes per network	
	5	Carrier density	Cm^{-3}
ν	5	Number of states	
ω	3	Angular frequency	Radians per second
Ω	3	Acoustic angular frequency	Radians per second
p	13	Probability	
	4	Momentum	Kilograms $\times$ meters/sec.
$\mathbf{p}$	4	Momentum vector	Kilograms $\times$ meters/sec.
P	5	Optical power	Watts
	8	Probability	
	13	Packet duration	Seconds
$\mathbf{P}$	5	Optical power density	Watts per meter^2
Π	5	Photon density	Meters^{-3}
ϕ	3	Sinusoidal phase angle	Radians
	13	Correlation function	
ψ	3	Spatial tilt of polarization ellipse	Radians
	4	Filter output crosstalk	
	5	Electric or magnetic field	See E or H.
	13	Unit amplitude of pulse of bandwidth W Hz	
q	5	Charge of the electron (1.6×10^{-19})	Coulombs
Q	8	Bit error rate function	
	10	Ratio of center frequency to bandwidth	
r	6	Extinction ratio	
	7	Electrooptic tensor	Radians per volt per meter
	8	Photon arrival rate	Seconds^{-1}

Symbol	Chapter introduced	Definition	Units, if not dimensionless
R	3	(Power) reflectance	
	8	Resistance	Ohms
	11	Ratio of output to input power	
R_{sp}	5	Spontaneous emission rate	Sec^{-1}
RIN	5, 9	Relative intensity noise	
$\mathcal{R}$	8	Responsivity	Amperes per watt
ρ	3	(Amplitude) reflectivity	
	8	Charge density	$Meters^{-3}$
S	3	Scattering matrix	
	4	Transmitted amplitude spectrum	Complex
σ	8	Standard deviation	
	11	Tap summing ratio	
t	3	Time	Seconds
T	8, 13	Bit duration	Seconds
	4	Power transfer function	Real
τ	3, 5	Time constant; lifetime	Seconds
	4	Acoustooptic access time	Seconds
θ	3, 4	Angle of incidence, reflection or transmission	Radians
$\mathbf{u}$	3	Unit vector	
Υ	5	Temperature	Degrees absolute ($^\circ K$)
v	3	Propagation velocity in medium	Meters per second
V	3	V-number	
	4	Acoustic velocity	Meters per second
	4, 8	Applied voltage	Volts
w	3	Beam waist diameter	Meters
	5	Laser active region width	Meters

Symbol	Chapter introduced	Definition	Units, if not dimensionless
W	13	Signal bandwidth	Hertz
x, y	4	Cavity lengths	Meters
X, Y	13	Signal amplitudes	Complex
XT	4	Crosstalk power	Watts
χ	3 6	Related to phase difference Amplifier noise enhancement factor	Radians
Z	3, 7 11	Interaction length Amplifier spacing	Meters Kilometers
$\langle\,\rangle$	8	Ensemble average	
$\bullet$	3	Dot product	

Index